NONLINEAR PHENOMENA IN STELLAR VARIABILITY

Edited by

M. TAKEUTI

Astronomical Institute, Tôhoku University,
Sendai, Japan

AND

J.-R. BUCHLER

University of Florida,
Gainsville, U.S.A.

Reprinted from Astrophysics and Space Science
Volume 210, Nos. 1–2, 1993

SPRINGER SCIENCE+BUSINESS MEDIA, B.V.

A C.I.P. Catalogue record for this book is available from the Library of Congress.

ISBN 978-0-7923-2769-1   ISBN 978-94-011-1062-4 (eBook)
DOI 10.1007/978-94-011-1062-4

Printed on acid-free paper

TABLE OF CONTENTS

III. MODELS

IAU COLLOQUIUM No. 134
NONLINEAR PHENOMENA IN STELLAR VARIABILITY

7 – 10 January 1992 held at Joyo Geibun Center, Mito, Japan

Scientific Organizing Committee

A. Baglin, M. Breger, J.-R. Buchler, W. Dziembowski,
Yu. A. Fadeyev, H. Mori, Y. Osaki, J. Percy, H. L. Swinney,
M. Takeuti (chairperson), W. Unno

Local Organizing Committee

H. Ando, T. Hamada (chairperson), K. Saijo, M. Takeuti, Y. Tanaka

The colloquium was held with 77 registered participants from 14 countries. It was scientifically sponsored by IAU Commission 27. The scientific program consisted of an opening talk, 20 invited reviews, 17 oral contributions, 29 posters, and a summary. The colloquium was held just after the disintegration of Soviet Union. Related with problems in the countries of the former Soviet Union, several astronomers did not succeed in participating the meeting. The papers sent by some of them to the Scientific Organizing Committee were presented to the participants as material for discussion. The titles and authors are listed below.

Yu. N. Efremov: The Cs Cepheids - overtone or first crossing?;

Larisa S. Kudashkina and Ivan L. Andronov: The multiperiodicities in the semi-regular variables;

Ivan L. Andronov: Autocorrelation function analysis of the rapid variability of the cataclysmic variables;

V. P. Arkhipova: Photometric evolution and the light oscillations of FG Sge in 1967-1991.

The colloquium was supported by several foundations and companies. We would like to express our thanks to the Commemorative Association for the Japan World Exposition (1970), the Science and Technology Promotion Foundation of Ibaraki, Fujitsu, Ltd., Hitachi Engineering Co. Ltd., Hitachi Tohoku Software Ltd., IBM Japan, Ltd., the Joyo Bank, the Joyo Geibun Center, the Mito Shinkin Bank, Mitsubishi Electric Corporation, NEC Corporation, Rikei Corporation, and the Astronomical Society of Japan for their support. We also express our thanks to Ibaraki University for its kind hospitality

T. Hamada and M. Takeuti

Astrophysics and Space Science **210**:vii, 1993.
© 1993 *Kluwer Academic Publishers.*

PREFACE

The nonlinear theory of oscillating systems brings new aspects into the study of variable stars. Beyond the comparison of linear periods and the estimate of stability, the appearance and disappearance of possible modes can be studied in detail. While nonlinearity in stellar pulsations is not a very complicated concept, it generally requires extensive and sometimes sophisticated numerical studies. Therefore, the development of appropriate computational tools is required for applications of nonlinear theory to real phenomena in variable stars.

Taking trends in variable star studies into consideration, the International Astronomical Union organized a colloquium for the nonlinear phenomena of variable stars at Mito, Japan in 1992. The colloquium served to give an overview of the new frontiers of variable star studies and to encourage further development of this field. The colloquium covered the fundamental theory, interesting observational facts, and the numerical modeling.

The publication of the proceedings was somewhat delayed since one of the editors, M. T., was overwhelmed by administrative work. We are sorry that the excellent reviews of Drs. H. Mori, M. Sano, and K. Makishima cannot be found in the proceedings. We also miss the summary given by Dr. W. W. Dziembowski. Throughout the editing procedure Dr. Y. Tanaka of Ibaraki University kindly helped us. Because of the unfortunate delay of the publication, the significance of several papers may be affected. Even so, we believe that the papers are useful to variable star researchers because of their scientific importance.

The editors wish to express their thanks to the editorial board of *Astrophysics and Space Science* and to Kluwer Academic Publishers for their willingness to publish the proceedings.

October 1993   M. Takeuti and J.-R. Buchler

Astrophysics and Space Science **210**:   1993.
© 1993 *Kluwer Academic Publishers.*

INTRODUCTION

M. TAKEUTI

Astronomical Institute, Tôhoku University, Sendai 980, Japan

Abstract. A historical sketch of the nonlinear theory of variable stars is outlined briefly. The main break-through came from the hydrodynamic study of stellar pulsation. From a theoretical point of view, coupling needs to be discussed more carefully. The impact of new opacities on the astrophysical problems is also discussed.

1 Nonlinear Phenomena in Variable Stars

In the classic textbook written by Rosseland, nonlinear phenomena in pulsating stars are described by inharmonic and relaxation oscillations. The former is the deviation of stellar oscillations from sinusoidal curves, and the latter involves the variability of U Gem stars and shock phenomena.

Astrophysical nonlinear phenomena, which will be studied in the present colloquium, differ from classical nonlinear phenomena. Trends in the studies of pulsation theory changed in recent decades. The strongest influence was the development of electronic computers. Hydrodynamic stellar models then succeeded in showing the features of pulsating stars. Hydrodynamic simulations applied to accretion discs also succeeded in showing details of the variability in cataclysmic variables. Even though such remarkable developments in the studies of stellar variability have occurred, several questions, such as the period-ratio of double-mode cepheids, the complex behaviour found in the δ Scuti stars, and the nature of irregularity observed in various giant stars and cataclysmic variables still remain. Irregular features found in hydrodynamic simulations have not yet been explained theoretically.

In the studies of the irregular nature of stellar pulsations, Baker, Moore and Spiegel pointed out the importance of nonlinear studies for one-zone stellar models. They suggested paying attention to recent developments in nonlinear dynamics which stressed the universal importance of deterministic chaos. The application of modern nonlinear dynamics to stellar variability first concentrated on the study of the conservative case. Later, dissipative systems were investigated. This study brings out new aspects on the nonlinear problem. Period-doubling bifurcation and intermittency have been recognized in hydrodynamical models. The limit-cycles of self-exciting systems have been investigated along the lines of Krogdahl, and used to discuss the progression of bumps in classical cepheids. These systems are studied on electronic computers and interesting features such as period-doubling bifurcation into chaos and phase-locking have been found. These numerical experiments remind us of several important characteristics of nonlinear oscillation theory. Progress in the nonlinear studies of stellar variability is to be reviewed and summarized in the colloquium.

Astrophysics and Space Science **210**: 1–5, 1993.

2 Scale Length of Variability

Stellar variability is usually accompanied by irregular features. Real variability consists of two kinds of instability; one is global instabilities such as the pulsations of a star or an accretion disc; and the other is local instability such as solar flares. The timescale of the latter is related indirectly to the dimensions of the star or the disc, so that the activity seems sporadic compared to the overall phenomena. The observed irregularity can be the superposition of these two types of instability. It is natural to suppose that the variability decomposed into a few modes could be caused by the local instability.

Since a deterministic system can yield chaotic motions, apparently sporadic variability can be produced by global phenomena. This is a new idea which comes from nonlinear dynamics. Several studies have been done on nonlinear oscillators which are expected to show deterministic chaos. The nonlinear oscillations of conservative models were studied first. Then, dissipative models such as the Moore-Spiegel type, the Tanaka-Takeuti oscillator, which is a modified version of the Rössler oscillator, and coupled Krogdahl type oscillators were examined. These models succeeded in showing the period-doubling bifurcation into chaos, but are not directly comparable with observational results. The nonlinear oscillations of one-zone models studied by Saitou, Takeuti and Tanaka, who tried to compare their results with the RV Tau stars, are also one of such oscillators. This exercise has been performed on accretion discs, but there is no complete review of the subject.

3 Breakthrough

The first results directly comparable with observations came from a careful hydrodynamic simulation of cool Population II giant stars. These should correspond to the RV Tau stars. The models demonstrate the period-doubling bifurcation into chaos. Strictly speaking, the models are not complete enough to compare with observations because the models do not include the effect of convection. Convection may play an important role for the stars studied here. What is important is the appearance of the period-doubling bifurcation, both in the simplified model oscillators and complicated hydrodynamic models. It thus is confirmed that the nonlinearity of the excitation and dissipation mechanisms in pulsating stars can produce the deterministic chaos. At least, therefore, we may apply the results of nonlinear dynamics for dissipative systems to semi-regular variable stars.

The physical processes at work in the outer layers of pulsating giant stars include the effects of opacity changes on radiation and other dissipative processes such as cooling. The nonlinearity in the opacity first dams up and then releases the flow of radiation, and then the outer layers drive the increasing and decreasing oscillations of the star. The decrease of the surface gravity makes the amplitude of oscillation large there compared to the inner parts. This causes an increase in nonlinearity both in the excitation and dissipation mechanisms.

4 Intermittent Enhancement of Pulsations

There exists the question as to the reality of intermittent enhancement of pulsations in low surface-gravity stars. This phenomenon was found by Tuchman, Sack and Barkatt and then also found in the models of Fadeyev and his collaborators. Nakata demonstrated that it is real in hydrodynamic simulations and appeared with decreasing surface gravity. Aikawa investigated the details of physical processes in the outer layers and has found that the development of a strong shock may play an important role. The occurrence of such a strong shock may be related to the high opacity layers which dam up the radiation very efficiently. Unfortunately, there is no precise numerical investigation of the effect of convection on the intermittency, although convection can weaken the effect of strong opacity.

5 Coupling

Over more than a decade, coupled oscillations of self-exciting systems have been investigated by several authors. The strength of coupling is determined by two different factors; one is resonance and the other is the similarity of wave functions that is indicated by the coupling coefficients. Resonance can be studied by comparing linear periods, so that this possibility is easily discussed. When resonance works efficiently, the largest amplitude will be that of the lightest mode which is usually a higher mode, since the law of equipartition works for the energy of oscillation among the coupled modes.

The calculation of the coupling coefficients, which are essential for the study of non-resonant coupling, is more complicated than the comparison of the period ratio. The formulation of Buchler and Goupil is elegant but unfortunately difficult to apply to numerical evaluation because of its complexity. On the other hand, simplified formulations are not as complete.

In any case, oscillations become either synchronized or non-synchronized, depending on the strength of coupling. The observed period in the synchronized oscillation differs slightly with one of the linear periods. For non-synchronized oscillations, the observed periods can be different to the linear ones. We have to pay attention to the phase-locking where the period-ratio tends to the ratio of small integers close to the linear period-ratio. The phase-locking needs carefull examination in the study of double-period variability. We can see complex variability in δ Scuti stars and white dwarf stars. The light curves sometimes decompose into several sinusoudal curves, but the modes are not so stable. The transition of the energy of oscillations should be investigated to analyze these stars. It seems probable that we can obtain new results on the inner structure and evolutionary aspects by considering the modal coupling.

6   Yellow Supergiant Stars

The semi-irregular variability found in yellow supergiant stars is studied on return maps on which we plot the successive maximum brightness. The maps never show any one definite pattern, but some of them show an egg-shape pattern and others show a horse-shoe-like pattern. In the numerical studies of one-zone stellar models, the egg-shape appears in low surface-gravity models and the horse-shoe pattern, in high surface-gravity models. It should be pointed out, however, that the egg-shape pattern will also appear for the double-mode oscillations.

The return map of UU Her shows the eff-shape pattern. After the variability of UU Her is decomposed into approximately double-period oscillations, there is no reason to suppose the surface-gravity would be low. UU Her should be a massive star.

7   Accretion Discs

The numerical simulation of the variability of accretion discs have been performed to a high precision. The origin of the variability is not in simple oscillation as is the radial stellar pulsation. The oscillation theory is, however, still applicable to more complex variability of accretion discs. The variability of U Gem is explained as a relaxation oscillation, so that the movement will be approximated by Moore-Spiegel type oscillations. They have two divergent fixed points and one saddle point. Such a model may explain the overall light variation of the discs.

The irregular variability of X-ray sources should be another target for nonlinear study. It seems interesting to separate global and local variations. The disc should show the UU Gem-like variability which controls the mass-flux at the inner edge of discs. Moreover, local activity at the inner edge is also expected. X-ray irradiation from the sources near the central body keeps the mass-flux nearly constant, but flare-like activity will occur frequently in such a high temperature domain accompanied by the strong magnetic field. The problem is now to distinguish the type of variability in the observed X-ray variability.

8   Conclusions

Before finishing the Introduction, we have to stress the importance of the examination of past results using new opacities. The difference in the opacities does not affect stellar models so much, but does affect the pulsation properties. We will see these effects later in the colloquium.

Trends in computing systems are towards down-sizing. More compact, easier to handle computers are now available. Work stations and personal computers are the most useful tools to analyze observational data and to try the models. The exchanges of young astronomers among institutions is fruitful for using these tools. They will have a chance to learn the skills of computation and the details of programming codes. Such technological progress may change the style of variable star studies.

Finally, I should stress that the nonlinearity is never connected with only the large amplitude phenomena. The motion near the limit-cycle should be studied as well as nonlinear phenomena. We anticipate that the papers presented in this colloquium will study the various cases of nonlinear phenomena.

I. FUNDAMENTAL THEORIES

A DYNAMICAL SYSTEMS APPROACH TO NONLINEAR STELLAR PULSATIONS

J. R. BUCHLER

University of Florida, Gainsville, FL32611, USA

Abstract. Over the last decade we have seen the application of novel techniques to the old problem of nonlinear stellar pulsations. Together with numerical hydrodynamics this approach provides a more fundamental understanding of the systematics of the pulsational behavior. For weakly nonadiabatic pulsations, whether regular or multi-periodic, dimensional reduction techniques lead to amplitude equations and to a description in terms of modal interactions and resonances. In particular they shed new light on the bump progression in the classical Cepheids. In more dissipative stars numerical hydrodynamical modelling has uncovered the existence of irregular variability, both in radiative and in convective models. An application of modern dynamical systems techniques has shown that this behavior occurs according to well understood routes from regular to chaotic behavior. The mechanism is very robust and represents the first non *ad hoc* theoretical explanation of irregular stellar variability. Finally, we discuss how a comparison with observations of irregular variability shows the need for more suitable observations, on the one hand, and of better techniques of signal processing, on the other.

1. Introduction

There is hardly a need to stress the importance of the study of stellar pulsations. Almost all stars undergo some kind of pulsational phase during their lifetimes, and an understanding of the large variety of pulsational behavior poses a challenge to the astrophysicist. From an astronomical point of view, nonlinear pulsations yield information about the parameters of the stars which their static siblings do not reveal us. More generally, the study of stellar pulsations has greatly improved our understanding of stellar structure and evolution, of galactic evolution and especially of cosmology, where the variable stars have provided the pillar on which our knowledge of the Universe's distance scale rests. For the physicist the pulsating stars are intriguing giant natural heat engines. Because of the extreme conditions of density and temperature encountered in stellar interiors they provide an excellent testing ground for Physics. For example, it is a long-standing discrepancy between the predictions of stellar pulsation and stellar evolution that has stimulated (Simon 1982) a revisitation of the atomic physics calculations with a subsequent substantial change in the opacities. Finally, to the dynamicist, pulsating stars are of interest because they exhibit very characteristic low-dimensional behavior in spite of the quite complicated nonlinear hydrodynamical equations which govern their behavior.

The work on nonlinear stellar pulsations can be grouped into three categories, numerical hydrodynamical, simple modelling and nonlinear dynamics approaches. The first, the *numerical hydrodynamical approach* was pioneered

Astrophysics and Space Science **210**: 9–31, 1993.

by Christy in the mid 60s and has been the workhorse for nonlinear modelling. Although it clearly constitutes a brute force attack, it is also the
approach that will yield the most detailed and accurate description of the
pulsations. However, its obvious shortcomings are the difficulty of extracting
the underlying systematics in the metamorphosis of the generated light and
radial velocity curves when the stellar parameters are varied. In particular,
it is easy to model and obtain the Hertzsprung progression in the bump
Cepheids, but it is difficult to understand its origin and what governs its
presence and shape.

The second approach of constructing *simple models* is very pedestrian,
but is essential for developing physical intuition and guidance. In the case of
the famous linear one-zone model of Baker (1966) it yielded a much clearer
understanding the destabilization of vibrational modes through the effects
of the equation of state and of the opacity, the so-called γ and κ mechanism.
In the context of nonlinear pulsations, Baker, Moore and Spiegel (1966) suggested the use of such a model in the form of a simple set of three first order
ODEs. Their suggestion came as a result of a model oscillator that they
had constructed for studying overstability in a convectively unstable zone
(Moor and Spiegel 1966). This simple oscillator model which is a little known
contemporary of the now famous Lorenz oscillator (*e.g.*, Cvitanovich 1984),
like the latter exhibits a myriad of different types of behavior, including
chaotic oscillations. Buchler and Regev (1982) developed a simple one-zone
model of interest for the oscillations of stars with extended convective partial ionization regions. It turned out this oscillator was very similar to the
Moore-Spiegel oscillator as well. Buchler and Perdang (1979) introduced a
two-zone model to understand the thermal relaxation oscillations found in
the study of stars with thin burning shells. Barranco, Buchler and Livio
(1981) and Livio and Regev (1984) used a similar model for X-ray bursters.
Recently Tanaka and Takeuti (1988) introduced additional one-zone models
for stellar pulsations, and Saitou, Takeuti and Tanaka (1989) showed that
the famous Rössler attractor can be transformed into a model stellar oscillator (for a review cf. *e.g.*, Takeuti 1990). This is particularly interesting in
view of the chaotic pulsations encountered in the numerical hydrodynamical
modelling of W Vir stars which seem to have the topology of the Rössler attractor (Kovács and Buchler 1988b). Generally speaking, it is important to
realize that such simple model equations (3 first order nonlinear ODEs) can
have a variety solutions, from static, to regular periodic, to period-doubled
and to chaotic, depending on the values of the model parameters. It would
therefore be astonishing if the more complicated hydrodynamical modelling
were not also to produce this type of behavior. The general drawback of
this approach is that the predicted behavior is not robust to the addition of
further zones and that it is therefore not easily generalizable and improvable
in a systematic fashion.

Finally, the *dynamical systems approach* is complementary to numerical hydrodynamics in that it gives a natural framework within which to understand the sometimes overwhelming computer output. Its astrophysical origins go back to the Hamiltonian approach of Woltjer (1936, 1937, 1946). (Here we only note in parentheses, because we are concerned with dissipative systems, that this Hamiltonian approach has been continued and applied to Kolmogorov unstable systems by Perdang 1983). Papaloizou (1973), realizing the importance of dissipation, was the first to use asymptotic perturbation methods with which he studied the nonlinear pulsations of upper main sequence stars. He also drew attention to the essential role played by resonances. Vandakurov (1979) and Dziembowski (1980), with the help of averaging techniques used in plasma physics, studied resonant wave coupling, but did not consider nonlinear nonadiabatic effects. At the same time Takeuti and Aikawa (1980, 1981, 1985) employed another asymptotic perturbation technique, the method of harmonic balance. again with an adiabatic approximation, and they added *ad hoc* van der Pol nonlinear dissipation terms. Buchler, Yueh and Perdang (1977) and Barranco, Buchler and Regev (1982) applied a multi-time method, but used a quasi-adiabatic approach which turned out not to be very useful from a practical point of view. Although all these studies were not fully satisfactory in some way or other they gave rise to amplitude equations of very similar form. This is not astonishing in retrospect as there exists a very general, systematical formalism to derive such equations provided rather general physical assumptions are satisfied.

2. The Dynamical Systems Approach

The first basic assumption that underlies this approach is one of *weak nonlinearity* which allows us to describe the nonlinear behavior of the system in terms of *modes*. Let use denote the deviation from static equilibrium of the basic variables (radius, velocity and a thermal variable, *e.g.*, the temperature) by $z(t) = (\delta R, \ldots, \delta v, \ldots, \delta T, \ldots, \ldots)$, where δR and δv respectively are vectors or scalars in the nonradial or radial cases. When for practical purposes the model is discretized into N mass zones, each of these quantities then has N components. We can write the equations of hydrodynamics and radiative transport (in the Lagrangean description) in the very compact form

$$\frac{\partial z}{\partial t} = \mathcal{L}z + \mathcal{N}(z) \qquad (1)$$

where $\mathcal{N}(z)$ is the strictly nonlinear part of the right-hand side. Linearization with a time-dependence $\exp(\sigma t)$ leads to the eigenvalue problem

$$\mathcal{L}e_k = \sigma_k e_k \qquad (2)$$

with

$$\sigma_k = i\omega_k + \kappa_k \qquad (3)$$

The complex eigenvectors e_k, together with their dual adjoints, form a complete orthogonal set and the displacement can be expressed in terms of the amplitudes $\{a_k\}$ of all the modes, *viz.* $z(t) = \sum_k a_k(t)e_k$. Substitution into Eq. (1) then leads to an equivalent system of coupled equations

$$\frac{da_k}{dt} = \sigma_k a_k + \sum_{lm} n_{klm} a_l a_m + \sum_{lmn} n_{klmn} a_l a_m a_n + \cdots \qquad (4)$$

the sum extending over all modes and over all combinations of the $\{a_k\}$ and complex conjugate (*c.c*) amplitudes $\{a_k^*\}$. The nonlinear coefficients $n_{klmn..}$ are constructible from the operator $\mathcal{N}$ and from the complex eigenvectors e_k and their duals (*e.g.*, Buchler and Goupil 1984, Buchler 1985). At this stage such equations can represent Hamiltonian as well as dissipative systems.

2.1. THE GALERKIN APPROXIMATION

This approximation is nothing but a truncation both in the number of modes and in the number of nonlinear terms in system (4). For example, for a single mode one finds

$$\frac{da_1}{dt} = (i\omega_1 + \kappa_1)a_1 + (n_{111}a_1^2 + c.c. + 2n_{11-1}|a_1|^2)$$
$$+ (n_{111}a_1^3 + 3n_{111-1}|a_1|^2 a_1 + c.c) \qquad (5)$$

In general, a Galerkin approximation is not a good approximation unless a large number of modes are included. The problem is that the predicted behavior is not robust in the sense that it can depend sensitively on the number of modes. The Lorenz equations, for example, have their origin in a truncation in terms of 3 Fourier components and have become famous not for the physics they represent, but for their interesting solutions.

2.2. THE AMPLITUDE EQUATION FORMALISM

The fundamental assumption of this approach is the existence of a *slow manifold*. What this means is that the modes can be split into two groups, the *principal modes* characterized by $|\kappa_k/\omega_k| \ll 1$ and the *slave modes* for which κ_k/ω_k is negative and of order unity. The physical idea which underlies the dimensional reduction method is very simple. When a system is disturbed away from its static equilibrium it is only during a short transient time-interval that it rings with all the eignefrequencies, an interval during which the amplitudes of the slave modes decay away very fast. This is followed by a slow evolution in which the system is completely specified by the behavior of the amplitudes of the principal modes, *i.e.*,

$$z(t) = z(\{a_k(t)\}) = \sum_{k=1}^{p} a_k(t)e_k + quadratic\ terms + cubic\ terms + \ldots (6)$$

These amplitudes can be considered generalized coordinates which parametrize in the slow manifold. *i.e.*, the subspace of phase-space corresponding to the slow manifold in which the system evolves. The behavior of the principal amplitudes themselves is described by a system of ordinary differential equations, the *amplitude equations* whose general form is

$$\frac{da_k}{dt} = (i\omega_k + \kappa_k)a_k + g_k(\{a_k\}), \quad k = 1,\ldots,p \tag{7}$$

where p is the number of principal modes. The expressions for $g_k(\{a_k\})$, the *normal forms* as they are also called depend on the number of modes and on what resonances, if any, are present, and are generic (*e.g.*, Guckenheimer and Holmes 1983). *Which specific form one needs to use can therefore be decided on the basis of the linear spectrum.*

The amplitude equations (Eq. (7) and the nonlinear terms $g_k(\{a_k\})$) can be obtained in a number of ways. Coullet and Spiegel (1984) presented a very elegant and general method for deriving the amplitude equations, whereas, concomitantly, Buchler and Goupil (1984; *cf.* also Buchler 1985) used a more intuitive multi-time method and gave explicit expressions for the case of radial stellar pulsations.

Because of the basic assumption of weak nonlinearity for the principal modes we can factor out a rapidly varying term from the amplitudes, *viz.*

$$a_k(t) = \exp(i\omega_k t)\tilde{a}_k(t) \tag{8}$$

It is clear that expression (6) for $z(t)$ represents a multi-periodic function with the eigen-frequencies of all the principal modes. Actually, the higher order terms gradually modify these frequencies because of nonlinear effects without however destroying the multi-periodicity.

To summarize, the dimensional reduction method reduces the complicated partial differential system (1) or its infinite dimensional counterpart (4) to as system of p ODEs for only the p principal modes. This is schematically illustrated in Fig. 1. In the amplitude equation formalism the slow manifold, while of the same dimension as the space of the principal modes, curves into the space of the slave modes. In contrast, in the Galerkin approximation the system is restricted to move in the space of the modes which are being included. In addition, the coefficients in the amplitude equations are different.

We now examine the predictions of the amplitude equation formalism for the simplest case, *viz.* that of a single principal mode. The amplitude equation, truncated at the lowest nonlinear terms, is given by

$$\frac{da_\alpha}{dt} = (i\omega_\alpha + \kappa_\alpha)a + Q|a_\alpha|^2 a_\alpha \tag{9}$$

Introducing into Eq. (9) the slowly varying amplitude-modulus and phase, defined by $\tilde{a}_\alpha(t) = A_\alpha(t)\exp(i\dot{\phi}_\alpha(t))$ we obtain

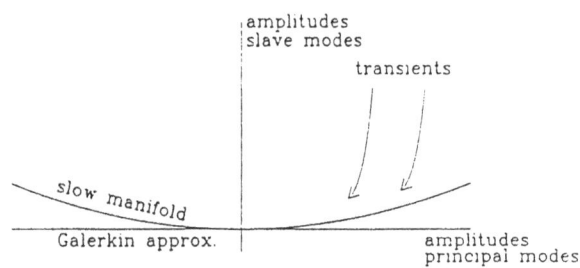

Fig. 1.   The slow manifold.

$$\frac{dA_\alpha}{dt} = \kappa_\alpha A_\alpha + Re(Q_\alpha)A_\alpha^3 \tag{10}$$

When $Re(Q_\alpha) < 0$, which is the usual situation in stellar pulsations, this equation has two *fixed points*, i.e., values for which $dA_\alpha/dt = 0$. The first one, $A_\alpha = 0$, represents the stable equilibrium model, whereas the second one, $A_\alpha^2 = -\kappa_\alpha/Re(Q_\alpha)$, represents a *limit cycle*, or pulsation of the stellar model with constant amplitude. The point $\kappa_\alpha = 0$ is the *Hopf bifurcation* point or, in the parlance of stellar pulsations, the *blue edge* for the sequence.

One also obtains an equation for the phase $\phi_\alpha(t)$, given by $d\phi_\alpha(t)/dt = Im(Q_\alpha)A_\alpha^2$ which for the limit cycle has the trivial solution $\phi_\alpha(t) = Im(Q_\alpha)A_\alpha^2$. When inserted into Eq. (8) it represents the nonlinear correction to the frequency, $\omega_\alpha^{NL} = \omega_\alpha + Im(Q_\alpha)A_\alpha^2$.

It is easily shown (*e.g.*, Buchler 1985) that the cubic saturation coefficient Q_α is a (specific) combination of the cubic term and products of the quadratic terms from Eq. (4), *viz.*

$$Q_\alpha = 3n_{\alpha\alpha\alpha-\alpha} + \frac{2}{i\omega_\alpha}\sum_k(-2n_{\alpha\alpha k}n_{k\alpha-\alpha} + n_{\alpha-\alpha k}n_{k\alpha\alpha}) \tag{11}$$

$$z(t) = (a_\alpha(t)e_\alpha + c.c.) + \frac{\alpha}{i\omega_\alpha}\sum_k\left(n_{k\alpha\alpha}a^2 - c.c. - n_{k\alpha-\alpha}|a_\alpha|^2\right)e_k \tag{12}$$

where the sums extend over *all* modes, principal as well as slaves.

These results are very general and are derived from the full dynamical system (Eq. 1 or Eqs. 4). It is perhaps instructive to compare these results to the Galerkin approximation. If a corresponding dimensional reduction is performed on the Galerkin approximation (Eq. 5) we obtain the same amplitude equation (Eq. 9), but the sums both for the cubic term and for the correction to z now reduce to a single term, $k = \alpha$. The Galerkin solution vector $z(t)$ therefore lies in the space spanned by e_α and its complex conjugate. In contrast, in the amplitude equation formalism the second order and higher terms in $z(t)$ acorrectly extend into all of phase-space (as illustrated in Fig. 1). Of course when more modes are kept in the Galerkin approximation, these modes then also appear in the sums and improve the solution.

3. Applications of the Amplitude Equation Formalism

A number of formal studies of amplitude equations have been made for a variety of physical situations (*e.g.*, Dziembowski and Kovács 1984, Takeuti and Aikawa 1980, 1981, Buchler and Kovács 1986a, b, 1987a, Moskalik 1985, 1986, Buchler and Goupil 1988, and many others). Such studies are very valuable because they elucidate the various types of behavior that can be expected, *e.g.*, fixed points, limit cycles and chaos, corresponding to stellar pulsations with, respectively, constant amplitudes, periodically modulated and erratically modulated stellar pulsations. In addition to the nature of the solutions they also allow an assessment of their stability. In particular, they show the effects of various types of resonances and the bifurcations that they can produce. It is important to note that often these results do not depend on the exact values of the parameters which appear in the amplitude equations, but are valid for broad ranges of values, lending quite general validity to such studies.

On the quantitative side, two types of studies have been made. Truly *ab initio* calculations in realistic stellar models have only been preformed for quadratic coefficients. Dziembowski (1982) and Dziembowski *et al.* (1985, 1988) computed the coefficients appropriate for a resonant condition between nonradial modes (a p mode coupled to two g modes) in δ Scuti stars. Takeuti and Aikawa (1980, 1981) computed the quadratic coupling coefficients in the case of a 2:1 resonance in classical Cepheid models. These last two studies approximated the exact (complex) linear eigenvectors by (real) adiabatic ones. The general expressions for the coupling coefficients in classical Cepheids were computed by Klapp, Goupil and Buchler (1985). The expressions are very complicated an extension to the cubic terms will almost certainly require the use of a symbol manipulation program.

The second, different approach consists of using numerical hydrodynamical computations of the pulsations to extract the values of the nonlinear coupling coefficients. This approach is perhaps less accurate, but it yields some useful results which we shall describe in the next sections. In particular, it allows a quantitative comparison of hydro-model sequences, it sheds new light on the important role played by resonances and it allows a search for specific pulsational behavior in stellar models. For example, the amplitude equations predict quite generally that in the presence of a 2:1 resonance a limit cycle sees its stability decreased, sometimes to the point of instability. Such general guidance from amplitude equations allowed Kovács and Buchler (1988a) to conduct a successful numerical hydrodynamical search for persistent beat pulsations.

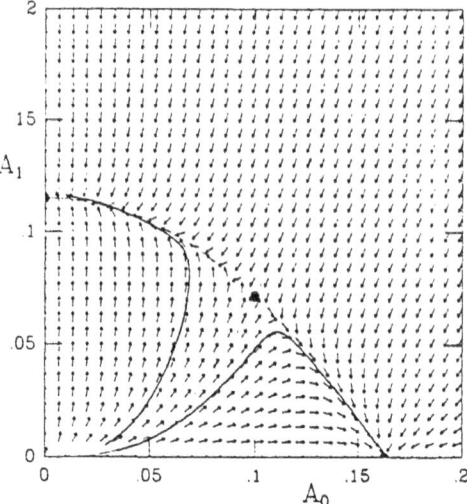

Fig. 2.   Flow field for nonresonant two-mode case; Separatrices: *thin lines*; Evolution of hydro-model: *thick lines*.

3.1. QUANTITATIVE STUDIES

The question arises how good the amplitude equation formalism is from a *quantitative* point of view. We shall now illustrate this for the case of a nonresonant RR Lyrae model with stellar parameters, $0.45 M_\odot$, $50 L_\odot$, 5900K (taken from Buchler and Kovács 1987a). The model is linearly unstable in the fundamental (0) and first overtone (1) and has stable limit cycles in these two modes (talking about a "limit cycle in a mode" is appropriate because the *nonlinear* limit cycles have a predominant projection onto the space spanned by their respective *linear* eigenvectors, and additionally their frequencies are also very close to the linear ones).

After disturbing the static equilibrium model we can follow its evolution with the numerical hydro-code. A frequency analysis shows that after a very short transient only two principal frequencies survive and that they are very close to the linear fundamental and first overdone frequencies. A time-dependent Fourier decomposition of, say the stellar radius $R_*(t)$, then yields the time-dependent amplitudes $A_0(t)$ and $A_1(t)$ and the phases $\phi_0(t)$ and $\phi_1(t)$. The amplitudes are plotted as thick lines in an amplitude-amplitude plot in Fig. 2 for two separate initializations of the hydro-code. One notes that in each case both amplitudes first increase before evolving toward a fundamental and first overtone limit cycle, respectively. This evolution will be considered the experimental "data". We will show that this somewhat unexpected behavior of the amplitudes is explained not only qualitatively, but also very well quantitatively by the apposite amplitude equations.

The amplitude equations for the *nonresonant two-mode* case to lowest order are given by

$$\frac{dA_0}{dt} = \kappa_0 A_0 + Re(Q_{00})A_0^3 + Re(Q_{01})A_1^2 A_0$$

$$\frac{dA_1}{dt} = \kappa_1 A_1 + Re(Q_{11})A_1^3 + Re(Q_{10})A_0^2 A_1 \qquad (13)$$

The integral curves of these equations depend on the 2 growth-rates and on the 4 nonlinear cubic saturation coefficients. These are to be determined by requiring a fit of the integral curves of the amplitude equations to the "data". Details of the procedure can be found in Buchler and Kovács (1987a). The structure of the 2-D phase-space is perhaps best seen from the flow field of Eqs. (13) (with the fitted coefficients) which is shown in Fig. 2. The thin solid lines represent the separatrices of the flow field. The system has four fixed points, the origin (static model) which is an unstable node point, two stable node points, corresponding to the two limit cycles and a saddle point (double-mode beat solution). The nonmonotonic behavior of the amplitudes is now easily understood: The double-mode fixed point exerts a strong attraction at first, but then repels the integral curves toward the limit cycles. The phases decouple again from the amplitudes as for the single mode case (Eq. 9) and they can be computed when the temporal behavior of the latter is known. We just mention here that their behavior is also very well described by the amplitude equations.

The location and stability of the limit cycles are easily obtained from the amplitude equations. Table 1 shows that they compare very closely with the values obtained from a numerical computation of the limit cycles with a relaxation hydro-code. The quantities $\lambda_1(\lambda_0)$ represent the Floquet exponents for the growth of the first overtone (fundamental) in the fundamental (first overtone) limit cycles, respectively.

TABLE I
Limit Cycle Characteristics

Relaxation code:	$A_0 = 0.166$	$\lambda_1 = -0.039\Pi_1$;	$A_1 = 0.117$	$\lambda_0 = -0.0093\Pi_0$
AE fit:	$A_0 = 0.163$	$\lambda_1 = 0.033\Pi_1$;	$A_1 = 0.115$	$\lambda_0 = -0.0098\Pi_0$

We conclude that despite the very complicated input physics (equation of state with H and He ionization, realistic expression for opacity) the hydrodynamical evolution of the stellar model takes place in a 2-D space (4-D when the decoupled phases are included). This evolution is quite accurately described by the system of two nonresonant amplitude equations for the fundamental mode and the first overtone, truncated at the lowest order (cubic) nonlinearities. In addition, because of the simplicity of these equations all its fixed points and their stability can readily be studied analytically (Buchler and Kovács 1986b).

3.2. EFFECTS OF RESONANCES

Resonances have a strong effect on the appearance and stability of the pulsations. A model is said to be resonant when there exists a relation of the form $\sum l_k \omega_k \approx 0$. ($l_k$ positive or negative integers) between the frequencies of the eigenmodes. In general, only the resonances of low order, i.e., for which $\sum_k |l_k|$ is small, can have an appreciable effect.

Since the two-mode resonances play a particularly important role we shall examine them in some detail. Let the resonance be characterized by the condition $n\omega_\alpha \approx m\omega_\beta$. The apposite amplitude equations are given by

$$\frac{da_\alpha}{dt} = \sigma_\alpha a_\alpha + Q_{\alpha\alpha}|a_\alpha|^2 a_\alpha + Q_{\alpha\beta}|a_\beta|^2 a_\alpha + P_\alpha a_\alpha^{*n-1} a_\beta^m$$

$$\frac{da_\beta}{dt} = \sigma_\beta a_\beta + Q_{\beta\beta}|a_\beta|^2 a_\beta + Q_{\beta\alpha}|a_\alpha^2|a_\beta + P_\beta a_\alpha^n a_\beta^{*m-1} \qquad (14)$$

The nonlinear coupling coefficients P_γ and $Q_{\gamma\delta}$ are complex in general.

Let modes α and β be linearly unstable and stable, respectively. Under these conditions, in the absence of a resonance these equations would reduce to Eqs. (13) and the system would have a single stable limit cycle corresponding to mode α.

We can distinguish three types of solutions depending on the value of m.

3.2.1. m=1, integer resonances: $n\omega_\alpha = \omega_\beta$

The characteristics of this type of resonance are that $A_\beta \neq 0$, always, and that the fixed points represent phase-locked (or synchronized) solutions, which are therefore singly periodic. These resonances thus cause a distortion of the light and velocity curves (e.g., bumps and shoulders). They have been found to play a major role in stellar pulsations.

The 2:1 resonance between the fundamental mode and the second overtone ($2\omega_0 \approx \omega_2$) has been found to be responsible for the Hertzsprung progression in the classical Cepheids. This was first conjectured by Simon and Schmidt (1976), but a full understanding had to await the development of the amplitude equation formalism. The observational low order Fourier phases both for the magnitude and for the radial velocity curves show a great deal of structure (e.g., Simon and Moffett, 1985 for the light curves and Kovács et al., 1990 for the radial velocity curves). It has been shown that cubic amplitude equations with terms describing the effects of the 2:1 resonance reproduce rather well this behavior of the Fourier phases (Kovács and Buchler 1989). Physically, the resonant, linearly stable second overtone gets entrained through its resonance with the fundamental mode. Because of phase-lock the solution remains periodic and the excitation of the second overtone manifests itself through the appearance of a secondary maximum on the light and radial velocity curves whose position varies with the proximity of the resonance. The amplitude equation formalism is the only formalism that is capable of explaining the Hertzsprung progression. which must be considered one of its major successes to date. We note that the same resonance is responsible for essentially identical progressions in BL Her model sequences (Buchler and Moskalik 1992).

The higher order. 3:1 resonance ($3\omega_0 \approx \omega_4$) occurs in low period classical Cepheids and in BL Her stars and gives rise to a deformation of the Fourier phases and amplitudes (Kovács and Buchler 1989. Buchler and Moskalik 1992). Moskalik and Buchler (1989) have analyzed the formal solutions of the amplitude equations, but a quantitative study has not been undertaken yet because this resonance overlaps with the wing of the 2:1 resonance and a fit involves too many terms to be practical.

3.2.2. m=2, half-integer resonances: $n\omega_\alpha = 2\omega_\beta$

The characteristics of this resonance are that they either have no effect on the pulsation (*i.e.*, $A_\beta = 0$) or they lead to a parametric excitation of the resonant overtone, depending on the parameters of the amplitude equations. In the latter case they lead to phase-locked period 2 pulsations because their period Π is given by $\Pi = n\Pi_\beta = 2\Pi_\alpha$). In other words these resonances affect the stability of the limit cycles. but not their appearance.

The 1:2 resonance has been found to be important in δ Scuti stars (Dziembowski 1982). Actually, he studied the 3-mode resonance between a p mode and tow g modes, but because the two g modes are required to have very similar frequencies, the 3-modes resonance has the same properties as the 2-mode resonance. Because of the large number of such resonances Dziembowski finds that amplitude saturation in these stars occurs through this resonance, rather that through the cubic terms; this has as a consequence a lower saturation amplitude. His numerical calculations of realistic stellar models yield amplitudes in agreement with observed ones. In an extension of this work Dziembowski *et al.* (1985, 1988) propose that the shortage of observed rapidly rotating variables is due to increased chances of resonances due to rotational splitting. and thus to unobservably low saturation amplitudes.

All numerical hydrodynamical studies of classical Cepheid model sequences found windows in which the pulsations displayed steady strictly periodically alternating cycles. This behavior has been trace to the 3:2 resonance ($3\omega_0 \approx 2\omega_1$) and the appropriate amplitude equations again give an analytical explanation of this behavior. Subsequently the same resonance was also found to be responsible for similar windows in BL Her models (Buchler and Moskalik 1992). In W Vir models a 5:2 resonance ($5\omega_0 \approx 2\omega_2$) is associated with period-doubling and the subsequent cascade of subsequent period-doublings to chaos which have been found in the radiative hydrodynamical models. This will be further discussed in §4.

3.2.3. m > 2, integer resonances:

As in the case $m = 1$ the fixed points of the amplitude equations represent phase-locked solutions with $A_\beta \neq 0$. These solutions are not as likely to be important because, generally speaking, the effects of a resonance decrease with the order. Such a criterion must be used with caution, however; the previous section showed that a relatively high order resonance (5:2) was found responsible for a period 2 bifurcation. So far no hydrodynamical calculations have shown any good evidence for the importance $m > 2$ resonances.

Generally speaking we still have a relatively poor *a priori* understanding of when a particular resonance is going to have an important effect on the pulsation.

3.2.4. Other Resonances

The 3-mode resonance $\omega_\alpha + \omega_\beta \approx \omega_\gamma$, for a while was thought to be the cause of beat behavior. Although this resonance can in principle give rise to beat behavior (*e.g.*, Takeuti and Aikawa 1980, Kovács and Kolláth 1988), in fact, no evidence thereof has been found in hydrodynamical models. Other resonances which may turn out to play an important role are the higher order 3-mode resonances $\omega_\alpha + \omega_\beta \approx 2\omega_\gamma$ and $\omega_\alpha + \omega_\beta \approx 3\omega_\gamma$ (*cf.* Kovács in this Volume).

3.2.5. Overlapping Resonances

In many real situations several resonances are sufficiently close so that their combined effect may give rise to very complicated behavior. For example, a particularly dramatic accumulation of resonances occurs in very low temperature BL Her models (Buchler and Moskalik 1992). The corresponding amplitude equations become very cumbersome, contain many coefficients and are very difficult to study. As our modelling of stellar pulsations progresses the next stage of refinement will require the consideration of resonance overlaps.

4. The Nature of Chaos in W Vir Stars

4.1. HYDRODYNAMICAL MODELS

The standard astronomical techniques are not adequate for understanding the behavior of W Vir model sequences. However techniques from nonlinear dynamics can be used to decipher the nature of the observed bifurcations, and amplitude equations help clarify the physical situation.

The hydrodynamical study of W Vir model sequences by Buchler and Kovács (1987b), Kovács and Buchler (1988b) found that several sequences

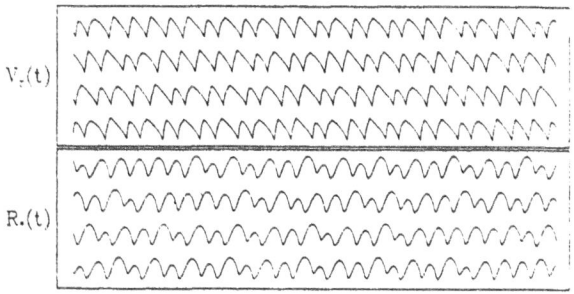

Fig. 3.   Temporal behavior of the radial velocity and of the stellar radius for the W Vir model.

of models undergo text-book successions of *period-doublings* (*cf.* Cvitanovich 1986, Bergé *et al.* 1986) as a control parameter. here T_{eff}, is gradually lowered. For higher luminosity sequences. on the other hand. the transition to chaos occurs through a *tangent bifurcation* (Aikawa 1987, 1988. Buchler. Goupil and Kovács 1987, Kovács and Buchler, 1988). Since this transition to chaos can happen after just a few a few period-doublings (Aikawa 1990) this seems to imply that the bifurcation diagram folds backward. Instead of displaying the metamorphosis of the radius or radial velocity curves here, we refer the reader to the quoted papers (*cf.* also Aikawa's review in this Volume, or Buchler 1990).

Here we merely examine one such W Vir model, with stellar parameters $M = 0.6M_{\odot}$, $L = 500L_{\odot}$, $T_{eff} = 4200K$ and a Pop. II composition of $X = 0.745$, $Z = 0.005$. The model comes from a sequence which has undergone a typical period-doubling cascade to chaos and its inverse chaotic undoublings. Figure 3 displays four short consecutive stretches of the temporal behavior of the stellar radius $R_{*}(t)$ (radius of the outermost zone) at the bottom and of the radial velocity $V_r(t)$ at the top. The fluctuations are much more apparent in the radius than in the velocity. Referring to the former we notice occasional intervals over which the oscillation is almost singly periodic and other intervals over which it exhibits strong RV Tau-like behavior.

The Fourier power spectrum, taken over 400 pulsations, is exhibited in Fig. 4. It displays a remarkably sharp peak at the fundamental frequency. but has a strong sub-harmonic structure, indicative of something interesting. Had the spectrum been obtained with gapped data beset with observational noise one might have concluded that the oscillation is periodic. However an O-C diagram, shown in Fig. 5, displays fairly large phase variations. up to 10%.

Perhaps the most powerful technique of the dynamicists is the phase-space reconstruction of an attractor. It allows one to construct a topologically equivalent structure from a *single* temporal signal: barring pathological cases, this can be any variable, observationally. for example the magnitude

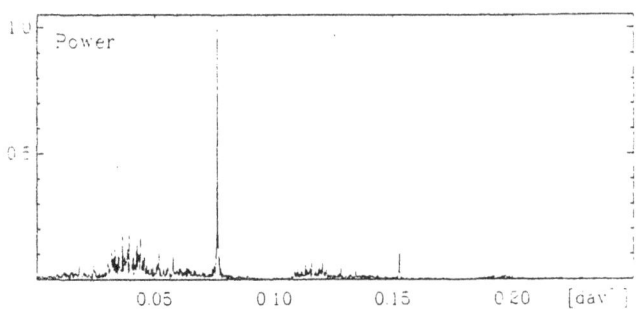

Fig. 4.   Power spectrum for W Vir model of Fig. 3.

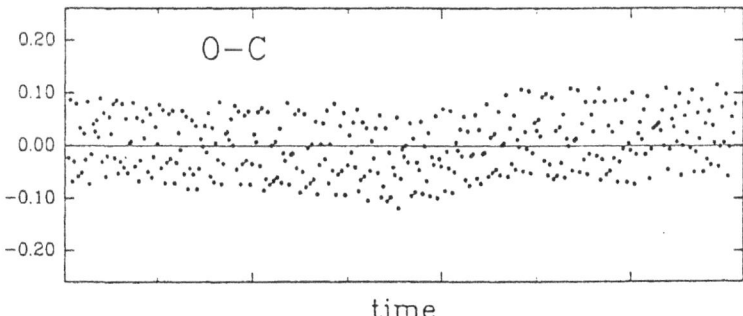

time

Fig. 5.   O-C diagram for W Vir model of Fig. 3.

or the radial velocity. For the numerical hydrodynamical output we find it convenient to use the stellar radius. While a mathematical theorem says that any n-D dynamic can be embedded in a (2n+1)-D space, in practice it has been found that many physical systems are not pathologicl and an n-D embedding space is sufficient. Indeed, while a limit cycle (single loop) is embeddable in 2 dimensions (it can be represented by 2 first order ODEs), a period 2 cycle (double loop) clearly cannot occur in a 2-D space because this would require an intersection point for the trajectory (which is not allowed by the uniquencess theorem of ODEs). A period-doubling therefore implies that the dimension of phase-space must be greater that 2 and that the reconstruction must at least be made in a 3-D embedding space.

In Fig. 6 we display such a 3-D phase-space reconstruction of the dynamic obtained by plotting the triplets of values, $\{R(t), R(t+\tau), R(t+2\tau)\}$ with τ equal to 40% of the pulsation "period". The signal shows a remarkably tight structure, strongly reminiscent of the *Rössler attractor*. (We note that a multi-periodic signal would have given rise to a space-filling structure, *cf.* *e.g.*, Fig. 9 in Kovács and Buchler 1988b). The Rössler attractor of course arises in a 3-D embedding space since it is generated by a set of 3 ODEs. It is therefore of interest to attempt to compute the dimension of our attractor. The Grassberger-Procaccia correlation method yields a low dimension, but

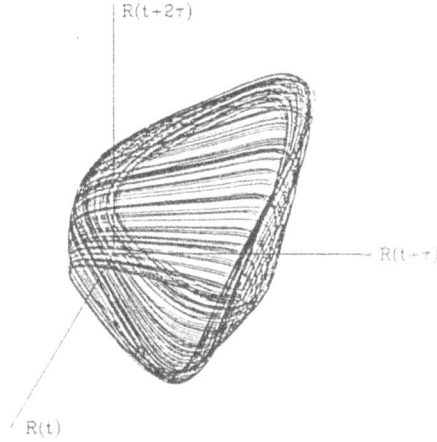

Fig. 6. 3-D phase-space reconstruction for W Vir model of Fig. 3.

is rather inaccurate and beset by uncertainties. An analysis based on a dynamical systems prediction method (Serre and Buchler 1992) also gives a low dimension, probably between 3 and 5. While the exact value of the embedding dimension is not certain, there seems to be little doubt that it is small.

We note in passing that there are many other useful techniques for exhibiting low dimensional structure, such as Poincaré sections, 1-D return maps, Lyapunov exponents, entropy methods, *etc.*

Two important questions arise. First, since the dimension of phase-space must be increased, in our modal description this necessitates the *excitation of another mode.* In the Rössler attractor this additional mode is necessarily a real, nonoscillatory mode. We will show that in our case, in contrast, it is oscillatory rather than secular. The second question concerns the robustness of period-doubling. Indeed, it occurs in many sequences of models and with a very different code (Aikawa 1987). It even survives when the heat flux includes time-dependent convection (Glasner and Buchler 1990).

A clue as to the origin of period-doubling comes from the classical Cepheid models. First of all, every sequence of classical Cepheid models (Buchler, Moskalik and Kovács 1990) has windows of strictly period 2 behavior in some range of T_{eff}, and a linear stability analysis of the models reveals a correlation of this behavior with the location of the half-integer resonance $3\omega_0 \approx 2\omega_1$. Second, a period-doubling bifurcation indicates that one of the Floquet coefficients $F_k = \exp(i\Phi_k + \lambda_k)$ must cross the negative real axis at -1, or that the Floquet phase, Φ_k passes through π. Since in the lowest approximation, $\Phi_k = \omega_k \Pi_0 = 2\pi \Pi_0 / \Pi_k$, we may anticipate that the bifurcation is associated with a half-integer resonance condition $\Pi_0 / \Pi_k = n/2$,

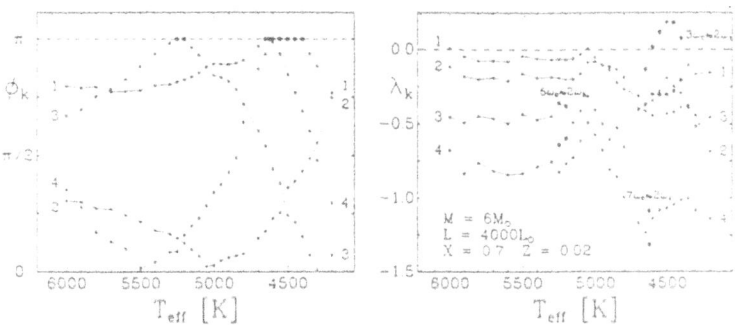

Fig. 7.   Low order Floquet phases and coefficients for classical Cepheid sequence.

where n is integer. The bifurcation itself, however, is strictly nonlinear, and to uncover its true nature we need to turn to numerical hydrodynamics and to the amplitude equation formalism.

It is found that the results of the Floquet stability analysis of the limit cycles for a sequence of Cepheid models (sequence I of Moskalik and Buchler 1990, 1991) display very characteristic behavior, namely bubbles in the Floquet exponents and concomitant plateaux in the phases, which are connected with the first overtone. Figure 7 presents the phases Φ_k and exponents λ_k of the Floquet coefficients for the lowest few modes as a function of the control parameter T_{eff}. (How this mode identification is carried out is described in Moskalik and Buchler 1991). Again it is seen that the bubbles and plateaux are associated with the linear modes which are in a half-integer resonance condition, as indicated in Fig. 7. One notes, however, that only for the first overtone does the Floquet exponent λ_1 pierce the stability boundary ($\lambda_1 = 0$) and give rise to period 2 behavior. As we have seen the classical Cepheids are weakly dissipative and the conditions for the validity of the amplitude equation formalism are satisfied. From Eqs. (14) it follows that the proper resonance terms to be added to Eqs. (13) are $P_0 a_0^{\varkappa n-1} a_k^2$ and $P_k a_k^{\varkappa n-1} a_0$, respectively. Moskalik and Buchler (1990) have shown that with the simplifying assumption of $Q_{01} = 0$ and of constant nonlinear coupling coefficients along the sequence as the resonance is traversed, the fixed points of these equations can be obtained analytically.

The amplitude equations predict that the Floquet coefficient for the resonant overtone, and the corresponding phase and exponent, behave as shown in Fig. 8. This is exactly the behavior seen in the Floquet coefficients of the numerical hydro-models.

For the W Vir models, on the other hand, we are beyond the range of validity of the amplitude equation formalism with $e.g.$, $\kappa_0 \Pi_0 \approx 0.20$. In order to obtain the cascade of period-doublings and chaos it would be necessary to introduce non-"normal" terms in the amplitude equations which

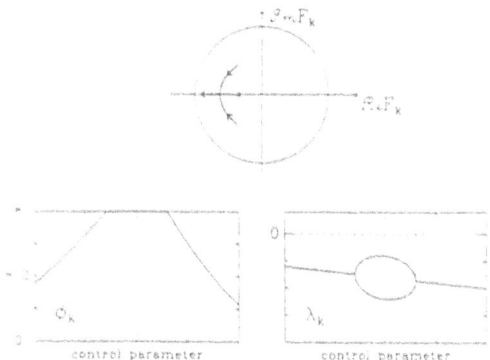

Fig. 8. Schematic behavior of the Floquet coefficient near the instability.

couple the oscillation to the modulation which now varies on the same time-scale. At the present time there is no known procedure for selecting such non-normal terms among all the nonlinear terms. Still. the Floquet analysis of the model sequences displays the same general type of behavior as for the classical Cepheids (Fig. 2 of Moskalik and Buchler 1991) and the same association of the bubbles and plateaux with the half-integer resonances. In these models it is the resonance $5\omega_0 \approx 2\omega_2$ that is responsible for the parametric excitation of the overtone. Now, however. inside the most important unstable bubble further period-doubling bifurcations and a transition to chaos occur. Still, it is the first period-doubling bifurcation which increases the dimension of phase-space and allows the subsequent period-doubling cascade to chaos to occur. We also note that a set of two coupled oscillators, designed to undergo a half-integer resonance, display the same type of period-doublings and chaos, including tangent bifurcations, as are observed in the hydrodynamical sequences.

We conclude that the bifurcation to period 2 behavior occurs because of the parametric excitation of a vibrational overtone which is brought about by a half-integer resonance. It is this association with a resonance which makes the occurrence of period-doubling so robust. Indeed, the locations of the resonances generally vary slowly and smoothly as the stellar parameters are varied, and they are not very much affected either by the numerical methods.

4.2. OBSERVATIONS

Traditional astronomical methods are well suited for analyzing periodic and multi-periodic signals, but they are inadequate when the purpose is to detect low-dimensional nonlinear behavior, and to determine its nature. It is true that almost 40 years ago small alternations were observed for the long period W Vir stars ($15-35^d$) in globular clusters by Arp (1955) in the light curves, and subsequently confirmed by Wallerstein (1958) in the radial velocities; similarly Lloyd-Evans (1971) reports such behavior in SZ Mon. Unfortunately, these investigations were not pursued, perhaps for want of suitable techniques of analysis and theoretical motivation. The observed alternations are very gentle in contrast to those found in the more luminous, longer period RV Tauri stars, and they appear very similar to the alternations in our hydrodynamical models. A systematic observation of long period W Vir stars has exciting prospect of confirming the period-doubling scenario.

Recently some attempts have been made to search for chaotic behavior. In the stellar context we note the work of Goupil et $al.$ (1988) on white dwarfs, of Canizzo et $al.$ (1990) and of Yanagida et $al.$ (in this Volume) on long-period variables and of Saitou, Takeuti and Tanaka (1989) on semiregular stars. The only really thorough study is that of Kolláth (1990) who, by applying a myriad of techniques to the study of 150 years of data on R Sct gives credible evidence for low dimensional chaos in this star.

Generally, the application of modern analyses are hampered by large observational noise and by erratically timed and gapped observations ($cf.$ Baglin in this Volume). Phase-space reconstructions require observations at equal time intervals, or at least data which can be interpolated as such with good accuracy. In order to make theoretical progress we have to convince the observers for the need of long-term observational programs of specific stars with good phase coverage, possibly multi-site to avoid too many gaps. Because of their relatively short periods and of the irregularities W Vir stars seem ideal candidates for that purpose.

5. Stochastic Effects

So far we have assumed that the model around which we expand Eq. (1) is truly static. In reality, the stellar interior is not quiescent, but undergoes complicated convective or turbulent motions. The latter can be considered to be the result of the nonlinear interaction of a large number of convective modes (of an originally truly static model), in static on average and treat the fluctuations as stochastic noise. We then substitute

$$\frac{\partial z}{\partial t} = \mathcal{L}(t)z + \mathcal{N}(z,t) + \Xi(t) \tag{15}$$

for Eq. (1), where the function $\Xi(t)$ represents additive noise and the time-dependence in $\mathcal{L}$ and $\mathcal{N}$ is parametric noise.

The amplitude equation formalism can be generalized to handle this situation (Stratonovich 1965, Buchler, Kovács and Goupil 1992) provided that, in addition to weak nonlinearity and weak dissipation, we assume (a) that the correlation time of the noise, τ_c is much smaller than the time-scale of modulation of the principal amplitudes, $i.e.$, $\tau_c \ll \tau_m \sim 1/\kappa_0$ and (b) that the stochastic processes are stationary, a reasonable assumption. Generally, τ_c is expected to be smaller than the period, therefore *a fortiori* smaller than τ_m. The result is a *Fokker-Planck equation* for the *probability distribution of the amplitudes* $w(\{a_k\})$.

If we look at the asymptotic case when the distribution has become stationary, and if, for illustration, we consider the 2 mode nonresonant case with additive noise only, the distribution $w(A_0, A_1)$ satisfies

$$\frac{\partial}{\partial A_0}\left[\left(\kappa_0 A_0 + Re(Q_{00})A_0^3 + Re(Q_{01})A_1^2 A_0 + \frac{C_0}{A_0}\right)w - C_0\frac{\partial w}{\partial A_0}\right]$$
$$+\frac{\partial}{\partial A_1}\left[\left(\kappa_1 A_1 + Re(Q_{11})A_1^3 + Re(Q_{10})A_0^2 A_1 + \frac{C_1}{A_1}\right)w - C_1\frac{\partial w}{\partial A_1}\right] = 0 \ (16)$$

where C_k is related to the spectral density of the noise.

It is well known that noise can have a critical effect on the behavior of a system near bifurcation points (*e.g.*, Moss and McClintock 1989). It is therefore of interest to explore how the modal selection problem is affected by noise, in particular whether it can generate beat behavior (Buchler and Kovács 1992; *cf.* also Kovács in this Volume).

Here we just give an example of how noise affects the behavior of an RR Lyrae model similar to that treated in section 3.1 (in particular Fig. 3). For a very low noise level the distribution has sharply spiked peaks at the fixed points of the noiseless amplitude equations. As the noise strength increases these peaks broaden and concomitantly move away from the axes. This leads to what is called precursor-noise in which the fluctuations are of the same order as the average amplitude of the mode. As pointed out by Kovács (in this Volume) the small, slowly varying overtone amplitudes detected by Walraven *et al.* in AI Vel could well be an example of such a stochastic excitation.

Noise can not only affect the appearance of the pulsation, it can also provoke qualitative changes in the nature of the pulsation. In Fig. 9 we show the distribution function for four increasing noise-levels (*a* to *d*, from left to right, top to bottom). Clearly, the peaks correspond to the limit cycles of the noiseless case. Although the peaks occur at a finite distance from the axes, *i.e.*, although both amplitudes are nonzero on average, the amplitude of the secondary mode remains of precursor type throughout. This situation therefore does not correspond to a true beat behavior. One notes, however, that sufficiently large noise can cause the disappearance of one of the two original states.

A more interesting situation comes about when the two linearly unstable modes are coupled to a large number of linearly stable modes which are stochastically driven. It then becomes possible for noise to convert the two peaks (single-mode limit cycles) into saddle points (unstable double-mode cycles) and the original saddle point into a peak. It is remarkable that the

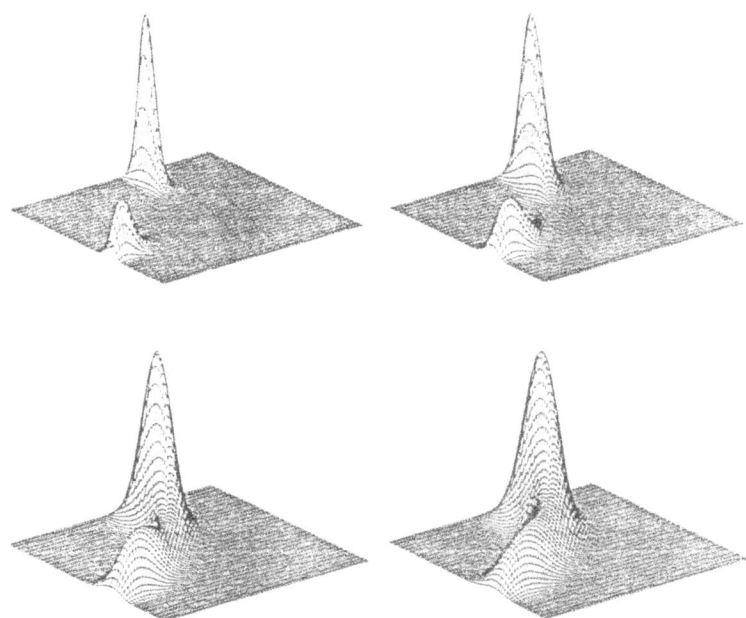

Fig. 9.   Probability densies for nonresonant two-mode case; origin at the leftmost corner; A_1 and A_2 axes upward and downward, resp.; increasing noise level from left to right and top to bottom.

primary amplitudes of this double-mode solution can be made to have arbitrarily small fluctuations when the number of secondary modes is made
very large (Buchler and Kovács 1992). Numerical hydrodynamical tests with
additive noise confirm the predictions of the Fokker-Planck treatment.

The astrophysical applications of this formalism are still in their infancy
and have interesting potential applications both in radial and nonradial stellar pulsations. In particular, it is well known that *parametric noise* which we
have not considered here can have very dramatic effects on the bifurcations
of the models (Moss and McClintock 1989).

6. Prospects

The application of techniques from dynamical systems has provided a new
outlook on the problem of nonlinear stellar pulsations. The amplitude equation formalism not only gives a more basic understanding of the nature of
the pulsations in terms of bifurcations, it also clarifies the role played by
specific resonances. However, much more remains to be learned about the
relative importance of resonances and especially about the consequences of
their overlaps. It has also become clear that new tools of data analysis are
necessary to understand not only the behavior of numerical hydrodynamical

models, but also of observational data. It is hoped that the new theoretical developments will stimulate an observational effort specifically geared toward detecting an analyzing nonlinear effects.

We have not even touched on some areas in which dynamical systems techniques have also been applied, or where they are expected to make an impact. For example, the problems of the interaction between the pulsation and rotation, magnetic fields, and convection are clearly going to receive increase attention. Astrophysical disks are expected to be another active area of application. Perhaps one of the most exciting topics is the study of spatio-temporal structures, their formation, propagation, interaction and destruction with applications to convection, turbulence, magneto-hydrodynamics and dynamo theory. Finally, another topic in nonlinear science is that of patterns and fractal structures for which astrophysics is destined to be a fertile ground.

Acknowledgements

It is a great pleasure to acknowledge the contributions of my long-time collaborators Géza Kovács, Pawel Moskalik and Marie-Jo Goupil. This work has been supported by NSF (grant AST89-14425), by the Pittsburgh Supper-computing Center, and by an RCI grant through IBM and the NE Regional Data Center at the University of Florida.

References

Aikawa, T.: 1987, *Astrophysics and Space Science* **139**, 281.
Aikawa, T.: 1988, *Astrophysics and Space Science* **149**, 149.
Aikawa, T.: 1990, *Astrophysics and Space Science* **164**, 295.
Arp, H. C.: 1955, *Astronomical Journal* **60**, 1.
Baker, N. H.: 1966, in *Stellar Evolution*, eds. R. F. Stein and A. G. W. Cameron (New York: Plenum Press), p. 333.
Baker, N., Moore D. and Spiegel E. A.: 1966, *Astronomical Journal* **71**, 845.
Barranco, M., Buchler, J. R., and Livio, M.: 1981, *Astrophysical Journal* **242**, 1226.
Barranco, M., Buchler, J. R., and Regev, O.: 1982, *Astrophysics and Space Science* **84**, 463.
Bergé, P., Pomeau, Y. and Vidal, C.: 1986, *Order Within Chaos* (N. Y.: Wiley).
Buchler, J. R.: 1985, in *Chaos in Astrophysics*, NATO ASI Ser. C, Vol. 161, eds. J. R. Buchler, J. M. Perdang and E. A. Spiegel (Dordrecht : Reidel), p. 137.
Buchler, J. R.: 1990, in *Nonlinear Astrophysical Fluid Dynamics*, eds. J. R. Buchler and S. T. Gottesman, *Ann NY Acad Sci*, 617, p.17.
Buchler, J. R. and Goupil, M. J.: 1984, *Astrophysical Journal* **279**, 394.
Buchler, J. R. and Goupil, M. J.: 1988, *Astronomy and Astrophysics* **190**, 137.
Buchler, J. R., Goupil, M. J. and Kovács, G.: 1987, *Phys. Letters* A**126**, 177.
Buchler, J. R. and Kovács, G.: 1986a, *Astrophysical Journal* **303**, 749.
Buchler, J. R. and Kovács, G.: 1986b, *Astrophysical Journal* **308**, 661.
Buchler, J. R. and Kovács, G.: 1987a, *Astrophysical Journal* **318**, 232.
Buchler, J. R. and Kovács, G.: 1987b, *Astrophysical Journal, Letters to the Editor* **320**, L57.

Buchler, J. R. and Kovács, G.: 1992, *Physica D*, submitted.

Buchler, J. R., Kovács, G. and Goupil, M. J.: 1992, *Astronomy and Astrophysics* submitted.

Buchler, J. R., Moskalik, P. and Kovács, G.: 1990, *Astrophysical Journal* 351, 617.

Buchler, J. R. and Moskalik, P.: 1992, *Astrophysical Journal* 391, (in press, May).

Buchler, J. R. and Perdang, J., 1979, *Astrophysical Journal* 231, 524.

Buchler, J. R. and Regev, O.: 1982, *Astrophysical Journal* 263, 312.

Buchler, J. R., Yueh, W. R., and Perdang, J.: 1977, *Astrophysical Journal*, 214, 510.

Canizzo, J. K., Goodings, D. A. and Mattei, J.: 1990, *Astrophysical Journal*, 357, L31.

Coullet, P. and Spiegel, E. A.: 1984, *SIAM J. Appl. Math.*, 43, 776.

Cvitanovich, P.: 1984, *Universality in Chaos*, (Bristol: Adam Hilger).

Dziembowski, W.: 1980, in *Nonradial and Nonlinear Stellar Pulsations*, eds. H. A. Hill and W. A. Dziembowski (Springer), p. 22.

Dziembowski, W.: 1982, *Acta Astr.* 32, 148.

Dziembowski, W. and Kovács, G.: 1984, *Monthly Notices of the RAS* 206, 497.

Dziembowski, W. and Krolikowska, M.: 1985, *Acta Astr.* 35, 5.

Dziembowski, W., Krolikowska, M. and Kosovitchev, A.: 1988, *Acta Astr.* 38, 61.

Glasner, A. and Buchler, J. R.: 1990, in *The Numerical Modelling of Nonlinear Stellar Pulsations; Problems and Prospects*, ed. J. R. Buchler, NATO ASI Ser. C302 (Kluwer, Dordrecht), p. 109.

Guckenheimer J. and Holmes P.: 1983, *Nonlinear Oscillations, Dynamical Systems and Bifurcation Theory*, (Springer, NY).

Goupil, M.-J., Auvergne, M. and Baglin, A.: 1988, *Astronomy and Astrophysics*, 196, L13.

Klapp, J., Goupil, M.-J. and Buchler, J. R.: 1985, *Astrophysical Journal*, 296, 514.

Kollath Z.: 1990, *Monthly Notices of the RAS* 247, 377.

Kovács, G. and Kolláth, Z.: 1988, in *Multimode Stellar Pulsations*, eds. G. Kovács et al. (Kultura: Budapest), p. 33.

Kovács, G. and Buchler J. R.: 1988a, *Astrophysical Journal* 324, 1026.

Kovács, G. and Buchler J. R.: 1988b, *Astrophysical Journal* 334, 971.

Kovács, G. and Buchler J. R.: 1989, *Astrophysical Journal* 346, 898.

Kovács, G., Kisvarsányi, E. and Buchler, J. R.: 1990, *Astrophysical Journal* 351, 606.

Livio, M. and Regev O.: 1984, *Astronomy and Astrophysics* 148, 133.

Lloyd, Evans, T.: 1971, *Observatory* 91, 159.

Moore, D. and Spiegel E. A.: 1966, *Astrophysical Journal* 143, 871.

Moskalik, P.: 1985, *Acta Astr.* 35, 229.

Moskalik, P.: 1986, *Acta Astr.* 36, 333.

Moskalik, P. and Buchler J. R.: 1989, *Astrophysical Journal* 341, 997.

Moskalik, P. and Buchler J. R.: 1990, *Astrophysical Journal* 355, 590.

Moskalik, P. and Buchler J. R.: 1991, *Astrophysical Journal* 366, 300.

Moss, F. and McClintock P. V. E.: 1989, *Theory of Continuous Fokker-Planck Systems*, Vols. I and II, Cambridge University Press, (Cambridge, UK).

Papaloizou, J. C. B.: 1973, *Monthly Notices of the RAS* 162, 143; *ibid.* 162, 169.

Perdang, J.: 1983, *Solar Physics* 82, 297.

Regev, O. and Buchler, J. R.: 1981, *Astrophysical Journal* 250, 776.

Saitou, M., Takeuti, M. and Tanaka, Y.: 1989, *Publications of the ASJ* 41, 297.

Serre, T. and Buchler, J. R.: 1992, *Astronomy and Astrophysics* submitted.

Simon, N. R.: 1982, *Astrophysical Journal, Letters to the Editor* 260, L87.

Simon, N. R.: 1986, *Astrophysical Journal* 311, 305.

Simon, N. R. and Lee, A. S.: 1981, *Astrophysical Journal* 248, 291.

Simon, N. R. and Moffett, T. J.: 1985, *Publications of the ASP* 97, 1078.

Simon, N. R. and Schmidt, E. G, 1976, *Astrophysical Journal* 205, 162.

Stratonovich R. L.: 1965, *Topics in the Theory of Random Noise*, Vols. I and II, (New York: Gordon and Breach).

Takeuti, M.: 1985, *Astrophysics and Space Science* 109, 99.

Takeuti, M.: 1990, in *The Numerical Modelling of Nonlinear Stellar Pulsations; Problems and Prospects*, ed. J. R. Buchler. NATO ASI Ser. C302 (Kluwer, Dordrecht), p. 121.

Takeuti, M. and Aikawa T.: 1980. *Monthly Notices of the RAS* **192**, 697.

Takeuti, M. and Aikawa T.: 1981, *Sci Rep Tôhoku Univ.* Eighth Ser. 2, 106.

Tanaka, Y. and Takeuti M.: 1988, *Astrophysics and Space Science* **148**, 229.

Vandakurov, Yu. V.: 1979, *Sov Astr* **23**, 421.

Wallerstein, G.: 1958. *Astrophysical Journal* **127**, 583.

Waltjer, J.: 1936, *Bull. Atr. Inst. Netherlands* **282**, 17: 1937, *ibid.* **303**, 193; 1946, *ibid.* **370**, 125.

PATTERNS OF APERIODIC PULSATION

E. A. SPIEGEL

Astronomy Department, Columbia University, New York, NY 10027, USA

Abstract. Techniques for deriving amplitude equations for stellar pulsation are outlined. For the simplest such equations with multiple instabilities, the derivation of a map for the patterns of pulsation phases is described. This map gives the time between two successive maxima of pulsation in terms of the time between the previous pair, under suitable conditions. The phase differences can be regular, chaotic or hyperchaotic.

1. Introduction

My aim in this paper is to sketch how we can try to understand chaos in stellar pulsation through simple models. The results are still a long way from insight into the bursting and intermittency that we see in many cosmic bodies from the sun to the galaxies, but I hope it is a beginning. The story has evolved slowly, starting with a three paragraph abstract (Baker et al., 1966) published some time ago. I reproduce the first paragraph of that paper in the Appendix. Though different in approach, the present paper is offered as a substitute for a complete paper, which was not written. Nor was the equation on which the work was based published. Nevertheless, I did set out to respond to skeptics who were unwilling to believe that a simple model could produce such complexity or that this could represent real stellar behavior. This paper is a summary of that answer, achieved with the help of many. It describes how to go from the primitive pulsation equations to an algebraic description of chaotic phase variations of pulsation, essentially completely analytically.

In retrospect, it must be admitted that our original derivation of a third-order differential equation for pulsation theory was rough, for all our brash confidence in it. However, we now know how to derive such equations by more formally correct procedures, and I will review that aspect of the problem here before describing how to go on to the reductions that permit analysis of the behavior of their solutions. Such analysis is the part of the theory that has not been discussed in pulsation theory as yet, as for as I know. It shows how even a pulsation that looks periodic can contain chaotic phase variations, as was implied by numerical integrations by Baker and M. C. Depassier on the nonlinear one-zone model. Their theoretical O-C diagrams, reported in a lecture at Columbia in seventies, showed the kind of mild chaos that can now be understood by asymptotic methods. Such results are robust, as is shown in the work of several groups using simple equations to model stellar pulsation theory, and as you will see from other papers in this volume. In particular, the work summarized by Buchler makes possible detailed comparison between the numerical solutions of the full pulsation equations and of the simplified systems.

For those who are already willing to accept the relevance of simple equations, either as models for complicated behavior or as asymptotically correct limits of the full equations. there may be no need to review the derivations. Such readers can simply skip right to the analysis showing how chaos arises. starting with §5. There they will see how to use singular perturbation theory to obtain results that can complement the usual numerical exploration of chaotic solutions. Indeed. I do not describe numerical results at all. Instead I make my points through asymptotic methods that lead to simple algebraic recursion formulae for pulsational phase shifts.

2. Amplitude Equations

In this section I want to recall when and why a description of stellar pulsation in terms of simplified models may be possible. Let us think about the equations of stellar radial pulsation in general from, using $U(\mathcal{M}_r, t)$ to designate the column vector whose components are the usual dependent physical variables of pulsation theory, $\rho, T, r, \dot{r}, \ldots$ and choosing $\mathcal{M}_r$, the mass within radius r, as the independent variable. We describe the form of the equations for radial pulsation as

$$\partial_t U = \mathcal{F}(U, \partial_{\mathcal{M}_r}). \tag{1}$$

A hydrostatic state of the system, U_0, satisfies the condition

$$\mathcal{F}(U_0, \partial_{\mathcal{M}_r}) = 0. \tag{2}$$

We suppose that solutions fulfilling this condition are distinct and we presume that we may ignore the slow changes caused by stellar evolution. These could be incorporated in the treatment at the cost of some complications, but I leave them out for simplicity and as an illustration of the main point of this section.

In considering the dynamics in the neighborhood of one of these static solutions, we introduce the disturbance vector, $u = U - U_0$. We can then generally rearrange (1) into the form

$$\partial_t u = Lu + N(u). \tag{3}$$

where L and N are linear and nonlinear operators whose structure will generally depend on U_0. This dependence may often be expressed in terms of control parameters that are functionals of the static mode, such as surface gravity and helium abundance.

We first look at the linear theory, which begins with the omission of the terms involved in N. Then we may seek solutions of the form $u =$

$a(t)\Phi(\mathcal{M}_r)$. For many of the solutions of the linear theory. the time dependence is simply like $\exp(\lambda t)$ where λ is an allowed value of the characteristic value problem

$$\mathbf{L}\Phi = \lambda\Phi. \tag{4}$$

In general, λ is complex and we write $\lambda = \mu + i\omega$ where μ and ω are real. We restrict ourselves to the case of mild instability, in which no values of μ much greater than zero. The assumption limits the strict range of applicability of our results. Nevertheless, they do provide us with a locally accurate view of the possible behavior and permit its rapid qualitative exploration.

Near to the critical situation, in which the parameters of the problem are carefully chosen so that the allowed values of λ have either $\mu = 0$ or $\mu < 0$, we distinguish between slow modes ψ_n with $|\mu|$ small and fast modes ϕ_m with $\mu < 0$ and of order unity. We assume that there are a few (N) show modes and an infinite number of fast ones. Postponing the issue of completeness to the next section, we decompose the solution vector into normal modes as

$$\mathbf{u}(\mathcal{M}_r, t) = \sum_{n=1}^{N} a^n(t)\psi_n(\mathcal{M}_r) + \sum_{m=1}^{\infty} b^m(t)\phi_m(\mathcal{M}_r). \tag{5}$$

On introducing this expansion into (3) and projecting out the coefficients, we get equations of the general form

$$\frac{d}{dt}\begin{pmatrix} \mathbf{a} \\ \mathbf{b} \end{pmatrix} = \begin{pmatrix} \mathcal{M} & 0 \\ 0 & \mathcal{K} \end{pmatrix}\begin{pmatrix} \mathbf{a} \\ \mathbf{b} \end{pmatrix} + \begin{pmatrix} \mathbf{F}(\mathbf{a},\mathbf{b}) \\ \mathbf{G}(\mathbf{a},\mathbf{b}) \end{pmatrix}, \tag{6}$$

where the $N \times N$ submatrix $\mathcal{M}$ has as proper values all the λ with small $|\mu|$ and the infinite matrix $\mathcal{K}$ has the paper values corresponding to the rapidly decaying modes. We denote the amplitude vectors (a^n) and (b^m) as $\mathbf{a}$ and $\mathbf{b}$.

For a moment, think of a^n and b^m as the abundances of chemical or nuclear species in some well-mixed medium and (6) as the rate equations for the various reactions. The rapidly reacting species (b^m) go quickly to equilibrium, but any changes in the equilibrium state are slow since they are dictated by the instantaneous abundances (a^n) of the slow reactors. So $\mathbf{b}$ is a function of $\mathbf{a}$ and the evolution equation of the system becomes an equation in terms of $\mathbf{a}$ alone. This reasoning, familiar in chemical kinetics, applies to normal modes of the pulsation problem as well, and it is known as center manifold theory (Carr, 1981). It is the same idea as we used when we assumed at the outset that the star is in equilibrium, even though we know that it is slowly evolving.

Formally then, we have $\mathbf{b} = \mathcal{B}[\mathbf{a}]$, and, on introducing this into (6), we get the "amplitude equation"

$$\dot{\mathbf{a}} = \mathcal{M}\mathbf{a} + \mathbf{g}(\mathbf{a}), \tag{7}$$

where $g(a) = F(a, \mathcal{B}[a])$. Since the stable modes would simply die out if they were not driven by the nonlinear couplings to the slow modes, we understand that $\mathcal{B}$ and $\partial_a \mathcal{B}$ vanish for $a = 0$.

The form of $\mathcal{M}$ comes from the linear theory and we take this up next. The specific details of g can be derived from the full problem so it is typically rather complicated to calculate it in numerical detail, but we can work out its general form, once $\mathcal{M}$ is known. In this respect, the situation is like fluid dynamics. We know the form of Navier-Stokes equations quite well, and we do not recalculate the viscosity every time we wish to use the equations for a different particular fluid. In this spirit, we can derive the derivation of the general equations that come up in all nonlinear instability problem, irrespective of details. I have summarized these matters before (Spiegel, 1985), so I shall recall only the main ideas in the next two sections.

3. The Linear Problem

To clarify the physics of ordinary vibrational instability, Baker (1966) isolated the key terms describing the process in his one-zone model. He started from the partial differential equations, or p.d.e.s. for radical motion. By finite differencing, in fact, he reduced these p.d.e.s to ordinary differential equations for the case with only one radial zone. We extended his derivation to the nonlinear case and I set out to write the present review by taking the published abstract of that extension (Baker et al., 1966) as the abstract for this paper. That did not quite work so, inspired by the fifteenth century rabbi who wrote "In may end is my beginning," I have put the first paragraph of the '66 abstract in the appendix.

In seeking to explain aperiodic pulsations, we set out to find a third order nonlinear equation like the one that had come up in overstable convection theory (Moore and Spiegel, 1966). I will not recall our physically motivated derivations here, but shall show how to proceed in the applied mathematical way to get similar results.

The radial oscillations of a spherical star obey an equation of the form

$$\partial_t^2 r = -\mathcal{G}_r + 4\pi r^2 \frac{\partial p}{\partial \mathcal{M}_r} \tag{8}$$

where $\mathcal{G}_r = GM_r/r^2$, $d\mathcal{M}_r = 4\pi r^2 \rho dr$ and p and ρ are pressure and density. Through the pressure and the equation of state, this equation is coupled into the heat equation which contains the luminosity at radius r. Since the latter is a first-order equation, we arrive at a system of third order in time. In the one-zone approximation, one removes the radial dependence from the right hand sides by coarse finite differencing and obtains a third-order ordinary differential equation, or o.d.e. That kind of equation can be derived systematically from the full equations by introducing a suitable choice of critical conditions, as we shall indicate.

Baker's linear third-order equation for the radial displacement, r', has three parameters as coefficients and, if we look for temporal behavior of the perturbations of the form $r' \propto \exp(\lambda t)$, these three coefficients appear in the equation of λ:

$$\lambda^3 = \alpha\lambda^2 + \beta\lambda + \gamma. \tag{9}$$

In the same way, there will be parameters in the complete models that can be tuned to bring on various instabilities. As I have said, we will work near to the critical situation of marginal stability since that makes things tractable analytically. There is no need to do that in numerical studies, once we have seen the kind of behavior we want to look for.

The simplest situation is the one in which one mode passed through marginality as we tune a parameter. That is, for real $\lambda = \mu$, μ passes through zero as the control parameter passed through a critical value. In (9), this critical situation occurs for $\gamma = 0$ with appropriate values of α and β, but we could just as well consider μ itself as the control parameter; this may be harder to do in practice, but it makes the principles easier to think about. The nonlinear outcome is an approach to a new steady state, as in the familiar Landau equation.

We could also have a simple passage to instability when, for a complex pair of proper values $\lambda = \mu \pm i\omega$, μ passes through zero. For ω bounded away from zero, we again pass through marginality by tuning a single parameter, expressed most simply as μ itself, and we find overstability. I will not dwell on this case either, for it is familiar and gives rise to limit cycles in the nonlinear regime.

To achieve a richer behavior, we can tune two parameters. For example, we may arrange to have two values of μ, for distinct conjugate pairs of oscillatory modes, pass through criticality together. Or we can tune both the μ and the ω for a single conjugate pair. In either case, there is a double zero for λ at criticality. When we move slightly off criticality, we need to control two parameters to lift this degeneracy since the two values of λ need two parameters to be fully characterized. The simplest description can be extracted from (9) when β and γ tend to zero for $\alpha < 0$. To study that case, we divide (9) by $\lambda - \alpha$ and, for small λ/α, find to leading order that $\lambda^2 + p\lambda + q = 0$ gives the proper values of the slow modes, where $p = (\gamma + \alpha\beta)/\alpha^2$ and $q = \gamma/\alpha$. The nonlinear amplitude equation associated with this case must be a second-order differential equation.

The point of this example is to suggest how an equation like (9) can in turn be similarly extracted from the linear theory when there are three show modes. Suppose that in the full problem there exists some equation $T(\lambda) = 0$ for all the proper values of λ, even if it is known only numerically.

In the neighborhood of a critical point for N modes, I can always do the same trick on T to get an Nth-order polynomial whose roots give me the λs for the slow modes near to marginality (Coullet and Spiegel, 1983). The calculation of the coefficients in this polynomial may need to be numerically assisted in real situations, but, near to criticality, such an equation is valid and does not rely on rough arguments for its justification.

In the tricritical case, (9) we need the freedom to control three parameters in order to make three modes become slow at once. I will not enumerate the diverse possibilities of three mode instabilities (see Tresser, 1984) but will concentrate on the situation where has a triple zero at marginality.

As we see from the previous section, our linear problem has the form

$$\partial_t \mathbf{u} = \mathbf{L}\mathbf{u}. \tag{10}$$

At marginality, the normal modes are null vectors such that $\mathbf{L}\Phi = \mathbf{0}$. However, since $\mathbf{L}$ is not self-adjoint, it will sometimes not have as many null vectors as there are zeros of λ. In this case, the conventional normal modes do not constitute a complete set and we need to add some more basis modes.

To see this, we consider three matrices:

$$\begin{pmatrix} 0 & 0 & 0 \\ 0 & 0 & 0 \\ 0 & 0 & 0 \end{pmatrix}, \quad \begin{pmatrix} 0 & 1 & 0 \\ 0 & 0 & 0 \\ 0 & 0 & 0 \end{pmatrix}, \quad \begin{pmatrix} 0 & 1 & 0 \\ 0 & 0 & 1 \\ 0 & 0 & 0 \end{pmatrix}. \tag{11a,b,c}$$

Matrix (a) has three characteristic vectors associated with the proper value zero, (b) has only two and (c) has only one. The critical conditions they describe each have triple zeros for λ, but they correspond to different dynamical situations: (a) arises in the Lotka-Volterra equations (or predator-prey equations) with three species, (b) figures in the Lorenz equations, and (c) is the one we chose, both in our convective and pulsational examples, so I will stay with that one here, even though it is the rarest of the three cases.

To get a complete set in case (c), we need to have two additional slowly growing solutions of the linear problem. That is provided by the general slow solution of (10), $\varphi_1(\mathcal{M}_r) + t\varphi_2(\mathcal{M}_r) + \frac{1}{2}t^2\varphi_3(\mathcal{M}_r)$ where $\mathbf{L}\varphi_1 = \varphi_2$, $\mathbf{L}\varphi_2 = \varphi_3$ and $\mathbf{L}\varphi_3 = 0$. Hence the transpose of matrix (c) is a matrix representation of $\mathbf{L}$ at criticality. (I put the transpose in (11) since that is what operates on the vector a^n in the amplitude equation.)

Slightly off criticality we want the proper values of λ to be the roots of (9). Hence, we need to modify matrix (c) so that it includes the three parameters α, β, γ of that equation when they are nonzero. There are many ways that are satisfactory of doing this and one standard choice, or normal form, is the following.

In the neighborhood of criticality we write the slowly evolving part of the solution as $\mathbf{u} = \sum_{n=1}^{3} a^n(t)v_n(\mathcal{M}_r)$. In order to fulfill (10) we then demand that

$$\dot{a} = \mathcal{M}a \tag{12}$$

where $\mathcal{M}$ is an appropriate three by three matrix. That is, we need to choose a matrix whose proper values are the ones we have found to represent our situation, either numerically or otherwise. In particular, for our present example, we may use the Jordan-Arnold form (Gilmore, 1981)

$$\mathcal{M} = \begin{pmatrix} 0 & 1 & 0 \\ 0 & 0 & 1 \\ \gamma & \beta & \alpha \end{pmatrix}. \tag{13}$$

This leads to the cubic (9) and it reduces to the matrix (11c) at criticality.

Near to the tricritical point for three slow modes, we have reduced the linear theory for the amplitudes of the triplet of slow modes to (12)–(13). This illustrative case is the one whose structure coincides with the one-zone model but the procedure is general. Now nonlinear terms must be brought into the theory to represent the coupling to stable modes, which inhibits the continued growth of the unstable modes. The situation has an analogy to particle physics. The damped modes are like virtual particles. When the slow modes begin to develop, they excite stable, or "virtual," modes that otherwise would die off quickly. Typically, the effect of these excited modes is to modify, or renormalize, the effective growth rates of the unstable modes through nonlinear terms that keep the amplitudes of the unstable modes from growing indefinitely. That is the content of the nonlinear generalization of (12).

4. The Pulsational Amplitude Equation

As we discussed in §2, the nearly marginal modes satisfy a nonlinear version of (12), which we write as

$$\dot{a} = \mathcal{M}a + g(a), \tag{14}$$

where g is strictly nonlinear. The question now is, what is the form of g? Remarkably, this is fixed by our choice of $\mathcal{M}$, which is why we gave it so much attention in the previous section.

It is simplest to obtain g at marginality and that provides an adequate approximation for many purposes. The center manifold theorem applies at marginality and it says that

$$u(\mathcal{M}_r, t) = v(\mathcal{M}_r, a(t)). \tag{15}$$

This is the usual starting point of the Bogoliubov method of asymptotic theory and it offers a revealing way to look at and develop nonlinear stability theory (Coullet and Spiegel, 1981, 1983). The time dependence of the solutions is carried by the internal variable a which gyrates in its own space according to (14). So we can turn (3) into

$$\mathcal{L}\mathbf{v} = \mathbf{N}(\mathbf{v}) - \mathbf{g} \cdot \partial_\mathbf{a} \mathbf{v} \tag{16}$$

where

$$\mathcal{L} = \mathcal{M}\mathbf{a} \cdot \partial_\mathbf{a} - \mathbf{L}. \tag{17}$$

For convenience in writing powers of the amplitudes, we introduce the notation $a^1 = A$, $a^2 = B$, $a^3 = C$. For example, at criticality, the first term in (17), a scalar operator, is

$$\Sigma := \mathcal{M}\mathbf{a} \cdot \partial_\mathbf{a} = B\partial_A + C\partial_B. \tag{18}$$

This is a representation of the angular momentum operator $\mathbf{J}_+$ of quantum mechanics for the case of particles of spin one. Moreover, the second term in $\mathcal{L}$ has $\mathcal{M}$ in its representation, and that is the matrix representation of $\mathbf{J}_+$. So both parts of $\mathcal{L}$ have similar structures, one operating on functions of position ($\mathcal{M}_r$) and the other operating on functions of the internal variable (**a**). Hence $\mathcal{L}$ itself is a representation of $\mathbf{J}_+$ (Spiegel, 1985), so it is an annihilation operator.

We are thinking about a star in conditions that are close to those described by $\alpha = \beta = \gamma = 0$, when the instability is weak. So we expect that the components of **a** are not too large and that Taylor series in those components may be used. The procedure then is to make Taylor expansions of **v** and **g** in terms of monomials $A^k B^l C^m$. The linear solution satisfies $\mathcal{L}\mathbf{v}_1 = 0$, hence we readily find $\mathbf{v}_1 = \mathbf{a} \cdot \psi = A(t)\varphi_1 + B(t)\varphi_2 + C(t)\varphi_3$. For each higher order, we have to solve an equation of the form

$$\mathcal{L}\mathbf{v}_S = \mathbf{I}_S - \mathbf{g}_S \cdot \psi, \tag{19}$$

where $S = k+l+m$ and $\mathbf{I}_S$ is a function of $\mathbf{v}_{S-1}, \mathbf{v}_{S-2}, \dots$ and $\mathbf{g}_{S-1}, \mathbf{g}_{S-2}, \dots$.

In solving this sequence of inhomogeneous equations order by order, we need to group the possible terms in the "wave functions" $\mathbf{v}_S$ into multiplets of terms that are not mixed by the action of $\mathcal{L}$. The terms within each multiplet are transformed into one another by the action of $\mathcal{L}$, as we just saw, for the effect of **L** on the φ_n. However, the last member of the multiplet is annihilated and the first one is not regenerated. For any multiplet, if there is a term in $\mathbf{I}_S$ that corresponding to this first member, it cannot be generated by operation with $\mathcal{L}$. Hence, a solution of the perturbation theory can be produced only if that term can be eliminated by a suitable term in $\mathbf{g}_S$. This requirement dictates the terms that must occur in $\mathbf{g}_S$ and that is how **g** is determined, up to some arbitrary gauge choices. The coefficients of the terms in $\mathbf{g}_S$ are reminiscent of Clebsch-Gordan coefficients.

These asymptotic methods thus permit a systematic derivation of the amplitude equations (Arneodo et al., 1982, 1985ab), however there are many equivalent ways to write them. We can arrange the results so that **g** has only a third component. Of these, four are quadratic terms, six are cubic terms, and so forth. One way to express the outcome is

$$\dddot{A} - f(A, \dot{A}, \ddot{A})\ddot{A} - g(A, \dot{A})\dot{A} - h(A)A = 0. \tag{20}$$

where

$$f(A, \dot{A}, \ddot{A}) = \alpha + \alpha_1 A + \alpha_2 A\ddot{A} + \alpha_3 A^2 + O(A^3), \tag{21}$$

$$g(A, \dot{A}) = \beta + \beta_1 A + \beta_2 \dot{A} + \beta_3 A^2 + \beta_4 A\dot{A} + \beta_5 \dot{A}^2 + O(A^3), \tag{22}$$

$$h(A) = \gamma + \gamma_1 A + \gamma_2 A^2 + O(A^3). \tag{23}$$

In this equation, we see that the nonlinear terms act to saturate or renormalize the growth rates of the linear theory when the amplitudes become large enough. There are a few too many of these nonlinear terms to keep track of and, in most studies, special choices of the coefficients in them have been made to limit the complications. In any case, when we are near enough to tricriticality, the dominant nonlinear term is either A^2 or A^3, depending on symmetries (Arneodo, et al., 1985b).

In the study of overstable convection that I referred to Moore and Spiegel (1966), the instability caused by the α term was turned off by the choice $\alpha < 0$. So the nonlinear terms that protect against that instability did not play a qualitatively significant role and were not needed. That was why we got by with $f = \alpha$. As to the choices of g and h, we kept them arbitrary for a time, but specialized to a simple choice for specific calculations. Some writers have assumed that this choice also represented the equation referred to in the 1966 abstract about stellar pulsation theory (Baker et al., 1966), perhaps on account of our loose description of it.

The equation referred to in the Appendix is a nonlinear one zone model, and though it is a special case of (20), it is not included in the possibilities offered by (21)–(23). To bring (21)–(23) into that form, we need one more step, Padé resummation. That is, if we rewrite g and h as rational functions, we get an equation like that derived for the one zone case by qualitative arguments. I will not write that out here. But I would like to add that the earlier study of overstable convection was also motivated by an interest in solar oscillations. We suspected that sound waves were subject to the same kind of convective overstability as rotating and magnetic convection (Chandrasekhar, 1961), hence sought a generic model of convective overstability.

5. Pattern Maps

I have indicated how, for a star with N slowly evolving modes, we may derive an Nth order o.d.e. by asymptotic methods. In general, these give an amplitude equation of the form

$$\dot{\mathbf{a}} = \mathcal{M}\mathbf{a} + \mathbf{g}(\mathbf{a}). \tag{24}$$

The simplest case for stellar pulsation arises with a complex conjugate pair of oscillatory modes going unstable. In that case, (24) corresponds to two equations. but they are complex conjugates of each other. So it really describes a first-order equation for a complex amplitude, hence a two dimensional phase space is involved. With more slow modes, (24) becomes of higher order. A common situation has two oscillatory modes and, in the richest example that I will refer to, we have a four dimensional phase space. For illustration, I will sometimes refer to a case of intermediate complexity, an asymptotic version of the third order system already discussed at length. However, the results I describe in this section apply also to the more familiar case of pulsation theory with two pairs of oscillatory modes.

Rather than speak of the general third-order system, I will refer to the asymptotic limit that it takes as we approach tricriticality. That limit can be found by careful application of the Poincaré-Linsteadt method for $N = 3$. Whether we do this for the general pulsation equations or for the amplitude equation (14), we get the same answer (Arneodo et al., 1985b) and this can be expressed either by specifying

$$
\mathbf{a} = \begin{pmatrix} A \\ B \\ C \end{pmatrix}, \quad \mathbf{g} = \begin{pmatrix} 0 \\ 0 \\ A^2 \end{pmatrix}, \quad \mathcal{M} = \begin{pmatrix} 0 & 1 & 0 \\ 0 & 0 & 1 \\ \gamma & \beta & \alpha \end{pmatrix}. \tag{25}
$$

or simply writing it out as

$$
\dddot{A} - \alpha \ddot{A} - \beta \dot{A} - \gamma A = A^2. \tag{26}
$$

In either form, I will use this special case for illustration.

For this particular equation, we may refer to a theorem of Shil'nikow to conclude that, for suitable choices of the parameters, (26) will have an infinite number of unstable periodic solutions. This is, in effect, the statement that there will be chaotic behavior as the system wanders from one to another periodic solution (Baker, Moore and Spiegel, 1971). To see this explicitly, the arguments of Shil'nikov may be used to extract a Poincaré map for the problem to good accuracy in terms of elementary functions (Arneodo et al., 1985). From that we see that there are sound asymptotic arguments that show that the solutions of the stellar oscillation equations may be chaotic if there exist static tricritical states. The same kind of discussion can be made for any number of other polycritical situations.

Another approach to studying the behavior of the solutions of the amplitude equations is suggested by the observation of Coullet and Elphick (1987) that the effective particle method of nonlinear field theory may be used on (26) to obtain results very like those following from the Shil'nikov arguments. An advantage of this alternative approach is that the results come out in terms of measurable quantities. In view of this, it has seemed worthwhile to systemize this procedure for general use (Elphick et al., 1990a), though it has been so far used mainly for studying p.d.e.s.

In studying the behavior of the amplitude equations, it is useful to isolate the fixed points and the homoclinic orbits of the system. The former are solutions with $\dot{\mathbf{a}} = \mathbf{0}$ and, in particular, we focus on $\mathbf{a} = \mathbf{0}$. A homoclinic orbit is one that is biasymptotic to a fixed point, that is, on this orbit, the system approaches the fixed point as $t \to \pm\infty$. A picture of $A(t)$ for a homoclinic orbit of (10) is shown in the figure. Both the fixed point and the homoclinic orbit are solutions with infinite period.

For the example in the figure, we have selected the parameters in the system so that the linear theory in the neighborhood of the origin has one monotonically growing solution, $\lambda = \mu_1 > 0$, and a damped oscillation, $\lambda = \mu_2 \pm i\omega$, $\mu_2 < 0$. Associated with each μ_i, there is a proper vector. Starting from initial conditions near the origin of phase space, the solution takes off along the vector associated with μ_1 with A growing like $\exp(\mu_1 t)$. After a time (during which it may take a turn or two around the fixed point with $A = -\gamma$) it spirals back in toward the origin coming into the plane formed by the two proper vectors associated with μ_2. In the figure, we see the exponential rise and the oscillatory decay suggested by our choice of proper values of λ.

I shall call the solution shown in the figure a principal homoclinic orbit since it corresponds to one turn around the other fixed point of (26). Secondary homoclinic solutions can also be found in which the solutions make several loops around the other fixed point before returning to the origin. Those solutions look somewhat like the figure, except that the peak has fine structure. The secondary homoclinic orbits may be considered as closely spaced groupings of the primary homoclinic orbit and need not be treated separately for our purposes. When solutions of (26) are obtained numerically, they typically show series of pulses, which may be regularly or irregularly spaced according to the parameter choices. The pulses have a characteristic shape that is approximated by the principal homoclinic orbit of the figure.

The general case, (24), is much the same. If there are only a couple of real modes in the problem, both the rise and the fall of $A(t)$ must be monotonic. However, for some examples with coupled oscillatory modes, we can expect oscillations in both the rise and decay. All these details can make a difference to the appearance of the signal (Elphick et al., 1990b). When there are even more modes in the problem, there can be several fundamental homoclinic orbits and the plot thickens. Not much calculation has been done in those cases, and I will say no more of them.

Suppose that we are in a low-dimensional situation with a simple homoclinic solution $A = H(t)$. The phase of this pulse is arbitrary. That is, if $H(t)$ is a solution of (26) or (24), so is $H(t - \tau)$, where τ is a constant. In fact, τ

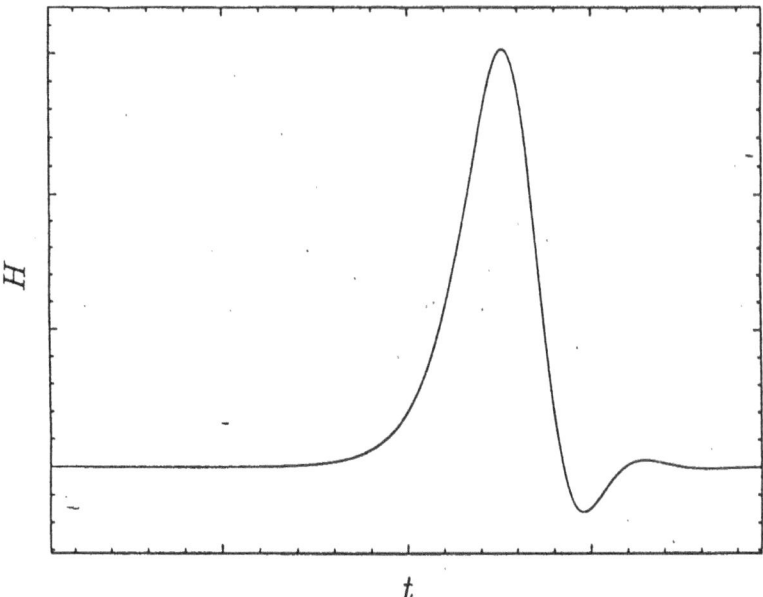

Fig. 1.   A(t) for a homoclinic orbit of (26), adapted from Elphick et al., (1991).

is the parameter of the group of transformations that shift the origin of the time and the presence of this group means that the solution is indifferent to the value of τ. Let us choose the phase so that $H(t - \tau)$ has its maximum at $t = \tau$. We are going to look for a more general solution with a sequence of pulses at times τ_n for $n = 1, 2, \ldots$. But before writing this down, we need to note a few things.

First, we have to decide at which level we want to work. We can study the behavior of just $A(t)$ or go to the full amplitude vector $\mathbf{a}$. Indeed there are times when we might wish to look for waves in the original p.d.e.s. of stellar pulsation theory and work with them to describe the spatio-temporal nature of the full solutions. For the present illustration, I shall stick with $A(t)$, as it is the simplest thing to do.

Then we need to decide something about the parameter range. The relative magnitudes of the rise and decay times of the pulses, as measured by μ_1 and μ_2 are important. Let $\mu_1 = \mu$ and $\mu_2 = -\mu\xi$, where μ and ξ are positive. In this discussion, I shall assume that ξ is of order unity; if it were not, the results would be different. Thus there are many other possible behaviors, which complicated (or enriches, if you prefer) these problems.

Finally, suppose that the mean spacing between pulse times is T. Then we can define a key parameter of the theory:

$$\varepsilon = e^{-T\mu}. \tag{27}$$

Throughout this discussion, I assume that $\varepsilon \ll 1$, that is, that the pulses do not overlap significantly. This is a situation often perceived in numerical studies; other situations may arise, but they are not so easily characterized.

When the pulses in the solutions are distinct, they generally look like $H(t)$, which we may use a building block for a series of pulses. That is, we can look for a solution of the form

$$A(t) = \sum_m H(t - \tau_m) + \varepsilon \mathcal{R}(t, \varepsilon), \qquad (28)$$

where the total number of pulses may be infinite, and $\varepsilon \mathcal{R}$ is the error made in trying to use a linear superposition of solutions in a nonlinear problem. The form of this error term expresses the belief that, as the spacing of the pulses increases, the error diminishes.

The procedure now is to base a perturbation expansion on (28) for small ε. We develop $\mathcal{R} = \sum_k \varepsilon^k \mathcal{R}_k$ and get a series of linear equations for the $\mathcal{R}_k$. We want to find finite solutions for the $\mathcal{R}_k$, and especially for $\mathcal{R}_0$, so that the error term, $\varepsilon \mathcal{R}$, should tend to zero as $\varepsilon \to 0$. However, if we proceed naively, this hope is not realized. That is, if we simply choose a set $\{\tau_m\}$ we find that $\mathcal{R}_0$ diverges — the perturbation theory is singular. Even though the phase of the simple homoclinic orbit is arbitrary, the phases of individual pulses cannot be chosen independently of each other. Up to a constant phase shift for the whole solution, the choice of the $\{\tau_m\}$ is not free and must be determined from the equations.

As is usual in singular perturbation problems, we broaden our outlook and let τ_k be a function of εt. Then, when we look at the problem of calculating $\mathcal{R}_0$, we find that the condition that this be possible in finite terms is a condition on $\dot{\tau}(\varepsilon t)$, where we use the dot to imply derivative with respect to argument. This solvability condition for $\mathcal{R}_0$ may be obtained in the normal way on multiplying by a null vector of the adjoint linear operator as spelled out in the references I have cited. The matter can be resolved in simple terms when certain simplifying approximations are included. We focus on the neighborhood of the mth pulse and make the approximation that only the interactions with its nearest neighbors, the $(m \pm 1)^{th}$ pulses, matter. Then, we get coupled equations of motion for all the τ_m.

We consider (24) in two, three or four dimensions, where there are at most two conjugate pairs of proper values of $\mathcal{M}$, say $\mu_1 \pm i\omega_1$ and $\mu_2 \pm i\omega_2$. As before, $\mu_1 = \mu$ is positive and $\mu_2 = -\mu\xi$, with $\xi \sim 1$. For a three-dimensional equation like (26), we can have only one of the two frequencies nonzero but, more generally, both the rising and the decaying edges of the pulse may be oscillatory if there are enough slow modes in the original problem. For pulse shapes with either oscillatory or monotonic rise and decay, we find (Elphick et al., 1990)

$$\dot{\tau}_m = \zeta_1 e^{-\mu \Delta_{m+1}} \cos(\omega_1 \Delta_{m+1} + \theta_1) + \zeta_2 e^{-\mu \xi \Delta_m} \cos(\omega_2 \Delta_m + \theta_2), \qquad (29)$$

where

$$\Delta_m = \tau_m - \tau_{m-1}, \tag{30}$$

and the ξ_i and θ_i $(i = 1, 2)$ are constants that depend on the pulse shape.

This is the same kind of result that you would get for solitary waves described by p.d.e.s with translational invariance. That leads to the interpretation of (29) as an equation of motion for the mth pulse under the influence of its nearest neighbors. In this dynamics, εt is the time and τ is like position. The pulses have no inertia in this example, so the velocity is given directly by the forces. Since the problem is one-dimensional, the force does not show an algebraic dependence on the separation but it does have an exponential cutoff, or range. There is also the interesting effect of the ripples in the force law coming from the oscillatory character of the modes. This permits bound states of pulse pairs to form at differing phase separations.

According to this system of equations of motion, the pulses tend to move into a locked pattern with constant velocity, $\dot{\tau}_m = V$, so that (29) turns into a *pattern map* that gives a deterministic relation between the phase difference of a pair of successive pulses and that of the previous pair. Once the system is relaxed into this uniformly progressing state, the possible phase patterns can be diverse (Elphick et al., 1990b). If $\omega_1 = \omega_2 = 0$, the pulses will be uniformly spaced, as in the case of a second-order system. When both ω_1 and ω_2 are nonzero, which is possible when we have a fourth order o.d.e. as a model, we can get hyperchaotic patterns (Glendinning and Tresser, 1985). Of course, when we deal with the kind of third-order systems suggested by (26), we can have oscillations on only one side of the pulse. For example, when the oscillatory tail is the leading one, with $\omega_1 = 0$ and $\omega_2 = \omega$, we get

$$Z_{m+1} = C - K Z_m^\xi \cos(\omega \ln Z_m - \Theta), \tag{31}$$

where

$$Z_m = e^{-\mu \Delta_m} \tag{32}$$

and C, K and Θ are constants involving the pulse shape and V. This is a map that gives the possible patterns of phases. Its fixed points correspond to periodic solutions with constant phased and the number of these tends to infinity as $C \to 0$.

6. Discussion

We have seen what can be done in situations near to polycriticality, where several modes can be simultaneously marginal. Such behavior, which I like to call competing instabilities (Spiegel, 1972), leads to amplitude equations of an order dictated by the number of significant parameters afforded by the available slow modes. For illustration, we have used a case of tricriticality where, according to how we unfold the singularity, we can get three monotonic modes or one monotonic and two oscillatory modes. Near such polycritical conditions, we can legitimately reduce the full pulsation problem described by p.d.e.s to one of o.d.e.s. The dimension of the phase space in which we must work has thus been reduced from infinity to a few.

Having found the amplitude equation, we need to analyze the diversity of behavior it may describe. We saw in the previous section that, by suitable choice of the parameters of the stability theory, we can describe a solution as a series of pulses in amplitude with a set of equations for the phases of the pulses. At first glance, this does not seem to be progress for cases where we have to analyze a long series of pulses. But in fact, since the phases do lock in rather quickly (see Elphick et al., 1990a), we may often go right to the limit where they have established their relative phases and we are led to an algebraic equation for the successive phase differences in the form of a map of the real line onto itself. In other words, we have gone from the full p.d.e.s of pulsation theory to an algebraic formula without needing any detailed calculations unless we want to evaluate the coefficients in the formula.

Of course, for this to be quantitatively good, we need to adopt specially chosen parameter ranges. Still, it does bring out asymptotically correct features of the content of the pulsation theory for those ranges. As those a little familiar with such maps will readily see from the sample result, (31), we can conclude that the pulsation equations give chaotic behavior arbitrarily closely to the onset of a triple instability. Such results are by now familiar in nonlinear stability theory and they naturally carry over to stellar pulsation.

However special they may be, these results do represent real behavior of the basic equations and may even show us the simplest form of the complex patterns that stellar pulsation may exhibit. They tell us that, with a few slow modes in the problem, otherwise regular pulsations can be expected to exhibit erratic phase variations. Moreover, in the simplest situations we have considered, these fluctuations are deterministic. In the example of (31), we can even see that there is a quantized aspect to the successive phase shifts as a result of the oscillatory character of some of the modes. The way that happens is that the pulses are locked in preferentially to alternate minima of the forces implied by (29). The existence of such minima in the forces is a consequence of the oscillatory tails of the pulses.

The implication of all this is that pulsational phase diagrams can be expected to exhibit the variety of behavior contained in a map like (31). The pulses can cluster into pairs, then into pairs of pairs, and so on through a full hierarchy of clusters of pulses, as in the usual period-doubling cascade. There can also be completely chaotic and hyperchaotic progressions in the phases. The pattern map (31) even describes intermittent, or bursts, behavior, but that is not so robust.

The question of the stability of phase patterns is a difficult one that is governed by the differential equations (29) and not by the map. This is too complicated to include in this sketch. Instead, I reluctantly leave the matter here, recalling an incident from the life of J. S. Bach. It is said that he had already retired for the night, though one of his children was still practicing on the harpsichord. When the child stopped abruptly, leaving a chord unresolved, Bach got of bed, went downstairs in his nightcap, resolved the dissonance on the harpsichord, and went back up to sleep soundly. You will understand how this tale has inspired the work described here. And just as Bach was up early next morning composing, we must now turn seriously to the question of how we may reach the next level of complexity in the patterns that the bursters or solar cycle lead us to expect from the theory.

One of the simplest things that we can do to increase the complexity allowed by the pattern map (31) is to relax the assumption that the field of pulses is so widely spaced. We then need to include the interaction with the next nearest (in time) pulses. Then (31) becomes a two-term recursion formula, implying a two-dimensional map. This makes for more complicated patterns that resemble a noisy version of (31). But these are still a far cry from the more erratic patterns of variation that we see, for example, in the X-ray bursters. Models for those, I believe, will have higher dimension than those I have discussed here. I hope that a sequel to this meeting will provide a forum to discuss them.

I am very grateful to the Air Force Office of Scientific Research whose support has permitted the continuation of this work for application to the solar cycle. And I am happy to thank Neil Balmforth for his careful reading of the manuscript.

Appendix

Appendix. The Original Abstract

Recently Baker (1966) devised a model for pulsational instability based on the dynamics of the layer in the star where the instability originates. The advantage of this model is that it leads to an ordinary third-order differential equation whose linearized form can be discussed quite readily. In the present work the model is used to study finite-amplitude oscillations. The equation governing the one-zone model is rather like the simpler third-order equation studied by Moore and Spiegel (1966) in connection with nonlinear overstability, and it exhibits the same kinds of phenomena. In particular it gives rise in different cases to aperiodic oscillations, relaxation oscillations and "stillstands" in the displacement curves, in addition to well-behaved periodic solutions.

References

Arneodo, A., Coullet P. H. and Spiegel, E. A.: 1982, "Chaos in a Finite Macroscopic System," *Phys. Lett.* **A92**, 368.

Arneodo, A., Coullet P.H., Spiegel, E. A. and Tresser, C.: 1985a, "Asymptotic Chaos," *Physica* **D14**, 327.

Arneodo, A., Coullet P. H. and Spiegel, E. A.: 1985b, "The Dynamics of Triple Convection." *Geophys. and Astrophys. Fluid Dynamics* **31**, 1.

Baker, N. H.: 1966, "Simplified models for Cepheid instability," in *Stellar Evolution*. eds. R. F. Stein and A. G. W. Cameron (Plenum Press).333–346.

Baker, N. H., Moore, D. W. and Spiegel, E. A.: 1966, "Nonlinear Oscillations in the One-Zone Model for Stellar Pulsation," *Astronomical Journal* **71**, 845.

Baker, N. H., Moore, D. W. and Spiegel, E. A.: 1971, "Aperiodic Behavior of a Non-Linear Oscillator," *Quart. Jour. Mech. Appl. Math.* **24**, 391.

Carr, J.: 1981, *Applications of Centre Manifold Theory*, (Springer-Verlag).

Chandrasekhar, S.: 1961, *Hydrodynamic and Hydromagnetic Stability*, (Oxford Univ. Press).

Coullet, P. H. and Elphick, C.: 1987, "Topological defect dynamics and Melniknov's theory," *Phys. Lett.* **A121**, 233.

Coullet, P. H. and Spiegel, E. A.: 1981, "A Tale of Two Methods," in *Summer Study Program in Geophysical Fluid Dynamics*, The Woods Holes Oceanographic Institution, ed. F. K. Mellor, p. 276.

Coullet, P. H. and Spiegel, E. A.: 1983, "Amplitude Equations for Systems With Competing Instabilities," *SIAM J. Appl. Math.* **43**, 774.

Elphick C., Meron, E. and Spiegel, E. A.: 1990a, "Patterns of Propagating Pulses," *SIAM J. Appl. Math.* **50**, 490.

Elphick, C., Meron, E., Rinzel, J. and Spiegel, E. A.: 1990b, "Impulse patterning and relaxational propagation in excitable media," *J. Theor. Biol.* **146**, 249.

Elphick, C., Regev, O., Ierley, G. R. and Spiegel, E. A.: 1991, "Interacting Localized Structures with Galilean Invariance," *Physical Review A: General Physics* **A44**, 1110.

Glendinning, P. and Tresser, C.: 1985, "Heteroclinic loops leading to hyperchaos," *Phys. Lett.* **A46**, 347.

Gilmore, R.: 1981, *Catastrophe Theory for Scientists and Engineers*, (Wiley and Sons, New York).

Moore, D. W. and Spiegel, E. A.: 1966, "A Thermally Excited Nonlinear Oscillator," *Astrophysical Journal* **143**, 871.

Spiegel, E. A.: 1972. "Convection in Stars. II. Special Effects," *Annual Review of Astronomy and Astrophysics* **10**, 261.

Spiegel, E.A.: 1985, "Cosmic Arrhythmias," in *Chaotic Behavior in Astrophysics*, eds. R. Buchler, J. Perdang and E. A. Spiegel (Reidel, Dordrecht), p. 91.

Tresser, C.: 1984, "Homoclinic orbits for flows in $\mathbf{R}^3$," *J. Physique* **45**, 837.

CHARACTERISATION OF THE DYNAMICS OF A VARIABLE STAR

M. AUVERGNE, A. BAGLIN and M. J. GOUPIL
*Observatoire de Paris-Meudon, DASAL, URA CNRS 335, 5 pl. J. Jassen
92195 MEUDON FRANCE*

Abstract. Tentatives to detect chaos in variable stars are presented. Constraints on the accuracy on individual measurements and length of the records, necessary to obtain defenitive conclusions, are discussed. Some directions for future improvements are sketched.

1. Introduction

Different facts favor the existence of chaos in pulsating stars:
- many of them have irregular pulsations (see i.e. Perdang 1985),
- laboratory experiment and theoretical works have shown that chaos exist in simple systems,
- extended hydrodynamical stellar models display chaotic radial pulsation in peculiar stages of evolution.

If chaos exists in stellar pulsations, it must manifest itself by a chaotic attractor, characterized by its major features:
- divergence of trajectories responsible for the long term unpredictability, as measured by Lyapounov exponents. Lyapounov exponents are probably the most useful indicators of a chaotic dynamic, but they have never been used (to our knowledge) in variable stars studies and they are not discussed in the following. Measurements of Lyapounov exponents on noisy data and limited time series are difficult as the positive exponents are often small and not easily distinguishable from zero.
- fractal structure at "small" scales, except for self-similar objects. But, the scale at which it appears in the reconstructed state space depends on the method of reconstruction (see §3) and is not known a priori.
- specific routes to chaos, as period doubling or tangent bifurcation. The routes to chaos have been extensively studied, especially in laboratory experiments in which it is possible to vary control parameters. For stars, global parameters, which could play this role, vary on the nuclear time scale, too long to be really useful.

2. The Data: Present and Future

The observables are radial velocity or brightness measurements in broad or narrow bands.

Astrophysics and Space Science **210**: 51–60, 1993.
© 1993 *Kluwer Academic Publishers.*

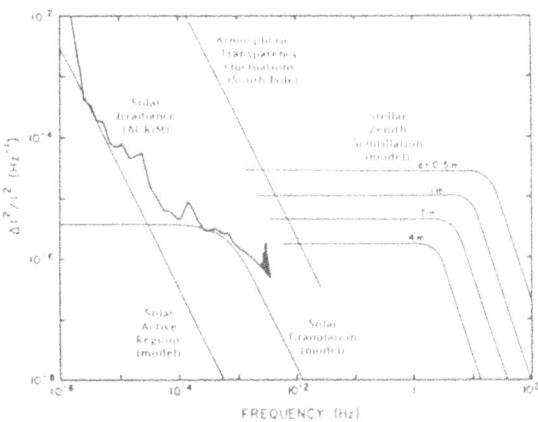

Fig. 1. Schematic Fourier spectra of scintillation and transparency fluctuations.

2.1. RADIAL VELOCITY

Radial velocity measurements are almost not affected by the atmospheric noise, at least in a first approximation. The accuracy is quite low $\sim 10^{-2}$. Improvements of the precision faces intrinsic limitations due to stellar atmosphere physics. Measuring velocities on one line is limited for most stars by the low photon flux or time resolution. Multiline measurements, as they are made by a Coravel type device, have to be performed on lines formed at the same level. The accuracy depends also strongly on the rotation and classical methods are limited to rotational velocity smaller than 20 km/s.

2.2. BRIGHTNESS

Brightness measurements are generally done from the ground, in more or less broad visible bandpass. The three independent sources of errors are photon noise, scintillation and transparency fluctuations. On figure 1 the Fourier spectra of the transparency and scintillation contribution are shown (Harvey 1988). The scintillation model assumes a wind velocity of 30 ms^{-1} at an altitude of 2 km. Let us remind that, to obtain a given accuracy, α on individual measurements, the noise level at all frequencies must be smaller than α. A typical spectrum of real brightness measurements of a constant star is given on figure 2.

2.3. PHOTOMETRIC TECHNIQUES AND THEIR ACCURACIES

(For a complete discussion see the paper of Young et al. 1991.) Photometry being the most common technique to study variable stars, we briefly review the various techniques and their accuracy.
- Classical one channel photometer: average accuracy about 10^{-2}.
    Differential photometry with such a device, gives a small time resolution

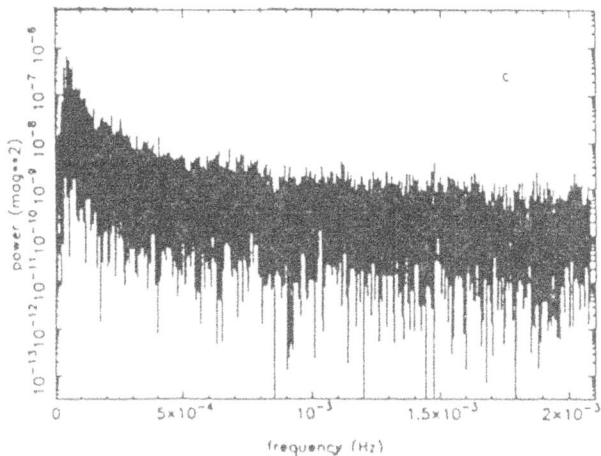

Fig. 2.   Fourier spectra of 232 hours of measurements of the assumed constant star HD
213 473, $V = 9.1$ (from Michel et al. 1992).

due to the regular telescope depointing from the target star to the comparison star. As the two stars are not very close, variations due to transparency
fluctuations are partly uncorrelated, unless for very good but rare nights.
- Multi channel photometer: average accuracy at most 10^{-3}.

The time resolution is better. The comparison stars are taken in the
telescope field of view, improving the correlation of the transparency fluctuations. The sky background can be easily monitored. But, particularly for
bright stars, the spectral types of the variable star and of the comparison
are often different and some chromatic effects in the atmospheric absorption
are present.
- CCD photometry: average accuracy a few 10^{-4}

This method is very promising: applied to a star cluster, several variables,
several comparisons and the sky background can be measured at the same
time. The limitation on the accuracy comes essentially from scintillation
(Frandsen 1992). But it produces images at a high rate and reduction procedures are heavier than for time series obtained with multi channel devices.
- Photometry from space.

It can be limited only by instrumental drift and photon noise because the
diameter of the collector remains small. The number of collected photons
outside the atmosphere is

$$N = 10^5 10^{-0.4V} d^2 \, \delta\lambda \, \mu$$

where d is the diameter of the entrance pupil, V the visual magnitude, $\delta\lambda$
the spectral bandwidth and μ the quantum efficiency of the photometer.
With $\delta\lambda \, \mu = 100$ and an exposure time of 30 s., the variance of the photon
noise is $A = 1/\sqrt{N dt}$. Lines of constant A are drawn on figure 3, in an

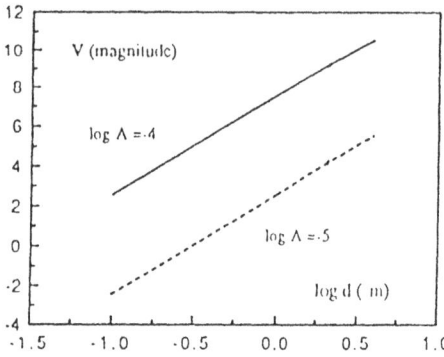

Fig. 3. Lines of constant A as function of the magnitude of the star and of the diameter of the collector.

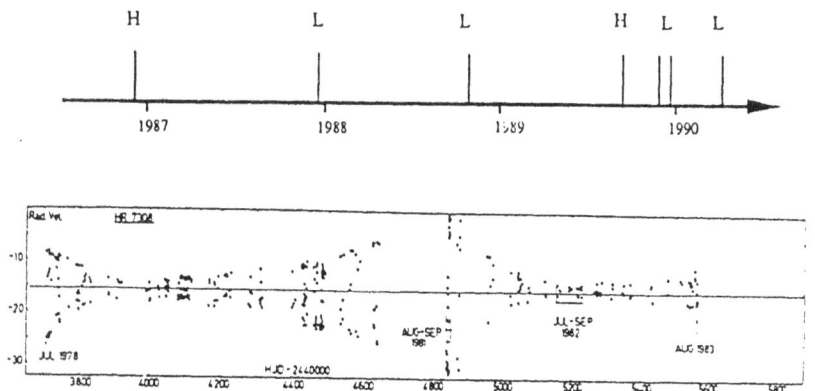

Fig. 4. a: Variations of the amplitude of the white dwarf G191-16 at different epochs from 1987 to 1990; H and L stands for high (0.2 Δm) and low amplitude (0.04 Δm); from Auvergne et al. 1990. b: Velocity curve of HR 7308 from Burki et al. 1982.

apparent magnitude - $\log d$ diagram. As they represent an unreachable limit for ground based observations they show that with a 2 meters telescope a relative precision of 10^{-4} cannot be achieved for a star fainter than 8.

2.4. TIME SCALES

To obtain a complete description of the dynamic one needs to record a large number of each characteristic time scale; but the long ones are not always known. For instance, very long time scale (several months to one year) exist in objects with short period. Two examples are shown on figures 4a and 4b. Long runs are necessary, favoring multisite campaigns or space missions. One of the best example is the monitoring of the Sun during 160 days by the IPHIR experiment on the Phobos mission.

3. Procedures to Detect Chaotic Behavior

For dynamical systems known only through a single observable $s(t)$ like stars, several methods exist to construct an attractor in a n dimensional space *equivalent* to the original one.

3.1. DELAY VECTOR

A theorem due to Takens (1981) states that a set (attractor) *equivalent* to the original one is obtained through the following algorithm known as the time delay method. Defining the set of vectors $x(t_i) = \{s(t), s(t + \tau), s(t + 2\tau), \ldots, s(t + (n - 1)\tau\}$, where n is the dimension of the Euclidian space used for the reconstruction, Takens shows that this process is an embedding i.e. that the original set and the embedded set are related through a diffeo-morphism, with the condition $n \geq 2D + 1$ where D is the dimension of the original set. This result is independent of the delay $\tau \neq 0$. However if τ is too small, $s(t) \sim s(t + \tau) \sim \ldots \sim s(t + (n - 1)\tau)$ and all points are close to the first bisector. Experience shows that a "good" value of τ is of the order of 30% of the characteristic time scale. This value has been related to the first minimum of the autocorrelation function.

3.2. SINGULAR VALUE ANALYSIS

The method known also as Karhunen-Loeve expansion is a decomposition of the signal on the basis of the eigenvectors of its correlation matrix. This linear transform determines the dominant components of the dynamic, but not necessarily those which govern the nonlinear part. Projection on the subspace generated by the eigenvectors associated with the largest eigenvalues gives also an embedding. Examples are given on figure 5. In some cases the components seem to correspond to physical parameters, temperature, velocity and radius, as in hydrodynamical models.

3.3. GEOMETRICAL CHARACTERISATIONS

Once, the attractor is reconstructed, Poincaré map and return maps provide information on its structure, hence on the underlying dynamics. Unfortunately, noise acts as a diffusive process in the embedding space.
-Poincaré map.

It contains a priori the same topological information as the flow and the same stability properties. Figure 6a show the Poincaré map of a simple model. Figure 6b is the map of the same model where one percent white noise, mimicking atmospheric transparency fluctuations, has been added. The fractal structure is partly swapped out as the attractor is broadened and the small scale structures are hidden. A statistical study of the density of points in the plane, allows to retrieve at least a part of the lost structure (Bijaoui 1974).

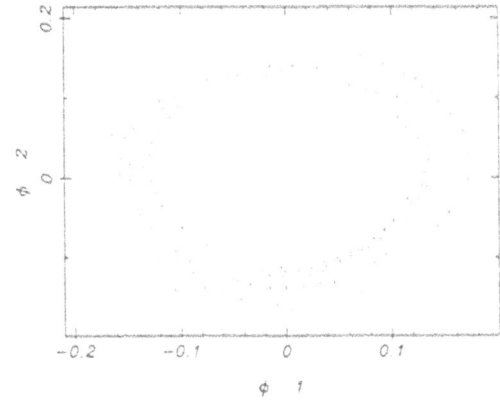

Fig. 5.   Projections on the two eigenvectors Φ_1 and Φ_2 corresponding to the two largest eigenvalues. The object is the white dwarf PG 1351+489. (Auvergne 1988)

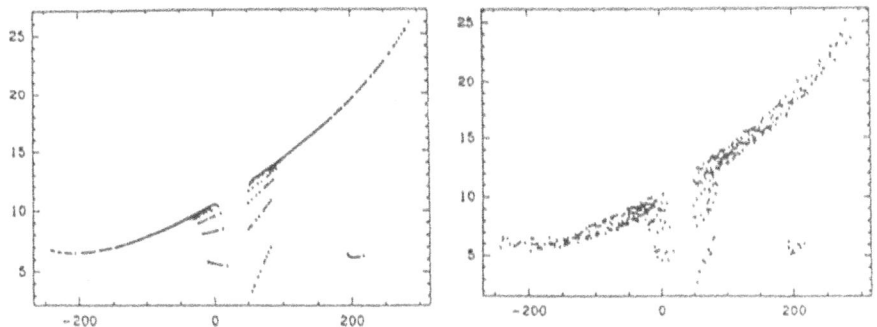

Fig. 6.   a: Poincaré map of a chaotic solution of a simple model (Auvergne and Baglin 1985). b: the same plus 1% noise.

-Return map.

A return map contains the same information than the Poincaré section, but it is generally difficult to use on astrophysical data as it seems to be more sensitive to noise than the Poincaré map. A complex first return map can arise from a high dimensional chaos. This is the case of a chaotic dynamic arising from a subcritical bifurcation of a subharmonic limit cycle. In this case a fourth return map displays a simpler one dimensional shape. Figures 7 and 8 display several first return maps from observations and from hydrodynamical calculations.

3.4. Dimension Computations

Roughly speaking, the dimension of a set is related to the amount of information needed to locate a point on it. Several "fractal dimensions" have been defined corresponding to the nature of the information one wants to

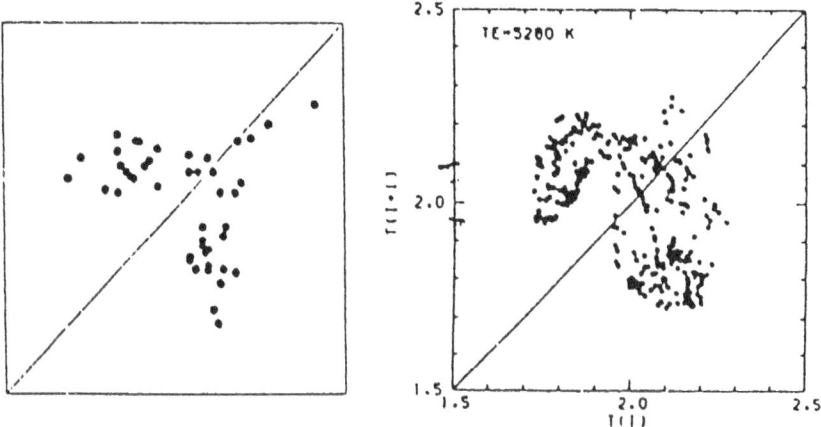

Fig. 7.   a: Return map of the semiregular variable S Vul, from Saitou et al. 1989. b: the same for the numerical simulation of a chaotic population II model, from Aikawa 1990.

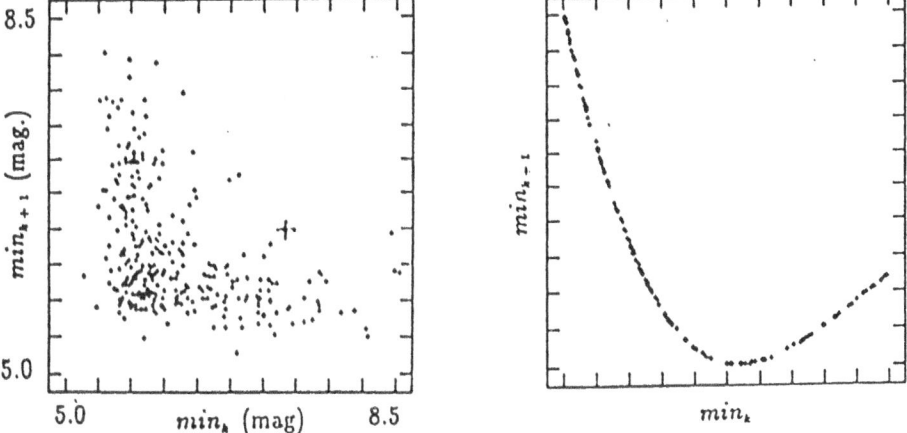

Fig. 8.   a: Return map of the RV Tauri star R Scutum, using light minima. b: return map of the Rössler attractor, from Kolláth 1990.

obtain (for a review on dimensions of chaotic attractors see i.e. Farmer et al. 1983). The most popular method to detect fractal structure is the correlation integral, defining:

$$C(l) = \lim_{N \to \infty} \frac{1}{N(N-1)} \sum_{i,j=1}^{N} \theta(l - ||x_i - x_j||)$$

where θ is the Heaviside function (Grassberger and Procaccia 1983). $C(l)$ counts the number of couples of points such that $l < |x_i - x_j|$, and for small values of l one can show that $C(l) \sim l^D$. Since the dimension of the attractor is not known, the dimension of the embedding space is also unknown. Therefore $C(l)$ is computed for increasing values of n. When the

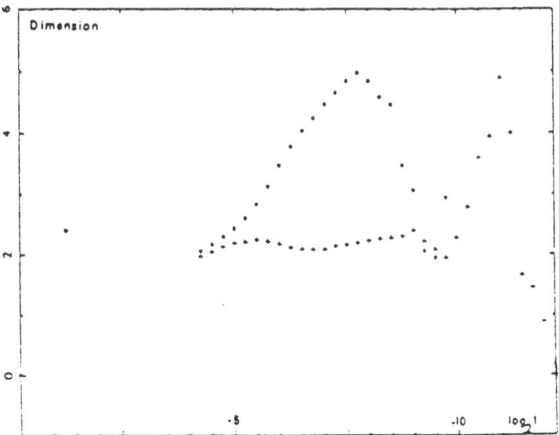

Fig. 9.   Slope of the correlation integral of a dimension 2.08 attractor without and with 1% white noise added.

slope of $C(l)$ becomes independent of n one has reached the dimension of the set.

But, as the scale at which the fractal structure appears is unknown (unless for self similar object), we are never sure to reach a noise free significant scale. This remark has not been always well understood and many published results are just artefacts. Figure 9 shows the noise influence on the correlation integral.

Several other important artefacts have been emphasized in the literature: the effect of a coloured noise by Osborne and Provenzale (1989) or the role of a low pass filter which, if the cutoff frequency is too small, increases the value of the computed dimension (Badii et al. 1988). Theiler (1986, 1990) and Smith (1987) have given constraints on the minimum number of points (as a function of the embedding dimension) to ascertain a satisfactory statistic on the attractor.

3.5. ROUTE TO CHAOS

Some simple dynamical systems, for instance the popular Rössler attractor, display (even in the chaotic state) subharmonic frequencies. This property is due to the phase coherence of such systems. Subharmonics have been found in several very different types of stars, like white dwarfs and Mira variables.

They show up naturally in Fourier spectra or on light curves by the alternation of different amplitude maxima or minima (Vauclair et al. 1989, Goupil et al. 1988, Buchler and Regev 1990). Figures 10 and 11 show power spectra of the Mira variable T UMa and of the white dwarf PG1351+489, with such subharmonics.

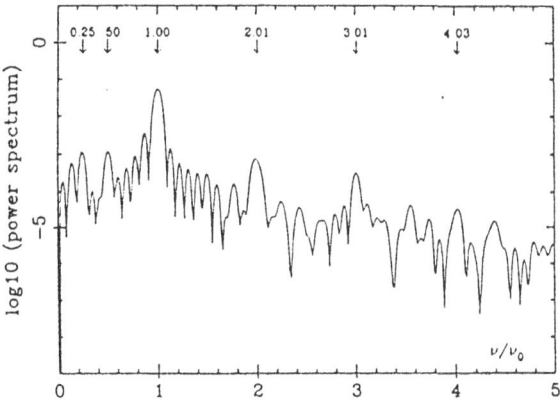

Fig. 10.   Power spectra of the light curve of T UMa (data from AAVSO) showing to peaks at $\nu/2$ and $\nu/4$.

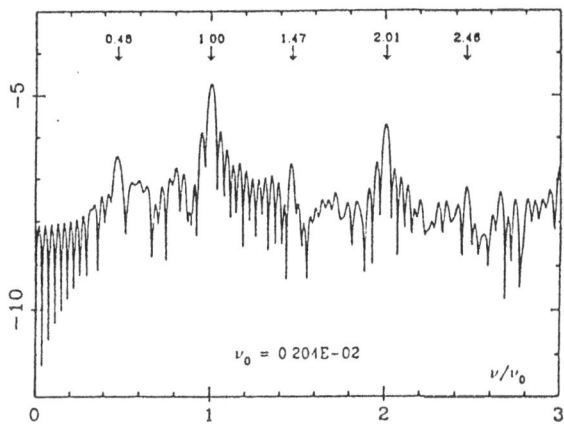

Fig. 11.   Power spectrum of PG 1351+459 (Goupil et al. 1988)

4. Conclusions

During the last ten years a lot of efforts have been devoted to the search of chaos in stars. Significant progresses have been made in the theoretical and computational domains (Aikawa 1987, Buchler et al. 1987, Buchler and Regev 1990) as well as in the time series treatment (Casdagli et al. 1990, Scargle 1992).

In real data some encouraging results have been obtained, but definitive conclusions can be derived only with extremely high quality data. In the most common case of photometric measurements from the ground, improvements will come essentially from the correction of transparency fluctuations, as the problem of the scintillation and photons noise is "simpler" to solve by increasing the telescope diameter.

Observations from space, though very costly, can be designed to be free from these difficulties. Two experiments, EVRIS and PRISMA, devoted to stellar variability detection are in preparation. They will produce, in a near future, low noise and very long time series, adequate for such studies.

References

Aikawa, T.: 1987, *Astrophysics and Space Science*, **139**, 218.

Aikawa, T.: 1990, *Astrophysics and Space Science*, **164**, 295.

Auvergne, M. and Baglin, A.: 1985, *Astronomy and Astrophysics* **142**, 388.

Auvergne, M.: 1988, *Astronomy and Astrophysics* **204**, 341.

Auvergne, M., Chevreton, M., Belmonte, J. A.. Vauclair. G., Dolez, N., and Goupil, M. J.: 1990, *7th European Workshop on White Dwarfs. NATO ASI Series.*, eds. G. Vauclair and E. Sion **336**, 167.

Badii, R., Broggi, G., Derighetti, B., Ravani, M., Cilberto, S., Politi, A., and Rubio, M. A.: 1988, *Physical Review Letters* **60**, 979.

Bijaoui, A.: 1974, *Astronomy and Astrophysics* **35**, 108.

Buchler, J. R., Goupil M. J. and Kovács, G.: 1987, *Phys. Lett.* **A126**, 177.

Buchler, J. R. and Regev, O.: 1990, in *The Ubiquity of Chaos*, ed. S. Krasner (American Association for the Advancement of Science, New York), p. 218.

Burki, G., Mayor, M., and Benz, W.: 1982, *Astronomy and Astrophysics* **109**, 258.

Casdagli, M., Des Jardins, D., Eubank, S., Farmer, J. D., Gibson, J., Hunter, N. and Theiler, J.: 1991, in *Applications of Chaos*, ed. Jong Kim (Wiley Interscience), in press.

Farmer, D., Ott, E., and Yorke, J.: 1983, *Physica* **D7** ,153.

Frandsen, S.: 1992, *private communication*.

Goupil, M. J., Auvergne, M., and Baglin, A.: 1988, *Astronomy and Astrophysics* **196**, L13.

Grassberger, P. and Procaccia, I.: 1983, *Physical Review Letters* **50**, 346.

Harvey, J. W.: 1988, *Advances in Helio- and Asteroseismology*, eds. J. Christensen-Dalsgaard and S. Frandsen (Reidel Publ. Co.), IAU symposium **123**, p. 497.

Kolláth, M.: 1990, *Monthly Notices of the RAS* **247**, 377.

Michel, E., Belmonte, J. A., Alvarez, M., Jiang S. Y., Chevreton, M., Auvergne M., Goupil M. J., Baglin A., Mangeney A., Roca Cortes T., Liu Y. Y., Fu J. N., and Dolez, N.: 1992, *Astronomy and Astrophysics* **235**, 139.

Osborne, A. and Provenzale, A.: 1989, *Physica* **D35**, 357.

Perdang, J.: 1985, in *Chaos in Astrophysics* NATO ARW C **161**, eds. J. R. Buchler, J. M. Perdang, and E. A. Spiegel. (D. Reidel Publ. Co., Dordrecht), p. 11.

Saitou, M., Takeuti, M., and Tanaka, Y.: 1989, *Publications of the ASJ* **41**, 297.

Scargle, J. D.: 1992, *preprint*.

Smith, L. A.: 1987, *Phys. Letters* **A133**, 283.

Takens, F.: 1981, *Dynamical Systems and Turbulence, Warwick, 1980*, Vol. 898 of *Lecture Notes in Mathematics* (Springer-Verlag, Berlin).

Theiler, J.: 1986, *Physical Review A: General Physics* **A34**, 2427.

Theiler, J.: 1990, *Physical Review A: General Physics* **A41**, 3038.

Vauclair, G., Goupil, M. J., Baglin, A., Auvergne, M., and Chevreton,M.: 1989, *Astronomy and Astrophysics* **215**, L17.

Young, A. T., Genet, R. M., Boyd, L. J., Borucki, W. J., Lockwood, G. W., Henry, G. W., Hall, D., Smith, D. P., Baliunas, S. L., Donahue, R., and Epand, D. H.: 1991, **103**, 221.

AN OVERVIEW OF STELLAR PULSATION THEORY

H. SAIO

Astronomical Institute, Tohoku University, Sendai, Japan
and
Max-Planck Institut für Astrophysik, Garching b. München, W-8046, Germany

Abstract. In this paper I will give overview of the stellar pulsation theory. Starting with basic equations I will discuss modal properties of oscillations and excitation mechanisms. I also mention briefly the effects of rotation.

1. Introduction

The pulsation is a global oscillation of a star. If the oscillation is spherically symmetric, it is called radial pulsation. One the other hand, if the oscillation does not have spherical symmetry, it is called nonradial oscillation or nonradial pulsation. Observationally, giant stars tend to pulsate radially, while main-sequence stars and degenerate stars tend to pulsate nonradially. (But there are exceptions, of course.)

Since the stellar pulsations are eigen-oscillations of the star, the pulsation frequencies contain information about the interior structure of the star. One of the classical applications of this fact is the well known period-luminosity relation of the Cepheids. The relation comes from the fact that the fundamental period is inversely proportional to the mean density of the star. More recently, the frequencies of nonradial pulsations of the sun and white dwarfs were used to infer their interior structure.

Although pulsations with finite amplitudes are nonlinear phenomena, the pulsation periods are well approximated by those from the linear analysis. Furthermore, the linear analysis tells us which star should pulsate because stellar pulsations are, in most cases, self-excited by vibrational instability (or overstability).

We start the discussion presenting the general basic equations for stellar oscillations. Then we discuss general properties of adiabatic pulsations and possible excitation mechanisms for stellar pulsations. And finally we mention briefly the effect of rotation on the pulsation frequencies.

The presentation of the material is intended for non-experts in the field. Since only basic facts are to be discussed, readers who wish to learn more details are recommended to consult a textbook; *e.g.*, Cox (1980) or Unno *et al.* (1989).

Astrophysics and Space Science **210**: 61–72, 1993.
© 1993 *Kluwer Academic Publishers.*

2. Basic Equations

2.1. EQUATIONS GOVERNING A SELF-GRAVITATING FLUID

The basic equations for the stellar pulsations are the equations of hydrodynamics with self-gravity. The continuity equation may be written as

$$\frac{\partial \rho}{\partial t} + \nabla \cdot (\rho \boldsymbol{u}) = 0, \tag{1}$$

where ρ is the density, and $\boldsymbol{u}$ is the fluid velocity caused by oscillation and
rotation in general. The momentum equation may be written as

$$\frac{\partial \boldsymbol{u}}{\partial t} + \boldsymbol{u} \cdot \nabla \boldsymbol{u} = -\frac{1}{\rho} \nabla p - \nabla \psi - \overline{\boldsymbol{V} \nabla \cdot \boldsymbol{V}} + \frac{1}{4\pi\rho} (\nabla \times \boldsymbol{B}) \times \boldsymbol{B}. \tag{2}$$

where p is the pressure, ψ the gravitational potential, $\boldsymbol{B}$ the magnetic field
and $\boldsymbol{V}$ the velocity of turbulent convection. The overbar means a small scale
average. Poisson's equation for the gravitational potential is written as

$$\nabla^2 \psi = 4\pi G \rho \tag{3}$$

with the gravitational constant G. The conservation of thermal energy may
be written as

$$\frac{dE}{dt} - \frac{p}{\rho^2} \frac{d\rho}{dt} = \epsilon - \frac{1}{\rho} \nabla \cdot (\boldsymbol{F}_R + \boldsymbol{F}_C), \tag{4}$$

where E is the internal energy per unit mass, ϵ the nuclear energy generation
rate per unit mass, $\boldsymbol{F}_C$ the energy flux carried by convection and $\boldsymbol{F}_R$ the
radiative flux which may be written as

$$\boldsymbol{F}_R = -\frac{4ac}{3\kappa\rho} T^3 \nabla T \tag{5}$$

by using a diffusion approximation, where a is the radiation constant, c the
speed of light and κ the radiative opacity.

For magnetic fields we may use the MHD approximation, in which the
evolution of magnetic fields is governed by the equation

$$\frac{\partial \boldsymbol{B}}{\partial t} = \nabla \times (\boldsymbol{u} \times \boldsymbol{B}). \tag{6}$$

To close equations we need an equation of state which gives a relation among
p, ρ and T, and the radiative opacity being expressed as a function of ρ, T
and chemical composition. Also, quantities related to convection $\overline{\boldsymbol{V} \cdot \nabla \boldsymbol{V}}$
and $\boldsymbol{F}_C$ must be given, which is the most complicated part of the theory
(see e.g., Unno and Xiong 1992 these proceedings). In most cases the effect
of convection is neglected.

2.2. Linearized Perturbation Equations

Let us consider a small perturbation around the equilibrium state. The displacement vector ξ is defined as $\xi \equiv r - r_0$, where the subscript 0 denotes a quantity at its equilibrium position. Then, the Lagrangian perturbation of velocity, δu may be written as

$$\delta u \equiv u(r_0 + \xi) - u(r_0) = \frac{d\xi}{dt} = \frac{\partial \xi}{\partial t} + u_0 \cdot \nabla \xi. \tag{7}$$

On the other hand, the Eulerian perturbation of velocity, u', is expressed as

$$u' \equiv u(r) - u_0(r) = \delta u - \xi \cdot \nabla u_0. \tag{8}$$

In the following part of this paper, we disregard magnetic fields for the sake of simplicity, and use spherical coordinates (r, θ, ϕ), in which the axis of $\theta = 0$ is equal to the axis of rotation. Furthermore, we assume that a possible velocity field in the equilibrium state is only due to a uniform rotation (i.e., $u_0 = \Omega \times r$) and that the equilibrium state is axisymmetric. The temporal and azimuthal dependence of the perturbed quantities may then be written as $\exp[i(\sigma t + m\phi)]$. The perturbation equations for eqs. (1)–(5) can be expressed as follows:

$$\rho' + \nabla \cdot (\rho \xi) = 0, \tag{9}$$

$$-(\sigma + m\Omega)^2 \xi + 2i(\sigma + m\Omega)(\Omega \times \xi) = -\frac{1}{\rho}\nabla p' + \frac{\rho'}{\rho^2}\nabla p - \nabla \psi' + (\overline{V \cdot \nabla V})'. \tag{10}$$

$$\nabla^2 \psi' = 4\pi G \rho', \tag{11}$$

$$i\sigma \rho T \delta S = \rho \epsilon \left(\frac{\delta \rho}{\rho} + \frac{\delta \epsilon}{\epsilon} \right) - \delta(\nabla \cdot F_R + \nabla \cdot F_C), \tag{12}$$

and

$$F'_R = F_R \left(3\frac{T'}{T} - \frac{\kappa'}{\kappa} - \frac{\rho'}{\rho} \right) - \frac{4acT^3}{3\kappa\rho}\nabla T', \tag{13}$$

where δ and the prime (') indicate Lagrangian and Eulerian perturbations, respectively. The entropy perturbation is denoted by S.

3. Adiabatic Oscillations of Nonrotating Spherical Stars

3.1. General Discussion

In many cases, the pulsation period of a star is much shorter than the thermal timescale of the envelope and therefore the nonadiabatic effects are small. In the adiabatic approximation, the governing equations become considerably simple because we can use the adiabatic relation

$$\frac{\delta\rho}{\rho} = \frac{1}{\Gamma_1}\frac{\delta p}{p} \tag{14}$$

instead of using the energy conservation and flux equations [Eqs. (12), (13)], where the adiabatic exponent Γ_1 is defined by $\Gamma_1 \equiv (\partial\ln p/\partial\ln\rho)_S$. [We also disregard the effect of convection in Eq. (10).] The governing equations for the perturbations in a nonrotating spherically symmetric star may then be written symbolically as

$$-\sigma^2\boldsymbol{\xi} + \mathcal{L}(\boldsymbol{\xi}) = 0, \tag{15}$$

where $\mathcal{L}$ is a Hermitian operator if the pressure vanishes at the stellar surface. (This is a good approximation for real and hence Hermiticity remains approximately true.) The explicit form of $\mathcal{L}$ is given in e.g., §15.2 of Cox (1980). Because $\mathcal{L}$ is Hermitian, all eigenvalues σ^2 are real, and eigenfunctions associated with different eigenvalues are orthogonal to one another; i.e.,

$$\int \boldsymbol{\xi}_i^* \cdot \boldsymbol{\xi}_j \rho d^3\mathbf{x} = 0 \qquad \text{if } \sigma_i^2 \neq \sigma_j^2. \tag{16}$$

Since σ^2 is real, the temporal behavior of the adiabatic perturbations is purely oscillatory when $\sigma^2 > 0$ or monotonic when $\sigma^2 < 0$ (dynamical instability).

We may write the term $\mathcal{L}(\boldsymbol{\xi})$ in Eq. (15) in the form

$$\mathcal{L}(\boldsymbol{\xi}) = f_1\hat{\mathbf{e}}_r + \nabla f_2, \tag{17}$$

where $\hat{\mathbf{e}}_r$ is the unit vector in the radial direction, and f_1 and f_2 consist of terms proportional to ξ_r or $\nabla_\perp \cdot \boldsymbol{\xi}_\perp$. Here $\boldsymbol{\xi}_\perp$ and $\nabla_\perp$ are defined as

$$\boldsymbol{\xi}_\perp \equiv \xi_\theta\hat{\mathbf{e}}_\theta + \xi_\phi\hat{\mathbf{e}}_\phi \quad \text{and} \quad \nabla_\perp \equiv \hat{\mathbf{e}}_\theta\frac{\partial}{\partial\theta} + \frac{\hat{\mathbf{e}}_\phi}{\sin\theta}\frac{\partial}{\partial\phi}.$$

Combining Eq. (17) with Eq. (15), we see that the angular dependence of ξ_r and $\boldsymbol{\xi}_\perp$ can be specified by using a single spherical harmonic $Y_l^m(\theta,\phi)$ as

$$\xi_r \propto Y_l^m(\theta,\phi) \quad \text{and} \quad \boldsymbol{\xi}_\perp \propto \nabla_\perp Y_l^m(\theta,\phi),$$

because a spherical harmonic $Y_l^m(\theta,\phi)$ is an eigenfunction of the operator $\nabla_\perp^2$, i.e.,

$$\nabla_\perp^2 Y_l^m(\theta,\phi) = -l(l+1)Y_l^m(\theta,\phi), \tag{18}$$

and hence

$$\nabla_\perp \cdot \boldsymbol{\xi}_\perp \propto l(l+1)Y_l^m.$$

Because of the above property of the spherical harmonics, the angular dependence of perturbed quantities can be expressed by a *single* $Y_l^m(\theta, \phi)$. Perturbed scalar variables are proportional to Y_l^m and the displacement vector is written as

$$\boldsymbol{\xi} = \left[\xi_r \hat{\mathbf{e}}_r + \xi_h \left(\hat{\mathbf{e}}_\theta \frac{\partial}{\partial\theta} + \hat{\mathbf{e}}_\phi \frac{1}{\sin\theta} \frac{\partial}{\partial\phi} \right) \right] Y_l^m(\theta, \phi) e^{i\sigma t}. \tag{19}$$

The governing equations are then reduced to differential equations of the radial coordinate only. (This property holds even for nonadiabatic oscillations; see e.g., Unno *et al.* 1989 §13.)

The effects of the angular dependence of eigenfunctions enter into the governing equations only through the terms proportional to $l(l + 1)$, which is related to the horizontal wave number of the oscillation, $\sqrt{l(l + 1)}/r$. Since the azimuthal order $m(-l < m < l)$ does not appear in the governing equations, the eigenvalues (*i.e.*, oscillation frequencies) are $(2l + 1)$-fold degenerate.

3.2. TOROIDAL DISPLACEMENTS

Inspecting Eqs. (15) and (17), we notice that these equations are also satisfied if

$$\sigma = 0, \quad \xi_r = 0, \quad \text{and} \quad \nabla_\perp \cdot \boldsymbol{\xi}_\perp = 0.$$

The last two relations are satisfied if we assume that displacements are toroidal; *i.e.*,

$$\boldsymbol{\xi} \propto \nabla_\perp \times (\hat{\mathbf{e}}_r Y_l^m). \tag{20}$$

The eigenfrequency is zero, because such toroidal displacements of a fluid do not cause any effect on the spherical equilibrium structure if there is no rotation or no magnetic field. Such modes of displacement are called trivial modes. [The oscillation modes whose displacement vectors have a form of Eq. (19) are sometimes called spheroidal modes.] If the star rotates, toroidal displacements describe Rossby waves with finite oscillation frequencies.

3.3. RADIAL PULSATIONS

Oscillations for $l = 0$ are spherically symmetric, *i.e.* radial pulsation. In this case Eq. (15) is reduced to

$$-\sigma^2 \xi_r - \frac{1}{r^4 \rho} \left(\Gamma_1 p r^4 \frac{d\xi}{dr} \right) - \frac{1}{\rho r} \left\{ \frac{d}{dr}[(3\Gamma_1 - 4)p] \right\} \xi_r = 0.$$

This equation with boundary condition ($\xi_r = 0$ at the center and $\delta p = 0$ at the surface) forms a Strum-Liouville type eigenvalue problem with the eigenvalue σ^2. Let the eigenvalue associated with the eigenfunction with n

nodes be represented by σ_n^2. The eigenvalues are ordered as $\sigma_0^2 < \sigma_1^2 < \sigma_2^2 <$
.... Thus the period of oscillation decreases as the number nodes increases, which is interpreted that the period is approximately the sound travel time between two adjacent nodes (Hansen 1972). From the above equation and the variational property of the eigenvalues the inequalities

$$(3\bar{\Gamma}_1 - 4)(-E_{grav}/I) > \sigma_0^2 > (3\bar{\Gamma}_1 - 4)4\pi G\bar{\rho}/3$$

can be derived, where $\bar{\rho}$ is the mean density, $\bar{\Gamma}_1$ an average of the adiabatic exponent Γ_1, E_{grav} the gravitational potential energy and I the moment of inertia of the star. (see §8.9, 8.10 in Cox 1980 for a derivation and discussion). If $\bar{\Gamma}_1 < 4/3$, at least the fundamental mode is dynamically unstable ($\sigma_0^2 < 0$). We note that since $(-E_{grav}/I)$ is proportional to the mean density of the star, the period of the fundamental mode $(2\pi/\sigma_0)$ is inversely proportional to the square of mean density, which states the period-mean density relation of pulsation.

3.4. p-MODES AND g-MODES

Let us now discuss the properties of nonradial pulsations. In order to make the discussion simple, let us disregard the Eulerian perturbations ψ' in this subsection. This approximation is called the Cowling approximation, and provides a good description of higher order modes. Using Eq. (19), and neglecting the Eulerian perturbation of gravitational potential, we obtain

$$\frac{1}{r^2}\frac{d}{dr}(r^2\xi_r) - \frac{g}{c_s^2}\xi_r + \left(1 - \frac{L_l^2}{\sigma^2}\right)\frac{p'}{\rho c_s^2} = 0, \tag{21}$$

$$\frac{1}{\rho}\frac{dp'}{dr} + \frac{g}{\rho c_s^2}p' + (N^2 - \sigma^2)\xi_r = 0, \tag{22}$$

where L_l and N are, respectively, the Lamb frequency and the Brunt-Väisälä frequency defined as

$$L_l^2 \equiv \frac{l(l+1)c_s^2}{r^2} \quad \text{and} \quad N^2 \equiv g\left(\frac{1}{\Gamma_1}\frac{d\ln p}{dr} - \frac{d\ln \rho}{dr}\right), \tag{23}$$

and g is the local gravitational acceleration and c_s the sound speed.

The two frequencies given in Eq. (23) play essential roles charaterizing noradial oscillations. This can be seen in a local analysis. Let us assume that

$$\xi_r, \quad p' \propto \exp(ik_r r),$$

where $|k_r| \gg 1$. Substitutuing these expressions into Eqs. (21) and (22), we obtain the dispersion relation

$$k_r^2 \simeq \frac{(\sigma^2 - L_l^2)(\sigma^2 - N^2)}{\sigma^2 c_s^2}. \tag{24}$$

The oscillation propagates in the radial direction when k_r is real. The dispersion relation shows that two types of oscillations are possible: one is called p-mode which is a propagating wave when $\sigma^2 > L_l^2$, N^2, and the other type is called g-mode which is a propagating when $\sigma^2 < L_l^2$, N^2. For extreme cases k_r^2 of p-modes increases as σ^2/c_s^2 increases, while k_r^2 of g-modes increases as $l(l+1)N^2/(\sigma^2 r^2)$ increases. The restoring force for the p-modes is the pressure force just as for the radial pulsations. Therefore the p-modes are the relatives to radial pulsations and are sound waves influenced by the gravitational field of the star. On the other hand, the restoring force for the g-modes is the buoyancy force, which works only for non-spherical symmetric perturbations.

The global oscillations of a star have a discrete spectrum of frequencies. In a simple stellar model such as a zero-age main-sequence model, the frequency range for p-modes is well separated from and higher than the frequency range for g-modes. The Lamb frequency for a given value of l monotonically decrease outward, while the Brunt-Väisälä frequency increases outward. At some zone in the star, these two frequencies have a same value, which approximately divides between the p-mode and g-mode frequency ranges (see Fig. 15.2 in Unno *et al.* 1989). The propagation zone of p-modes is in the envelope where the amplitude is large, while the propagation zone of g-modes is in the core. (However, the loci of propagation zones are opposite in white dwarfs.) The frequencies of p-modes increases as the number of nodes increases, while the frequencies of g-modes decreases as the number of nodes increases. Between the lowest order g- and p-modes ($n = 1$) for a given degree l (larger than one) there exists a mode called f-mode which has no node in the radial direction in the amplitude distribution. (See Fig. 17.2 in Cox 1980, or Fig. 14.1 in Unno *et al.* 1989. But nodes appear in a centrally concentrated model.)

As the central concentration of the star gradually increases with evolution, the Brunt-Väisälä frequency in the core and hence the frequencies of g-modes increase. When the frequency of a g-mode approaches and exceeds the frequency of the f- or a p-mode, the two frequencies undergo an 'avoided crossing' (see Fig. 15.7 of Unno *et al.*), because two different modes with a same l cannot have the same frequency. In an evolved star, the frequency range of g-modes overlaps the frequency range of p-modes. Any mode whose frequency is in the overlapping range has two propagation zones in the star; p-mode propagation zone in the envelope and g-mode propagation zone in the core. Although such a mode has dual character, the overall character may be distinguished by inspecting which propagation zone traps more pulsation energy.

In a highly evolved star, the Brunt-Väisälä frequency is extremely high in the core so that for any nonradial pulsation with a moderate frequency $N^2 \gg \sigma^2$ in the core. Then, as discussed above, the radial wave number is extremely high in the core, which means that the eigenfunction shows a rapid spatial oscillation in the core. Since the short wavelength of the spatial oscillation of the eigenfunction causes thermal dissipation of pulsation energy, it is difficult to excite a nonradial oscillation in a giant star.

For oscillations with a large radial order n, there exist asymptotic formulations for the frequencies (Tassoul 1980, see also Unno et al. 1989);

$$\sigma \simeq \pi(n + l/2) \left(\int_0^R \frac{1}{c} dr \right)^{-1} \tag{25}$$

for p-modes and

$$\sigma \simeq \frac{[l(l+1)]^{1/2}}{n\pi} \int_{r_a}^{r_b} \frac{N}{r} dr \tag{26}$$

for g-modes, where $N^2 > 0$ in a zone of $r_a < r < r_b$. We note that the separations of *frequencies* of high order p-modes is equidistant, while the separation of *periods* $(= 2\pi/\sigma)$ of high order g-modes is equidistant.

4. Excitation Mechanism

4.1. ENERGY EQUATION

In discussing the excitation of pulsations, we will use, for the sake of simplicity, the quasi-adiabatic approximation, in which the entropy perturbation is evaluated from Eq. (12) (energy conservation) using the adiabatic relation and adiabatic eigenfunctions in the right hand side of the equation. This approximation gives reasonable results only when nonadiabaticity is small.

After some manipulations using Eqs. (9)–(12) and (14) we obtain

$$\frac{dE_W}{dt} = \int_0^M \delta T \frac{d\delta S}{dt} = \int_0^M \delta T \delta \left[\epsilon - \frac{1}{\rho} \nabla \cdot (F_R + F_C) \right] dM_r, \tag{27}$$

where

$$E_W = \frac{1}{2} \int \left[(u')^2 + \left(\frac{p'}{\rho c_s} \right)^2 + \frac{g^2}{N^2} \left(\frac{p'}{\Gamma_1 \rho} - \frac{\rho'}{\rho} \right)^2 + \frac{\rho'}{\rho} \psi' \right] dM_r, \tag{28}$$

which represents the kinetic potential energies of the pulsation. [In deriving the above equations we disregarded the effect of rotation and convection in Eq. (10).] Equation (27) shows the change in pulsation energy caused by the interactions with nuclear energy generation rate and with the energy flux.

Integrating over one cycle of pulsation, we obtain an expression for the work integral of pulsation;

$$W = \oint dt \frac{dE_W}{dt}$$

$$= \frac{\pi}{\sigma} \int_0^M dM_r \left[\frac{\delta T_r}{T} \delta \epsilon_r - \frac{\delta T_r}{T} \delta \left(\frac{1}{\rho} \nabla \cdot F_R \right)_r - \frac{\delta T_r}{T} \delta \left(\frac{1}{\rho} \nabla \cdot F_C \right)_r \right], \tag{29}$$

where the temporal and angular dependence of the perturbed quantities are assumed as

$$\delta T = Re[\delta T_r Y_l^m e^{i\sigma t}], \qquad \delta \epsilon = Re[\delta \epsilon_r Y_l^m e^{i\sigma t}],$$

$$\delta \left(\frac{1}{\rho} \nabla \cdot F \right) = Re \left[\delta \left(\frac{1}{\rho} \nabla \cdot F \right)_r Y_l^m e^{i\sigma t} \right], \ldots$$

with $Re(\ldots)$ stands for the real part of the indicated quantity. The subscript r is attached to indicate that the quantity is a function of r only. All the quantities with subscript r are real since we are using the quasi-adiabatic approximation. When the work W is positive, the pulsation energy grows over one pulsation cycle, *i.e.* the pulsation is excited (overstable). Let us examine each term in the integrand of Eq. (29).

4.2. *Epsilon*-MECHANISM

The first term of the right hand side of Eq. (29) corresponds to the driving by the *epsilon*-mechanism. After some manipulations we may write

$$\int_0^M dM_r \frac{\delta T_r}{T} \delta \epsilon_r = \int_0^M dM_r \epsilon \left(\epsilon_R + \frac{\epsilon_\rho}{\Gamma_3 - 1} \right) \left(\frac{\delta T_r}{T} \right)^2, \qquad (30)$$

where $\epsilon_T \equiv (\partial \ln \epsilon / \partial \ln T)_\rho$ ($\sim 4 - 30$), $\epsilon_\rho \equiv (\partial \ln \epsilon / \partial \ln \rho)_T$ ($\sim 1 - 2$), and $\Gamma_3 - 1 \equiv (\partial \ln T / \partial \ln \rho)_S$ ($\sim 2/3$). Thus this term always gives a positive contribution to the work integral W. Physically, this term may be understood as follows: In the compression phase, temperature and hence nuclear energy generation rate are higher than in the equilibrium condition so that matter gains thermal energy. Therefore, the amplitude in the next expansion phase is larger than the previous one. In the expansion phase the nuclear energy generation rate is lower than the equilibrium value and hence the matter loses its thermal energy. Therefore, the amplitude in the next compression phase is higer than the previous one. In this way, the amplitude of pulsation gradually increases as pulsation goes on.

However, the amplitude of pulsation is so small in the nuclear burning region that the *epsilon*-mechanism is too weak to excite pulsation in most stars except very massive ($M \gtrsim 100 M_\odot$) main-sequence stars.

4.3. *Kappa*-MECHANISM

The second term on the right hand side of Eq. (29) may be written as

$$-\int_0^M \frac{\delta T_r}{T} \delta \left(\frac{1}{\rho} \nabla \cdot F_R \right)_r \sim \int_0^R \left(\frac{\delta T_r}{T} \right)^2 \frac{d}{dr} \left[\left(\kappa_T + \frac{\kappa_\rho}{\Gamma_3 - 1} \right) L_R \right] + \ldots (31)$$

where $\kappa_T = (\partial \ln \kappa / \partial \ln T)_\rho$ and $\kappa_\rho = (\partial \ln \kappa / \partial \ln \rho)_T$. This integral corresponds to the *kappa*-mechanism of driving. If radiative luminosity L_R is constant as in a radiative envelope, a region in which

$$\frac{d}{dr}\left(\kappa_T + \frac{\kappa_\rho}{\Gamma_3 - 1}\right) > 0 \tag{32}$$

helps to drive pulsation.

Let us discuss the reason for the *kappa* mechanism driving. When the above condition is satisfied, the opacity perturbation in the compressed phase of pulsation increases outward so that the radiative luminosity is blocked. Then, the zone gains thermal energy in the compressed phase. On the other hand the zone loses thermal energy in the expanding phase. As discussed previously for the *epsilon* mechanism, the pulsation energy tends to grow in this zone.

The *kappa* mechanism is responsible for pulsations of the stars in the cepheid instability strip. Also, it is recently found that the pulsation of the β-Cephei variables is excited by the *kappa* mechanism corresponding to the opacity peak caused by metal ions in a zone with a temperature of around 2×10^5K (e.g., Moskalik and Dziembowski 1992).

We note that when nonadiabaticity is very large the above excitation must be changed and complex phenomena occur (see Gautschy and Glatzel 1990).

4.4. EFFECT OF CONVECTION

The third term on the right hand side of Eq. (29) represents the effect of the perturbation of the convective flux. Since convection is turbulent, it is very difficult to evaluate this term correctly (see e.g. Unno and Xiong 1992). Moreover, the effect of convection also appears in the momentum equation (10), where the same difficulty applies. In most calculations the influence of convection on the stability of the stars is completely neglected. They give reasonable results for blue stars, where convection is weak, for example, the blue edge of the Cepheid instability strip and the β-Cephei variables. However, it is necessary to include the effect of convection in order to obtain the red edges of instability regions and to study the pulsations of red variables such as the Mira variables, where convection play an important role.

5. Effects of Rotation

For a rotating star the governing equations for adiabatic oscillations [Eqs. (9)–(11)] may be reduced to

$$-(\sigma + m\Omega)^2 \xi + 2i(\sigma + m\Omega)\Omega \times \xi + \Omega \times (\Omega \times \xi) + \mathcal{L}(\xi) = 0. \tag{33}$$

Although this equation is also correct for differentially rotating stars, we only consider the case of uniform rotation. Taking the scalar product with ξ^* and integrating over the whole volume we obtain

$$-(\sigma + m\Omega)^2 a + (\sigma + m\Omega)b + c = 0, \tag{34}$$

where

$$a \equiv \int_V \boldsymbol{\xi}^* \cdot \boldsymbol{\xi}\rho d^3\mathbf{x}, \qquad b \equiv 2i \int_V \boldsymbol{\xi}^* \cdot (\boldsymbol{\Omega} \times \boldsymbol{\xi})\rho d^3\mathbf{x},$$

and

$$c \equiv \int_V \boldsymbol{\xi}^* \cdot [\mathcal{L}(\boldsymbol{\xi}) + \boldsymbol{\Omega} \times (\boldsymbol{\Omega} \times \boldsymbol{\xi})]\rho d^3\mathbf{x}.$$

The quantities, a, b, and c can be proved to be real (Lynden-Bell and Ostriker 1967). Solving the above equation we obtain

$$\sigma + m\Omega = \frac{1}{2a}\left(b \pm \sqrt{b^2 + 4ac}\right). \tag{35}$$

For a non-rotating star ($\Omega = 0$), Eq. (35) gives

$$\sigma = \pm\sqrt{\frac{c}{a}} \equiv \pm\sigma_0.$$

We note that the sign of the frequency is physically not important because it changes only the phase of pulsation. When rotation is slow ($\sigma_0 \gg \Omega$), $4ac \gg b^2$. Then

$$\sigma = \pm\sqrt{\frac{c}{a}} + \frac{1}{2}\frac{b}{a} - m\Omega + O(\Omega^2) = \pm\sigma_0 - m\Omega(1 - C_{nl}) + O(\Omega^2).$$

Thus rotation completely lifts the $(2l + 1)$ fold degeneracy.

When the rotation frequency Ω is comparable to or larger than σ_0, which is expected for high order g-modes, eigenfunctions and eigenvalues are considerably modified from those for non-rotating stars and new features arise (see e.g., Ch. 6 in Unno et $al.$ 1989, or Saio and Lee 1991).

A special case is the toroidal displacements given in Eq. (20), for which $c = O(\Omega^3)$ and hence $\sigma_0 = 0$. For this case Eq. (34) leads to

$$\sigma + m\Omega = 0 \quad \text{or} \quad 2m\Omega/[l(l + 1)].$$

(The next terms are of the order Ω^3.) The latter frequency corresponds to the global Rossby waves (or planetary waves).

Acknowledgements

The author is very grateful to A. Gautschy for helpful discussions and his hospitality.

References

Cox, J. P.: 1980, *Theory of Stellar Pulsation* (Princeton Univ. Press, Princeton).

Gautschy, A. and Glatzel, W.: 1990, *Monthly Notices of the RAS* **245**, 597.

Hansen, C. J.: 1972, *Astronomy and Astrophysics* **19**, 71.

Lynden-Bell, D. and Ostriker, J. P.: 1967, *Monthly Notices of the RAS* **136**, 293.

Moskalik, P. and Dziembowski, W. A.: 1992, *Astronomy and Astrophysics* **256**, L5.

Saio, H. and Lee, U.: 1991, in *Rapid Variability of OB Stars: Nature and diagnostic Value*, ed. D. Baade, ESO Conference and Workshop Proceedings No. 36, ESO, Garching.

Tassoul, M.: 1980, *Astrophysical Journal, Supplement Series* **43**, 469.

Unno, W., Osaki, Y., Ando, H., Saio, H., and Shibahashi, H.: 1989, *Nonradial oscillations of stars* (2nd ed.) (University of Tokyo Press, Tokyo).

Unno, W., and Xiong, D. R.: 1992, in *Nonlinear Phenomena in Stellar Variability*, ed. J. R. Buchler and M. Takeuti, in press.

NON-LINEAR OSCILLATIONS AND BEATS IN THE
BETA CANIS MAJORIS STARS

A. S. BARANOV

Institute of Theoretical Astronomy of USSR Academy of Sciences,
10, Kutuzov Quay, St. Petersburg, 191187

Notwithstanding a great number of hypotheses, suggested for explaining superpositions of the light- and of the velocity variations of the β Canis Majoris stars, no one of these does it satisfactorily. Possibly it is due to an inadequate elaboration of the non-linearly oscillation theory. Analysis and critical evaluation of the existing hypotheses are given by Mel'nikov and Popov (1970). Our explanation consists in existence of close frequencies corresponding to various oscillation modes which are non-linearly interacting.

Equations of motion of an ideal incompressible fluid under condition of preserving the equilibrium figure symmetry with respect to the equatorial plane (lateral oscillations) have the form (Baranov 1988):

$$\dot{q}_{1,2} = \frac{\partial H}{\partial p_{1,2}}, \quad \dot{p}_{1,2} = -\frac{\partial H}{\partial q_{1,2}}, \tag{1}$$

where variable q_1 and q_2 are in the following way connected with semi-axes a and b of ellipsoid (the variable connected with the remaining semi-axis c of an ellipsoid has been already eliminated from the equation of motion): $\exp(q_{1,2}) = a \pm b$, the Hamiltonian $H = H(q_1, q_2, p_1, p_2, t)$, where t is time. We remind that the variables describing orientation of the figure don't enter into the Hamiltonian since impulses conjugated by it are combinations of invariants of motion: $p_{3,4} = \pm(C \mp L)/2$, where C and L are values of circulation and the moment of motion quantity. Aside from the indicated integrals of motion, the equations (1) allow the energy integral. It is also necessary to take into account the condition of the mass constancy.

Although the system of equations has a canonical form this circumstance is, in fact, not utilized. Nevertheless theorems of existence and uniqueness are almost automatically extended on our problem.

Now substitute into equation (1) expansions of coordinates and impulses from equilibrium state for which are assumed ellipsoids of revolution. The detailed calculations are omitted since they are given by Baranov (1988).

In the case of a stationary state (pure rotation) the equations (1) lead to the known Maclaurin formula.

In linearized approximation the equations (1) are again easily solved.

Finally, we compare the terms, containing ε^2 – the squares of deviations from equilibrium state in both parts of equations (1). Here we have to consider the difference of the true oscillations period from that of the

Astrophysics and Space Science **210**: 73–76, 1993.
© 1993 *Kluwer Academic Publishers.*

linearized oscillations. To this we introduce the new variable τ so that $t = (1 + \epsilon^2 h + \ldots)\tau$, where quantity h is introduced so that in terms of τ functions describing oscillations have a constant period. Such refinement of the calculation is not influencing all previous calculations.

The equations of motion (1) after some transformations are reduced to the form:

$$Q'' = -\sigma_p^2 Q + \lambda_1 A^2 + \lambda_2, \tag{2}$$

where $A^2 = \tilde{C}(1 + e \sin T)/J$, σ and σ_p are frequencies of lateral and pulsating oscillations, $\tilde{C}$ is an amplitude of lateral oscillations, $T = \sigma\tau$, e is eccentricity of the meridian cross-section of an ellipsoid, $J = 2B_{11} - \Omega^2/2$, in this connection we used in accordance with Chandrasekhar (1969) the notation

$$B_{ijk\ldots} = \alpha_1\alpha_2\alpha_3 \int_0^\infty \frac{ds}{\Delta(\alpha_i^2 + s)(\alpha_j^2 + s)(\alpha_k^2 + s)\ldots},$$

$$(\Delta^2 = (\alpha_1^2 + s)(\alpha_2^2 + s)(\alpha_3^2 + s), \quad \alpha_1 = a, \ \alpha_2 = b, \ \alpha_3 = c).$$

Ω is the angular velocity of a figure rotation. Values of the constants λ_1 and λ_2 are presented by Baranov (1988).

The equation (2) may be solved in a general form

$$Q = \lambda_2/\sigma_p^2 + u(\tau) + Q_1 \cos \sigma_p\tau + Q_2 \sin \sigma_p\tau,$$

where $u(\tau)$ is solution of the equation

$$Q'' + \sigma_p^2 Q = \lambda_1 A^2, \tag{3}$$

possessing the same period as the quantity A. Since we consider purely lateral oscillations then assume $Q_1 = Q_2 = 0$. The function $u(\tau)$ is sought in the form

$$u(\tau) = \alpha_1 + \alpha_2 \sin \sigma\tau \tag{4}$$

where α_1 and α_2 are still unknown functions. Substituting expression (4) into equation (3) after transformations which are omitted here, find:

$$\alpha_1 = \frac{\lambda_1 \tilde{C}}{\sigma_p^2 J}, \quad \alpha_2 = \frac{\lambda_1 \tilde{C} e}{J(\sigma_p^2 - \sigma^2)}.$$

The case $\sigma_p \neq \sigma$ was studied (Baranov 1988).

Phenomenon of various modes oscillations resonance for an ellipsoids of revolution explains a superposition of the light variation harmonics and the radial velocity of the β Canis Majoris stars. Equality of frequencies in the linearized problem is due to common properties of spherically symmetric problems (Bhagavantam and Venkatarayudu 1951, Vilenkin 1965) having as a consequence $2n + 1$ - multiple degeneration of frequencies (n is a principal index of a spherical harmonics as an angular part of the equation of oscillations). Various modes originate from one another by turns and a linear superposition. If the linearity is disturbed a superposition is no longer a lawful operation in physical sense. On the other hand, various modes influence each other and loose independence. Therefore, in non-linear analysis there appear typical resonance terms. The situation is in a certain degree analogous to the case of the asteroidal commensurability 1:1 if the asteroid orbit is a quasi-circular but is inclined to Jupiter's orbit. Let it be emphasized that in a non-linear case frequencies of oscillations slightly deviate from predicted values of linear theory at the expense of the final amplitude. If there exist synchronously some oscillation modes then the above mentioned displacement is, generally speaking, different for each of the modes.

Considerations of a qualitative pattern described is connected with one of the interesting and important properties of self-oscillating systems – the phenomenon of a forced synchronization which is sometimes called a caputure. At sufficiently small difference between a proper frequency of the system (frequency of lateral oscillations in this paper) and a frequency of an external force (whose role belongs here to pulsating oscillations) a stable periodic motion acquires the frequency of the latter. If in such manner, the difference $\sigma_p - \sigma$ is sufficiently small, then there takes place a synchronization of frequencies. The main problem of the theory is finding the value of the capture interval i.e. the value of that largest difference of frequencies at which capture still occurs, where as by further increase of difference between frequencies the capture does not take place and there appears a special regime related to presence in the system of quasiperiodic motion with two main frequencies which the one is the frequency of pulsating oscillations and the other – more or less changed frequency of lateral oscillations (regime of beats).

It is clear that possible is the superposition of the greater number of frequencies which in absence of the simple resonance correlations between these leads to the oscillations having still less regularities. Multifrequency oscillations in many aspects remind the stochastic processes in the sense that prediction of the further course of evolution encounters if not with principal then with substantial difficulties. In the above aspect the synchronization phenomenon and that of stochasticity are contrary. Emergence of synchronism leads to suppression of stochasticity and on the contrary development of stochasticity implies the lesser degree of the oscillations synchronism of separate parts of the system.

Emphasize the principal difference of our hypothesis from the point of view expressed elsewhere on an independent origin of various frequencies met in the model of oscillations of the β Canis Majoris stars (Chandrasekhar and Lebovitz 1962). On the contrary in our scheme one considers the oscillation frequencies in the linearized problem which differ in orientation only and therefore synchronous by their character.

We remind that we have considered the homogeneous case only but it clarifies many features of more general and complex structural models since degeneration with respect to symmetry does not depend upon the concrete law of a density change.

References

Baranov, A. S.: 1988, *Pis'ma v Astronomicheskij Zhurnal* **14**, 754. (In Russian).

Bhagavantam, S. and Venkatarayudu, T.: 1951, *Theory of Groups and its Application to Physical Problems* (Andhra University).

Chandrasekhar, S.: 1969, *Ellipsoidal Figures of Equilibrium*(New Haven).

Chandrasekhar, S and Lebovitz, N. R.: 1962, *Astrophysical Journal* **136**, 1105.

Mel'nikov, O. A. and Popov V. S.: 1970, *Pulsating Stars* (Moscow), p282. (In Russian).

Vilenkin H. Ja.: 1965, *Special Functions and the Group Representation Theory* (Moscow). (In Russian).

ONE ZONE MODELING OF IRREGULAR VARIABILITY OF STELLAR CONVECTIVE ENVELOPE

W. UNNO

Research Institute for Science and Technology, Kinki University, Kowakae,
Higasi-Osaka-shi, Osaka, 577 Japan

and

D.-R. XIONG

Purple Mountain Observatory, Academia Sinica, Nanjing, 6001,
Peoples' Republic of China

Abstract. One zone modeling of the irregular variability of red super-giants is intended with regard to the nonlinear coupling of finite amplitude pulsation with convection. The nonlocal mixing length is employed for the evaluation of the convective flux, the turbulent pressure and the turbulent power of temperature fluctuations. The radial pulsation and the Boussinesq convection are assumed for simplicity. The one zone is defined as the layer having the entropy maximum and the minimum at the bottom and at the top, respectively. The quasi-adiabatic approximation is consistent with this definition in fixing the zone to the same mass range. The spatial derivatives are evaluated under the assumption of homologous changes with the equilibrium homologous parameters. Then, a set of 6 simultaneous first order nonlinear ordinary differential equations are obtained as the one zone representation of the irregular variability of the convective envelope.

Key words: Irregular variability – Convection pulsation coupling – Nonlinear dynamical system – One zone modeling

1. Introduction

Stars are dynamically stable, and so are red giants. Because of deep convective envelopes, however, the stability of red giants is relatively weak, and the oscillation easily grows to a large amplitude with the significant driving kappa-mechanism. Saitou, Takeuti and Tanaka (1989) has shown in the one-zone treatment that the saturation of the diving power due to the large-amplitude dissipation leads to the semi-regular variability of yellow giants. An improved one-zone modeling for the irregular variability is intended in the present paper with full regards to the fundamental role of the convective envelope.

The radiation hydrodynamics has been formulated conveniently for the study of the nonlinear coupling between the convection and the pulsation (Xiong, 1977; 1989). For the qualitative study of the origin of the irregular variability, however, simplifying approximations such as the Boussinesq convection and the isotropic turbulence may well be introduced (Unno and Xiong, 1990: hereafter referred to as Paper 1). Perhaps, the intrinsic difficulty of the standard one-zone modeling in which the zone is fixed with respect to the mass element may be the Lagrangian shift of the zone of partial ionization which is crucial for the physics of the variability. In the present paper, we will define the one-zone boundaries instantaneously as the layer of the maximum and the minimum entropy. This definition of the one zone has the advantage of ensuring negligible shifts of the zone with respect to the mass element by the introduction of the quasi-adiabatic approximation which is suitable for the qualitative study.

The mathematical technique of the one-zone modeling is either setting boundary conditions as many as the number of different kinds of spatial derivatives or assuming homologous changes of variables with the equilibrium homology invariants. In the original one-zone modeling, Baker (1966) neglected all the spatial derivatives of the relative variations except for the relative luminosity change. In that case, the only boundary condition was the zero luminosity change at the lower boundary. Proper boundary conditions for the thick one-zone are approximately given by the asymptotic solutions of the linear pulsation in the upper atmosphere and the radiative interior (Okamoto and Unno, 1967). In the nonlinear treatment, however, the assumption of the homologous changes of variables within the convection zone is simpler and consistent with the one-zone modeling. In the latter case, however, since the equilibrium relative luminosity change is zero, the homologous change method is untenable for the luminosity variation and Baker's boundary condition should be employed.

In the present paper, nonhomologous changes due to the shock wave formation, for example, are disregarded. But, since the time scales of the oscillation and the convection can be comparable (Buchler, 1990) and convection contains intrinsic instability, there can be ample chance for the deep convection zone to oscillate chaotically. Further numerical studies are desired.

2. Equations of Pulsation-Convection Coupling

Equations governing radial pulsations in the convective envelope are given in Paper 1 as follows,

$$\bar{\rho} D_t(1/\bar{\rho}) = r^{-2} \, \partial_r(r^2 \bar{u}), \tag{1}$$

$$\bar{\rho} D_t \bar{u} = -\partial_r(\bar{p} + \bar{p}_t) + r^{-2}\partial_r(r^2 \mu_t \partial_r \bar{u}), \tag{2}$$

and

$$\bar{\rho}\bar{T} D_t \bar{S} = -r^{-2}\partial_r[r^2(\bar{F}_R + \bar{F}_C)] - \bar{\rho} D_t \bar{E}_t - \bar{p}_t(r^{-2}\partial_r)(r^2 \bar{u}) + \mu_t(\partial_r \bar{u})^2, \tag{3}$$

where overbars indicate the horizontal average (radial oscillation!),

$$D_t = \partial_t + \bar{u}\partial_r, \tag{4}$$

$$F_R = -C_p \kappa_R (\partial_r T) \quad \text{with} \quad \kappa_R = 4acT^3/(3C_p \bar{\kappa}\rho), \tag{5}$$

$$F_C = \rho(h' + u'^2/2)u_r'. \qquad h_t = E_t + (p_t/\rho) = (5/2)u'^2, \tag{6}$$

and μ_t denotes the turbulent viscosity given by

$$\mu_t = R_{\text{eff}}^{-1} \bar{\rho}(\overline{u'^2}/3)^{1/2} H_p, \tag{7}$$

R_{eff} being the effective Reynolds number($\approx 10[1 + (\tau_t/\tau_o)^2]$),and H_p the pressure scale height.

The convection quantities $\overline{F_C}$ and $\bar{p}_t$ representing the coupling of pulsation with convection in the above equations are governed by the following equations (see Paper 1 for the derivation),

$$(D_t + 2\mu_t k^2)(\bar{p}_t/\bar{\rho}) = (4/9)\{-\bar{p}_t[r^{-2}\partial_r(r^2 \bar{u})] + \alpha(g + D_t\bar{u})(\vec{F}_C - \vec{F}_K)\}, \tag{8}$$

$$[\bar{\rho}D_t + (\mu_t + \kappa_t + \kappa_R)k^2](H/\bar{\rho}) = -(2/3)H(\partial_r \bar{u})$$

$$+(2/3)\rho_T G(g + D_t\bar{u}) - \bar{p}_t[\partial_r(\bar{h} + \bar{h}_t) - \bar{\rho}^{-1}\partial_r(\bar{p} + \bar{p}_t)], \tag{9}$$

and

$$[\bar{\rho}D_t + 2(\kappa_t + \kappa_R)k^2 + (C_{p,T} + 1)\bar{\rho}(D_t\bar{T}/\bar{T})](G/\bar{\rho})$$

$$= -2(C_p\bar{T})^{-1}H[\partial_r(\bar{h} + \bar{h}_t) - \bar{\rho}^{-1}\partial_r(\bar{p} + \bar{p}_t)], \tag{10}$$

where

$$G = C_p\bar{\rho}\overline{T'^2}/\bar{T}, \tag{11}$$

$$\rho_T = -(\partial \log\rho/\partial \log T)_p, \qquad C_{p,T} = (\partial \log C_p/\partial \log T)_p, \tag{12}$$

$$\overline{F_K} = (1/2)\overline{\rho u'^2 u_r} = -(1/2)\kappa_t[\partial_r(\bar{p}_t/\bar{\rho})]. \tag{13}$$

The turbulent conductivity κ_t is taken to be equal to the turbulent viscosity μ_t, $\kappa_t = \mu_t$ (cf. Xiong and Chen, 1992).

3. The One-Zone Modeling

For irregular variability which depends critically on the parameters of the system, one-zone modeling is dispensable for overall qualitative studies. We define the one-zone convection model by the zone of decreasing entropy with height so that the entropy is maximum at the base and minimum at the top. The technique of the one-zone modeling lies in the replacement of the spatial derivatives with the corresponding ratios of the physical quantities. Thus, the only independent variable of the system will be the time t. To achieve this goal, we first transform the variables from the (t, r) coordinate system to the $(t, \bar{S})$ coordinate system to write down the above system of equations. We write s instead of $\bar{S}$ hereafter for brevity. We also employ the quasi-adiabatic approximation, $D_t \bar{S} = 0$, in the transformation except in the energy equation. Then, the time-derivative of a quantity Q, $D_t Q$, in the (t, r)-representation is simply described by $d_t Q$ in the (t, s)-representation. Also, since the convection zone in the above definition is occupied by the same material, the assumption of the homologous variation within the zone reduces the spatial derivative, $\partial_r Q$, to $[Q_r](Q/r)$ in which $[Q_r]$ denotes the homology invariant,

$$[Q_r] = \partial \log Q / \partial \log r. \tag{14}$$

The homology invariants are taken to be the same as the equilibrium value by assumption. For instance, we have

$$r^{-2}\partial_r(r^2 \bar{u}) = 3(\bar{u}/r), \tag{15}$$

$$\partial_r p = -V(p/r) \quad \text{with} \quad V = (GM_r\rho/rp)_{eq}, \tag{16}$$

and

$$\partial_r \bar{p}_t = -(1/3)[4 + (1 - \nabla)V](\bar{p}_t/r) \quad \text{with} \quad \nabla = (\partial \log T/\partial \log p). \tag{17}$$

An approximate relation, $r^2\overline{\rho u'^3} = const.$, has been used to derive the above $[(\bar{p}_t)_r]$ value.

In the energy equation, however, neither the quasi-adiabatic approximation nor the homology approximation are obviously applicable. Baker's boundary condition that the luminosity L is constant in time, L_{eq}, at the base of the zone (Baker, 1966; Saitou et al., 1990) so that

$$(\bar{\rho} r^2)^{-1}\partial_r L = 2(L - L_{eq})/m, \tag{18}$$

m being the mass of the zone, should be used. The application of the above procedures of the one-zone modeling to the set of equations given in the last section yields the required one-zone equations given in the following section.

4. Summary

The results are summarized below. Omitting overbars from the symbols, we obtain the following set of one-zone equations,

$$\rho d_t(1/\rho) = 3(u/r), \tag{19}$$

$$\rho d_t u = V(p/r) - V_t(p_t/r) - \rho g - k_0^2(\mu_t u/r^2), \tag{20}$$

where

$$V_t = (V/3)(1 - \nabla) \quad \text{and} \quad k_0^2 = (V/3)(2 + \nabla) - 4, \tag{21}$$

for the equilibrium model by the mixing-length theory, and

$$C_p[d_t T - \nabla_{ad}(T/p)d_t p]$$
$$= -2(L - L_{eq})/m - (3/2\rho)[d_t p_t + 5 p_t u/r] + (\mu_t/\rho)(u/r)^2, \tag{22}$$

where

$$L = 4\pi r^2 (F_R + F_C), \qquad F_R = -(16V\nabla/3)(acT^4/\bar{\kappa}\rho r), \tag{23}$$

$$F_C = H + F_K, \quad \text{and} \quad F_K = -(V\nabla/3)(\kappa_t p_t/r\rho). \tag{24}$$

The convection variables p_t, H, and G are governed by the following set of one-zone convection equations,

$$[d_t + 2\mu_t(\pi/l_e)^2](p_t/\rho) = (4/9)[-3p_t(u/r) + \rho_T(\rho/T)(g + d_t u)H], \tag{25}$$

$$[\rho d_t + (\mu_t + \kappa_t + \kappa_R)(\pi/l_e)^2](H/\rho) = -(2/3)H(u/r)$$
$$+(2/3)\rho G(g + d_t u) + p_t[C_p(T/r)V(\nabla - \nabla_{ad}) - 6V_t(p_t/\rho r)], \tag{26}$$

and

$$[\rho d_t + 2(\kappa_t + \kappa_R)(\pi/l_e)^2 + (C_{p,T} + 1)\rho(d_t T/T)](G/\rho)$$
$$= -2(C_p T)H[C_p(T/r)V(\nabla - \nabla_{ad}) - 6V_t(p_t/\rho r)], \tag{27}$$

where l_e denotes the mixing length ($\simeq$ scale height).

References

Baker, N.H.: 1966, in *Stellar Evolution*, eds. R.F.Stein and A.G.F. Cameron (Prenum Press, New York), p. 333.

Buchler, J.R.: 1990, in *Nonlinear Astrophysical Fluid Dynamics, Annals New York Acad. Sci.*, **617**, 17.

Saitou, M., Takeuti, M., and Tanaka, Y.: 1989, *Publications of the ASJ* **41**, 297.

Unno, W. and Xiong, D.-R.: 1990, in *Progress of Seismology of the Sun and Stars*, eds. Y. Osaki and H. Shibahashi (Springer-Verlag), p. 103.

Xiong, D.-R.: 1977, *Acta Astronomica Sinica* **18**, No.1, 26.

Xiong, D.-R.: 1989, *Astronomy and Astrophysics* **209**, 126.

Xiong, D.-R. and Chen, J.-J.: 1992, *Astronomy and Astrophysics*, in press.

ACCRETION DISK INSTABILITIES

S. MINESHIGE

Department of Physics, Ibaraki University, Mito 310, Japan

Abstract. Basic properties of accretion disk instabilities are summarized. We first explain the standard disk model by Shakura and Sunyaev. In this model, the dominant sources of viscosity are assumed to be chaotic magnetic fields and turbulence in gas flow, and the magnitude of viscosity is prescribed by so-called α model. It is then possible to build a particular disk model. In the framework of the standard model, accretion disks are stationary, but when some of the basic assumptions are relaxed, various kinds of instabilities appear. In particular, we focus on the thermal limit-cycle instability caused by partial ionization of hydrogen (and helium). We demonstrate that the disk instability model well accounts for the basic observed features of outbursts of dwarf novae and X-ray nova. We then introduce other kinds of instabilities based on the α viscosity model. They are suspected to produce time variabilities observed on a wide range of timescales in close binaries and active galactic nuclei.

1. Introduction

It is widely believed that accretion disks play important roles in a variety of astrophysical objects, including active galactic nuclei (AGN), close binary systems, such as cataclysmic variables (CVs) and X-ray binaries (XBs), and protoplanetary systems. Disk accretion is indeed a very efficient mean to convert rest-mass energy of infalling material into radiation energy; huge energy output is thus possible in AGN disks. Observed high-energy phenomena in XBs and AGN are often interpreted as the activities of disks or disk coronae. Protoplanetary disks provide places for planet formation. Moreover, jets or winds in AGN and protostars seem to be generated in the inner portions of the accretion disks.

Suppose a point mass star with gravitational potential, $\Phi = GM/r$ (where M is the mass of the central star and r is the distance from the star), embedded in gas clouds. If gas particles in ambient clouds are at rest with respect to the central star; namely, when clouds have no systematic angular momentum, gas particles will directly fall onto the central star due to the gravitational force. This is a case of spherical accretion.

If a gas particles in clouds are rotating around the star in the same direction, i.e., if a gas cloud has an angular momentum, on the other hand, gas cannot directly accrete onto the star, but it will rotate around the star on an eccentric orbit. The orbit of each gas particle is characterized by its semi-major axis, a, and its specific angular momentum, l. After a few rotation, each gas orbit is expected to be circularized by losing eccentricity through collision between particles. The formation of a circular ring or disk is therefore highly plausible. The radius of the circular ring, the circularization radius, is related to its specific angular momentum by,

Astrophysics and Space Science **210**: 83–103, 1993.
© 1993 *Kluwer Academic Publishers.*

$$r_{\text{circ}} = \frac{l^2}{GM}. \tag{1}$$

To sum up, the coexistence of a deep potential well and ambient gas clouds with an angular momentum with respect to the central potential well leads to the formation of a rotating ring or disk.

This is. however, a rotating disk and not an "accretion disk." For rotating gas to accrete onto the central star, each gas particle should somehow lose its angular momentum. We are now encountered with the fundamental issue in the theory of viscous accretion disks; viscosity provides a mechanism for angular momentum transport and heating of plasmas, but the ordinary molecular viscosity is too small to cause mass accretion in reasonable timescales (less than the age of the universe). We therefore require that disk plasmas be turbulent and/or full of chaotic magnetic fields. Turbulence and magnetic fields are dominant sources of viscosity. This is the basic assumption adopted in the theory of accretion disks (Pringle and Rees 1972; Shakura and Sunyaev 1973; Novikov and Thorne 1973; for a review, see Pringle 1981).

The plan of this review paper is as follows: §2, we explain the standard model (hereafter referred to as SS model). In this model, disks are assumed to be quasi-steady. However, by relaxing some of basic assumptions of SS model, we can derive various kinds of disk instabilities, which will manifest themselves as some observable effects. The most successful example as such is the thermal limit-cycle instability discussed in §3. This instability is caused by partial ionization of hydrogen and helium and is thought to trigger outbursts of CVs and LMXBs. Other kinds of instabilities are briefly reviewed in §4. Final section is devoted to summary.

2. Standard Disk Model

2.1. BASIC ASSUMPTIONS

The standard accretion disk model by Shakura and Sunyaev (1973) is constructed on the following assumptions:

(a) Gas in the disk is rotating around a central star on a circular orbit with the local Keplerian velocity given by,

$$v_\varphi = \sqrt{\frac{GM}{r}}, \tag{2}$$

and gradually moves inward with radial velocity, v_r,

$$v_r \ll v_\varphi, \tag{3}$$

by losing angular momentum. Here we use the cylindrical coordinate, (r, φ, z) with the z axis perpendicular to the disk plane and the r axis in the central

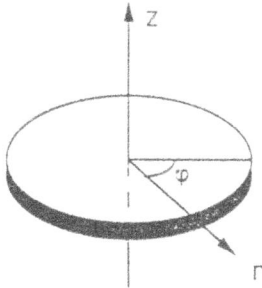

Fig. 1. Cylindrical coordinate, (r, φ, z), used in this review.

plane; $z = 0$ (see Fig. 1). If we define the viscous timescale and the dynamical timescale as

$$t_{\rm vis} \equiv \frac{r}{v_{\rm r}}, \quad t_{\rm dyn} \equiv \frac{r}{v_\varphi}, \tag{4}$$

we find from equation (3)

$$t_{\rm vis} \gg t_{\rm dyn}. \tag{5}$$

(b) The disk is in stationary state in the sense that the mass inflow rate,

$$\dot{M} = -2\pi r v_{\rm r} \Sigma, \tag{6}$$

(which is positive for inflow) is constant in space and time, where

$$\Sigma = \int_{-\infty}^{\infty} \rho dz, \tag{7}$$

is surface density.

(c) Disks are geometrically thin,

$$H \ll r, \tag{8}$$

where H is the half thickness of the disk. Typically we find $H/r \sim 1/30$. It is then reasonable to use one-zone approximation to describe the vertical structure of the disk, in which all the derivatives with respect to z are replaced by $1/H$. Furthermore, we can safely decouple two-dimensional equations (in r and z directions) into a set of two, separate, one-dimensional equations: those describing the radial (r) structure and those for the vertical (z) structure.

(d) Hydrostatic balance holds in the vertical (z) direction. This condition is under one-zone approximation written as,

$$\frac{P}{H} = \rho g_z, \tag{9}$$

where g_z is the vertical component of the gravity at the surface of disk $(z = H)$ due to the central star, and is

$$g_z \simeq \frac{GM}{r^2}\frac{H}{r} = \Omega^2 H, \tag{10}$$

ρ is the mean density of the disk, and Ω is the Keplerian angular frequency,

$$\Omega = \frac{v_\varphi}{r} = \sqrt{\frac{GM}{r^3}}. \tag{11}$$

(e) The disk is optically thick;

$$\tau \gg 1, \tag{12}$$

where

$$\tau = (\kappa_{es} + \kappa_{abs})\rho H, \tag{13}$$

and κ_{es} and κ_{abs} are opacities due to electron scattering and absorption, respectively. The emergent energy flux from a unit area of the surface, which determines the cooling rate of the disk, is then

$$F = Q^- = \frac{4\sigma T^4}{3\tau}, \tag{14}$$

where σ is the Stefan-Boltzmann constant and T is the mean disk temperature (usually represented by a value at $z = 0$).

(f) The $r - \varphi$ component of the shear stress tensor is prescribed as

$$\sigma_{r\varphi} = -\alpha P, \tag{15}$$

where local pressure P is given by the sum of gas pressure and radiation pressure;

$$P = P_{gas} + P_{rad}, \tag{16}$$

and α is the viscosity parameter ($\alpha < 1$). Viscosity in the accretion disk plays two important roles: angular momentum transport and heating of disk plasmas. The heating rate of the disk is then

$$Q^+ = \frac{3}{4}W\Omega, \tag{17}$$

where W is vertically integrated viscous stress,

$$W \equiv \int_{-\infty}^{\infty} \alpha P dz. \tag{18}$$

(g) Finally we assume energy balance. Potential energy of accreting gas is first converted to thermal energy via viscous process, and then is released as radiation. Therefore, local energy balance at each radius is given by

$$Q^+ = Q^-, \tag{19}$$

where heating is due to viscous heating [eq. (17)] and cooling is via black-body radiation [eq. (14)].

2.2. Stationary Relations

Stationary conditions for radial disk structure are derived from the following basic equations;

$$\frac{\partial \Sigma}{\partial t} = \frac{1}{2\pi r}\frac{\partial \dot{M}}{\partial r}, \tag{20}$$

$$\dot{M}\sqrt{\frac{GM}{r}} = 4\pi\frac{\partial}{\partial r}\left(r^2 W\right), \tag{21}$$

where the former equation represents mass conservation and the latter describes angular momentum transport.

By setting $\partial/\partial t = 0$ in equation (20), we have constant $\dot{M}$ and

$$F = \frac{3}{8\pi}\frac{GM\dot{M}}{r^3}\left[1 - \sqrt{\frac{r_*}{r}}\right], \tag{22}$$

where r_* is the radius of the inner edge of the disk and we used relations (19) and (21). The emergent local flux is hence independent of the magnitude of the viscosity, and is proportional to r^{-3} at $r \gg r_*$. The larger α is, the smaller becomes the disk mass, M_d, and so does surface density, Σ. For a given mass-input rate, however, the total energy emitted locally, F, and the total disk luminosity, L_d, are independent of α, because increase (decrease) in radial flow velocity caused by increase (decrease) in the α value totally compensates with the decrease (increase) in the disk mass.

Now we obtain expressions for physical quantities, such as T, H, Σ, W, τ, as functions of radius r for given parameters M, $\dot{M}$, and α.

Effective temperature of the disk, defined as

$$T_{\text{eff}} \equiv (F/\sigma)^{1/4} \propto r^{-3/4} \quad \text{at } r \gg r_*, \tag{23}$$

has radial dependence in accretion disks (see Fig. 2). For fixed M and r, T_{eff} increase as $\dot{M}$ increases.

2.3. Comparison Between Disk Models and Stellar Atmosphere Models

As we have seen previously, vertical disk structure is in many respects analogous to stellar atmosphere structure; e.g., gas layer is in hydrostatic balance, excess energy is carried towards the surface by radiation, and local energy balance holds. There are, however, some discrepancies between them:

(a) One single star has only one value of T_{eff} under spherically symmetric assumption, whereas T_{eff} in a stationary disk is radially dependent. Roughly speaking, an accretion disk is composed of ensemble of different types of stellar atmospheres with radially different T_{eff} and surface gravity g. (Note,

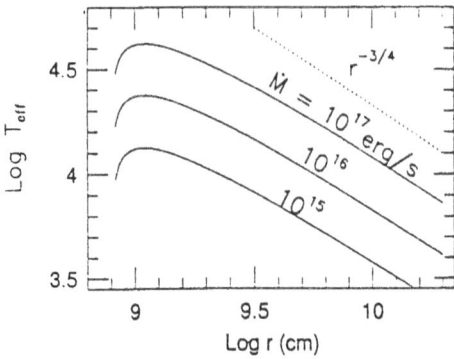

Fig. 2.   Effective temperature distribution of the stationary disks for given values of $\dot{M} = 10^{15}$, 10^{16} and 10^{17} erg s^{-1}. Here the mass of a central star is $M = 1M_\odot$.

however, disk structure is not exactly equal to that of stellar atmospheres; see below.) Disk spectra are thus composed of multi-color spectra.

(b) Energy generation in a star takes place only at a core. Hence energy generation rate is practically zero, $\epsilon = 0$, in stellar atmospheres. In contrast, viscous heating is everywhere available in a disk, so $\epsilon \neq 0$ in the disk structure (see eq. 17 and 18).

(c) Gravity is nearly constant in stellar atmospheres, whereas $g_z(z) \propto z$ in a disk.

3. Thermal Limit-Cycle Instability

Now we are at the position to discuss the stability of accretion disks based on the α model. We first describe the thermal limit-cycle instability, which has most successively linked the theory of time-dependent accretion disk models and the observed time variabilities in close binary systems.

3.1. Partial Ionization of Hydrogen in The Disk

It is well known that the partial ionization process of hydrogen in stars provides so-called κ-mechanism for stellar pulsation (see, e.g., a review by Saio 1992 in this volume). When plasma temperature is low enough for hydrogen to recombine, the opacity κ changes quite rapidly with change in temperature; $\kappa \propto T^a$ with $a = 5 - 10$. The outgoing radiation is thus efficiently blocked in a compressed phase of stellar pulsation because of enormous increase in κ in higher temperature regime compared with the value in an expanded phase, leading to overstable oscillation. Partial ionization process of hydrogen thus manifests itself as visible effects in stars.

The question then arises as to what visible phenomenon is expected in accretion disk when partial ionization process of hydrogen sets out. Rapid changes in opacity due to ionization and recombination of hydrogen and helium are very likely to produce some visible effects in disks as well, because hydrogen is most abundant in disks and the partial ionization process is a basic atomic process.

The answer to this question is the thermal limit-cycle instability, which yields sporadic light variation in disk luminosity. It is now widely believed that this mechanism causes outbursts of dwarf novae (DN), a subgroup of CVs, (Hoshi 1979; Meyer and Meyer-Hofmeister 1981; Cannizzo et at 1982; see also Smak 1984a for a review), and those of soft X-ray transients (or X-ray novae, XN), a subgroup of low mass X-ray binaries (e.g., Mineshige and Wheeler 1989; Mineshige et al. 1990).

3.2. S-Shaped Thermal Equilibrium Curves

To study time-dependent properties of unstable accretion disks, we need to vertically integrate the disk structure, because the presence of vertical convection is a key to understand the instability (Meyer and Meyer-Hofmeister 1981, 1982; Faulkner et at 1983; Smak 1982a; Mineshige and Osaki 1983). Note that one-zone approximation, introduced in §2, is valid only for radiative disks, where convective energy transport is negligible.

Basic differential equations for vertical structure are equations of continuity,

$$\frac{d\Sigma_z}{dz} = -2\rho, \tag{24}$$

hydrostatic balance,

$$\frac{dP}{dz} = -\rho g_z, \tag{25}$$

energy balance,

$$\frac{dF}{dz} = \frac{3}{2}\alpha P\Omega, \tag{26}$$

and energy transport,

$$\frac{d\ln P}{d\ln T} = \nabla_{\text{rad}} \quad \text{or} \quad \nabla_{\text{conv}} \tag{27}$$

with supplementary equations

$$\frac{d\tau}{dz} = -\kappa\rho, \tag{28}$$

$$\frac{dW}{dz} = -\alpha P, \tag{29}$$

where $\Sigma_z(z)$ is surface density integrated from the surface ($z = H$) to z, F_z is energy flux at z, and $\nabla_{\rm rad}$ and $\nabla_{\rm conv}$ are pressure gradient for radiative energy transport and convective energy transport, respectively. Convective gradient is calculated by the mixing-length theory (Cox and Giuli 1968; Paczyński 1969) and Rosseland-mean opacity, κ, is taken from the tables by Cox and Stewart (1970) and Alexander et al. (1983).

We integrate these equations from the surface at $z = H$ (where H is a parameter) with the surface boundary conditions,

$$\Sigma_z = 0, \quad T = T_{\rm eff}, \quad \dot{F}_z = \sigma T_{\rm eff}^4,$$

$$\tau_1 = 2/3, \quad {\rm and} \quad \kappa(\rho_1, T_{\rm eff})P(\rho_1, T_{\rm eff}) = \tau_1 g_z(H), \tag{30}$$

towards the center ($z = 0$). Here the last relation in eq.(30), derived from the combination of equations (25) and (28), gives a condition for density at the surface, ρ_1. The value of the scale height H is chosen so that the following central boundary conditions are to be met,

$$F_z = 0, \quad {\rm at} \quad z = 0. \tag{31}$$

For fixed values for r, α, and M, we get a family of solutions as functions of $\dot{M}$. Figure 3 schematically displays equilibrium relation between the mass flow rate out of a ring at r, $\dot{M}_{\rm out}$ (see Fig. 4) and Σ. The essence of the thermal instability is summarized in this S-shaped equilibrium curve. There exist three solutions for a certain range of Σ; $\Sigma_B < \Sigma < \Sigma_A$. Hydrogen is fully ionized in the high state (HII state), and it is neutral in the low state (HI state). These two branches are thermally and viscously stable (see, e.g., Mineshige and Osaki 1983), whereas the middle state, in which hydrogen is partially ionized, turns out to be unstable against thermal and viscous instabilities. When the mass input rate into a ring, $\dot{M}_{\rm in}$, falls onto a range, $\dot{M}_A < \dot{M} < \dot{M}_B$ (see Fig. 3), limit-cycle behavior of the disk unavoidably appears. The presence of such limit cycles was intuitively predicted by Osaki (1974).

Suppose the disk lying in the low state. Since $\dot{M}_{\rm in} > \dot{M}_{\rm out}$ in this state, Σ (and disk mass) should increase. The evolutionary track of such an unstable disk follows the low equilibrium line up to point B as indicated by an arrow in Fig. 3. Since there is no low state at $\Sigma > \Sigma_B$ and $Q^+ > Q^-$ on the right side of the S shape, disk temperature will increase. Hydrogen in the disk starts to get ionized and eventually the disk jumps to the HII hot state, where $\dot{M}_{\rm out} > \dot{M}_{\rm in}$ holds. The disk mass (or Σ) should decrease there, as the disk gas falls onto the central star. At point A, the thermal instability again takes place and the disk falls into the HI cool state. All the hydrogen in the disk is neutral there. Now one limit cycle is completed.

In summary, any accretion disks with mass-input rate in a range,

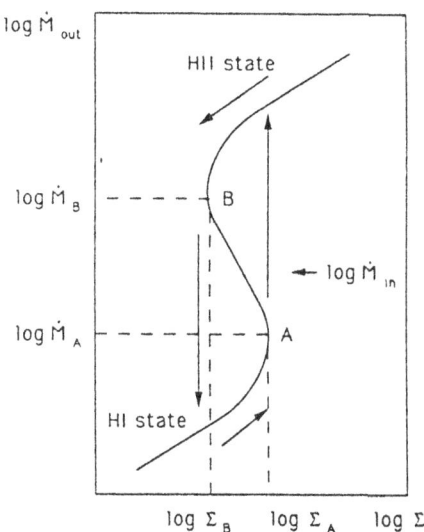

Fig. 3. Schematic thermal equilibrium curve in dwarf-nova disks. Hydrogen is fully ionized (HII) on the upper branch, partially ionized on the middle branch, and are neutral (HI) on the lower branch. The middle solution between points A and B is thermally unstable. When the mass input rate falls onto a range between $\dot{M}_A$ and $\dot{M}_B$, the disk alternates between two stable states, as indicated by arrows.

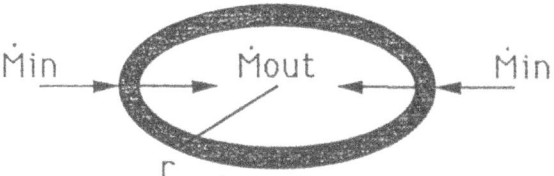

Fig. 4. Mass outflow, $\dot{M}_{out}$, from a ring at radius r and mass inflow, $\dot{M}_{in}$, into a ring.

$$\dot{M}_A < \dot{M}_{in} < \dot{M}_B, \tag{32}$$

inevitably produces L_d variation due to the relaxation oscillations of the disk between HI and HII states.

Smak (1982b) plotted $\dot{M}$ and r_d for some individual CVs, and found that those CVs with $\dot{M}$ satisfying the criterion (30) at $r = r_d$ exhibit DN outbursts, while those with $\dot{M} > \dot{M}_B$ have no outbursts and are thus classified as nova-like objects. This is a direct evidence for the disk-instability model for outbursts of DN. A similar plot was attempted for XN by Cannizzo et al. (1985), who confirmed the same result for X-ray binaries.

3.3. TIME EVOLUTION OF UNSTABLE DISKS

Local behavior of the unstable disk is characterized by the thermal limit cycles. Next, we need to simulate the global response of the local instability. The time-dependent basic equations for radial disk structure are equations

(20), (21), and vertically averaged energy equation;

$$\frac{C_p \Sigma}{2} \left(\frac{\partial T}{\partial t} + v_r \frac{\partial T}{\partial r} \right) = Q^+ - Q^- + \frac{H}{r} \frac{\partial (r F_r)}{\partial r}, \tag{33}$$

where F_r is the radial heat flux;

$$F_r = \frac{4acT^3}{3\kappa\rho} \frac{\partial T}{\partial r}. \tag{34}$$

They are diffusion type equations for gas (eqs. 20 and 21) and for heat (eqs. 33 and 34). By means of time dependent simulations, we can see how local instability is spatially propagated (Papaloizou et al. 1983; Smak 1984b; Mineshige and Osaki 1985; Cannizzo et al. 1986).

Note that this is a typical example of the dissipative structure (Nicolis and Prigogine 1977). We may find good examples of the dissipative structure in various contexts of astrophysical plasmas. Spatial pattern formation of three-phased interstellar medium is discussed by Tainaka et al. (1992) in this volume.

The results of transition wave propagation are displayed in Figure 5, which is taken from Mineshige (1987). Note that sharp peaks in the heating front in Σ diagram look like shock feature, but front speed is $v_F \sim \alpha C_s$ (where C_s is the sound velocity), and is subsonic (recall $\alpha < 1$; Meyer 1984).

The mass accretion rate at $r = r_*$, the V-magnitude, and the total disk luminosity (Mineshige and Wheeler 1989).

$$L_d = \int_{r_*}^{r_d} 2 \cdot F(r) 2\pi r dr, \tag{35}$$

is depicted in Fig. 6 for the case of X-ray nova (Mineshige and Wheeler 1989). The basic features of the theoretical light curves calculated based on the disk instability model are in good agreement with those of the observed X-ray and optical light curves of DN and XN.

Finally the same mechanism could operate in the disks in galactic nuclei and may be responsible for AGN phenomena (Mineshige and Shields 1990). It is also suggested that the thermal limit cycles cause FU Ori eruptions in protostar systems (Lin and Papaloizou 1985).

4. Other Disk Instabilities

In this section, we briefly introduce the fundamentals of other disk instabilities together with some observational implications.

4.1. INSTABILITIES IN RADIATION-PRESSURE DOMINATED DISKS

As long as the α prescription (15) is adopted, accretion disks are thermally and viscously unstable when radiation pressure, P_{rad}, dominates over gas

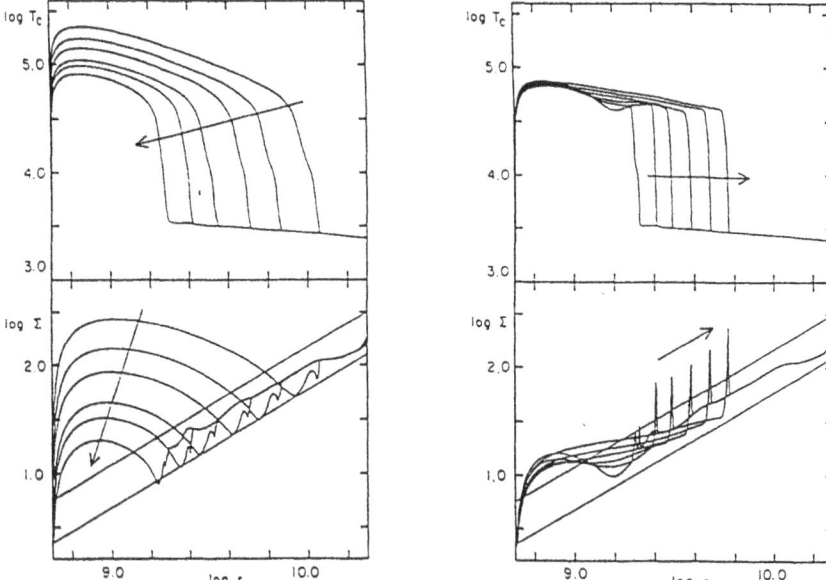

Fig. 5. Dynamics of transition wave propagation in dwarf novae. The left panels depict the time evolution in the central temperature (upper) and in surface density (lower) distribution in the disk in the decay phase, whereas the right panels display that in the rising phase.

pressure, P_{gas}; $P_{rad} > P_{gas}$ (Lightman and Eardley 1974; Shibazaki and Hoshi 1975; Shakura and Sunyaev 1976). The fate of these instabilities are, however, still open to question at this moment; Shapiro et al. (1976) argued that the instabilities may lead to the formation of two-temperature, optically thin disks, while the slim disk model by Abramowicz et al. (1988) asserts that a third branch is realized by efficient advective cooling in the relativistic disks, so that the disk can undergo limit cycles, similarly to the case of DN or X-ray binaries (see §3).

Let us examine the latter model in some details. The thermal equilibrium curve is schematically depicted in Fig. 7. In P_{rad} dominated disks, diffusion timescale t_{vis} becomes comparable to the thermal timescales t_{th}, where

$$t_{th} \equiv \frac{C_p \Sigma T}{Q^+} \sim \frac{1}{\alpha \Omega}. \tag{36}$$

We thus have $t_{vis} \geq t_{th}$, so the disk trajectory in the $(\Sigma, \dot{M})$ plane is somewhat distinct from the case of DN (cf. Fig. 3). Evolutionary paths are also indicated in Fig. 7 by arrows. When point C is reached, $\dot{M}$ increases by the onset of the thermal instability. In the meantime, the value of Σ decreases because of short viscous timescale. Detailed calculations have been performed

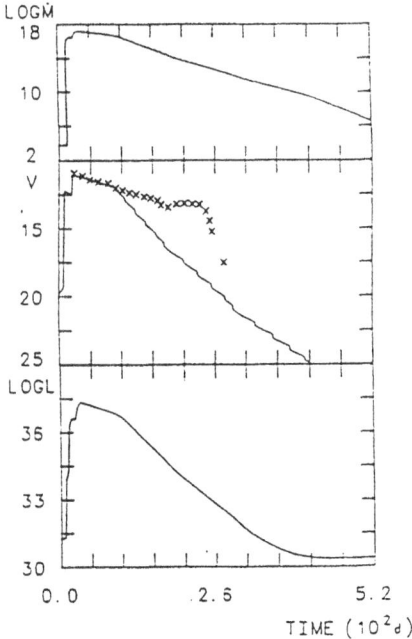

Fig. 6.   Theoretical light curves of X-ray nova outbursts. From top, the mass-accretion rate at the inner edge of the disk, the V-magnitude, and the bolometric luminosity. Crosses represent the observed V-magnitudes of the black-hole nova A0620-00 by Whelan et al. (1977).

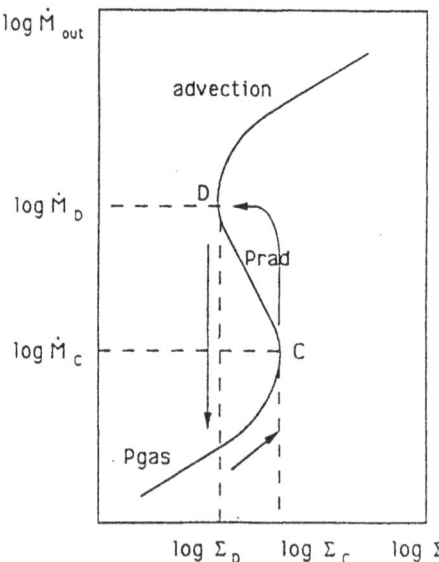

Fig. 7.   Schematic thermal equilibrium curve in the slim disk model. The disk is gas pressure dominated on the lower branch, while it is radiation pressure dominated on the middle branch. The upper branch is also radiation pressure dominated, but advective cooling is substantial on this branch, which is distinct from the middle branch. The expected trajectory is schematically displayed by arrows.

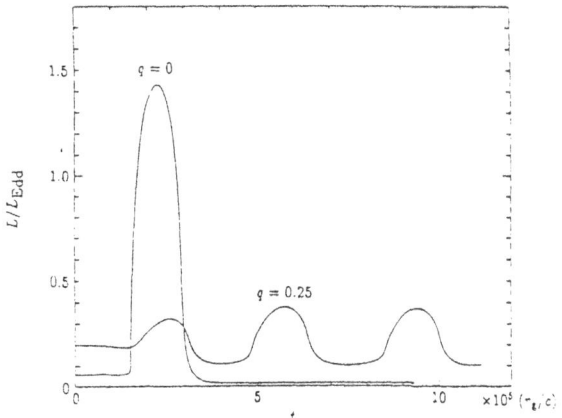

Fig. 8.   Light curves of radiation pressure-dominated disks (due to courtesy of Honma et at. 1991). Here the r_φ component of the shear stress tensor is prescribed as $\sigma_{r\varphi} = -\alpha P_{\mathrm{tot}} \beta^q$, where $\beta \equiv P_{\mathrm{gas}}/P_{\mathrm{tot}}$.

by Honma et al. (1991), who succeeded in reproducing limit cycle-type light curves as illustrated in Fig. 8.

There are some observational implications of these instabilities, which include quasi-periodic oscillation (van der Klis 1989), Type II bursts of X-ray binaries (Lewin et al. 1976), and AGN activities. For details of these observational features, see a review by Matsuoka (1992) in this volume.

4.2. INSTABILITIES IN OPTICALLY THIN DISKS

In optically thin disks, the following two process are substantial.

(a) Two-Temperature Plasmas

When a disk is optically thin, and is thus of low density, the disk plasma may be in a two-temperature state; electron temperature is less than proton temperature, $T_p > T_e$, by several orders of magnitudes. Reason for this can be explained as follows: In accretion process, a proton acquires more energy than an electron because of larger proton mass than the electron mass. Cooling of plasmas is, on the other hand, done mainly by electrons through bremsstrahlung and/or inverse-Compton cooling. In low density plasmas, coupling between protons and electrons is not strong if only Coulomb interaction is taken into account. Protons then cannot be cooled as quickly as electrons are.

By changing the efficiency of Coulomb coupling, the disk plasma may possibly alternate between a two-temperature state and a one-temperature state, thereby modulating hard X-ray intensities, although a specific mechanism for such transition is not yet known.

(b) e^+e^- Pair Generation

When the disk is optically thin, efficiency of plasma cooling is substan-

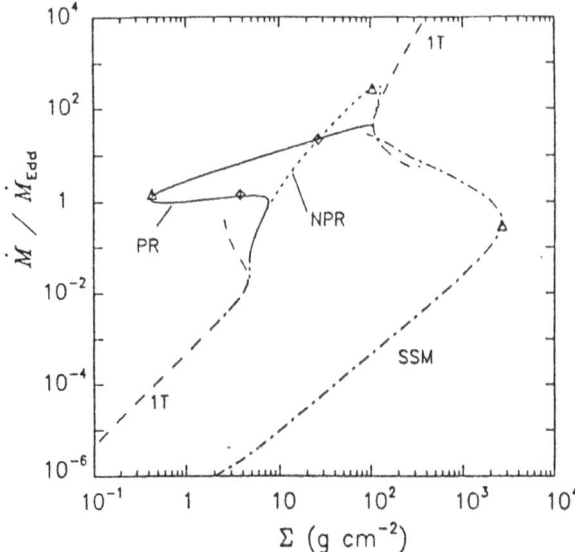

Fig. 9.   Thermal equilibrium curves of one-temperature and two-temperature disks.
One-temperature solutions are indicated by the dashed lines with a label "1T", whereas the
short-dash line labeled by NPR depicts the two-temperature solutions without e^+e^- production, and the solid line labeled by PR represents those with e^+e^- pairs. The dash-dot
line corresponds to the modified Shakura-Sunyaev solution.

tially reduced. This, in turn, increases plasma temperatures (both T_e and
T_p). When T_e increases up to the temperature corresponding to electron
rest-mass energy, $k_B T_e \sim 511$ keV (where k_B is the Boltzmann constant),
copious e^+e^- pairs will be generated.

   Thermal equilibrium curves of hot accretion disks are illustrated in Fig.
9 for a one-temperature disk and a two-temperature disk. In this figure,
the equilibrium solutions with and without e^+e^- pairs are indicated by the
solid line and the dashed lines, respectively. We find that some portions of
one-temperature, optically thin branches are thermally stable, while all the
two-temperature, optically thin branches are thermally unstable. We also
see that the equilibrium curves are substantially modified by the presence
of e^+e^- pairs. Moreover, the disk equilibrium structure can be more clearly
understood in the three dimensional $(\dot{M}, \Sigma, z)$ space, where $z = n_+/n_p$ is
the ratio of the number density of positrons to that of protons.

   In Fig. 10 we depict the cross-section of such three-dimensional equilibrium curves by fixing the value of Σ. The proton heating rate is equal to the
proton cooling rate on the thermal equilibrium curves (dashed line), while
pair generation rate balances with pair annihilation rate on the pair equilibrium curves (solid line). There are three pair-thermal equilibrium solutions
for a range of Σ. They are all thermally unstable, whereas only the middle one is unstable against perturbations in z. We now find a possibility of

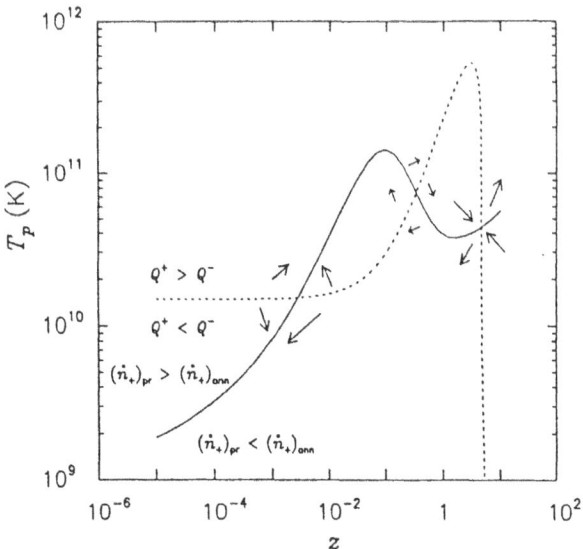

Fig. 10. Thermal equilibrium curve (by the short dashed line) and pair equilibrium curve (by the solid line) in the (z, T_p) plane. Three pair-thermal equilibrium solutions are found. Expected trajectories are also indicated by arrows. Limit cycles may be possible around the middle solution.

having limit-cycle behavior around the middle branch; When T_p increases, z will also increase, because pair generation is more efficient at higher temperature. The number of coolants (i.e., electrons and positions) thus increases, which leads to rapid increment in the proton cooling rate, thereby decreasing T_p. This, in turn, decreases the pair creation rate, resulting in decrease in the number of coolants and in proton cooling rate. The proton temperature again increases. Limit cycles as such may be responsible for rapid X-ray variabilities (flickering) in XB and AGN.

In conclusion, optically thin disks are thermally unstable (Pringle et al. 1973), unless a disk plasma remains in a one-temperature state or unless e^+e^- pair production is taken into account (Kusunose and Mineshige 1992).

4.3. IRRADIATION-INDUCED INSTABILITIES

So far we considered as a heating source viscous heating only. In the case of X-ray binaries or active galactic nuclei (AGN), however, X-ray illumination of the outer portion of the disk by a central star or by the inner disk could dominate over viscous heating (SS73; Begelman et al. 1983). In such cases, overstable oscillation in gas flow may be induced by the irradiation-induced instability (Meyer and Meyer-Hofmeister 1984; Shields et al. 1986; Schwarzenberg-Czerny 1989).

When $\dot{M}_{in}$ increases, mass accretion rate through the inner edge of the disk, $\dot{M}_{acc}$, should also increase on the viscous timescales (see Fig. 11).

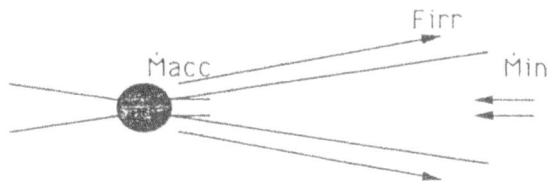

Fig. 11. A schematic view of an irradiated accretion disk. The mass accretion rate from the outer portions of the disk $\dot{M}_{\mathrm{in}}$ is largely influenced by the presence of strong irradiating flux, F_{irr}.

Enhancement in $\dot{M}_{\mathrm{acc}}$ is likely to reinforce irradiating flux, F_{irr}, thereby causing further increase in $\dot{M}_{\mathrm{in}}$. The irradiation thus works as positive feedback. As long as $\dot{M}_{\mathrm{in}}$ and $\dot{M}_{\mathrm{acc}}$ are kept large, gas in the outer portions of the disk will eventually be depleted. This will drastically reduce mass accretion rate and F_{irr}. We then have overstable oscillation under a certain condition (Shields et al. 1986). Chaotic behavior may also be produced (Bell et al. 1991).

4.4. RADIAL PULSATIONAL INSTABILITY

The pulsational instability in stars is now well-known phenomenon, but little is known about radial pulsational instability in accretion disks. Kato (1978) was the first to investigate radial pulsations of accretion disks. Along this line, Bulmenthal et al. (1984) performed more detailed analysis.

In contrast to the stars, in which self-gravity force is balanced with pressure gradient force, centrifugal force, $r\Omega^2$, is balanced with the gravity due to a central star, $g = GM/r^2$, in accretion disks. When radial perturbation is added to a ring in equilibrium, the radius of the ring will oscillate around the equilibrium value because of epicyclic motion of orbiting gas elements.

The oscillation amplitudes will increase with time when α viscosity model is used. The amplifying mechanism is explained by two processes: the thermal process and the dynamical process. The thermal process is analogous to the ϵ mechanism for stellar pulsation (see a review by Saio 1992 in this volume). If energy generation rate, ϵ, increases in a compressed phase, oscillation will acquire energy. Unlike the stellar case, in which energy generation is only available at the core where oscillation amplitudes are small, this mechanism really works well in the accretion disks, in which viscous heating is everywhere available.

Disk pulsation also acquires energy through the work done on a rotating ring by viscous stress. This is the dynamical process. As a result, disks are overstable against radial axisymmetric perturbations under a variety of conditions (Blumenthal et al. 1984). The results of nonlinear numerical simulations (cf. Okuda and Mineshige 1991) confirm the presence of global overstable modes (see Okuda and Mineshige 1992 in this volume). They are suspected to cause QPOs in CVs and low-mass X-ray binaries.

4.5. NON-RADIAL OSCILLATIONS

Disks are also unstable against non-axisymmetric perturbation. Kato (1983) discussed non-radial oscillations of accretion disks with $m = 1$ mode (where m is the azimuthal wave number) in Newtonian potential. Long-term V/R variation of Be stars may be explained by this sort of one-armed oscillation (see Okazaki 1992 in this volume).

Kato (1989) further considered the non-radial oscillations in relativistic disks. He used the pseudo-Newtonian potential and found by the linear analysis some marginally overstable modes, in which horizontal motion is essential. One-armed oscillations of the disks in X-ray binaries is suggested to cause low frequency noise (LFN). For more details, please refer to the original papers.

4.6. TIDAL INSTABILITY

Finally, we discuss the disk instabilities in more realistic potentials. The standard SS model assumes axisymmetric potential, and so neglect non-axisymmetric effects, such as the tidal effect by a companion star, which could be important in outer portions of the disk in close binaries. When the mass ratio, $q = M_2/M_1 < 1/4$, where M_2 and M_1 are masses of a companion star and of a compact star, respectively, the orbital motion of a test particle around a central star becomes unstable at large radii (Paczyński 1974). Two-dimensional simulations demonstrate that the disk shape then becomes elliptical, and its semi-major axis gradually rotates in the inertial frame in the same direction as that of the orbital motion (Whitehurst 1988; Hirose and Osaki 1990). When the elongated rim of the disk is passed by the companion star, the disk luminosity is periodically enhanced due to excess energy dissipation by the tidal torque. The period of such light variation is slightly longer than the orbital period by a few percents. This is believed to explain so-called superhump light variations observed in superoutbursts of SU UMa stars, a subgroup of DN, and possibly in some XBs.

When superhump light variation is observed, it is possible to roughly estimate the mass of a compact star, M_1, from the empirical relation between M_1 and the superhump and orbital periods. Using this method, Mineshige et al. (1992) conclude that the compact components in X-ray novae, GS2000+25 and GS2023+338 (V404 Cyg), are very likely to be black holes instead of neutron stars.

A unified model for ordinary outbursts and superoutbursts in SU UMa system is discussed by Osaki (1989) (see also Osaki et al. (1992) in this volume.)

5. Summary

We have briefly explained recent development of the theoretical studies on accretion disks, with emphasis on the disk instabilities. As we have seen previously, the current understanding of the accretion disks largely depend on the α viscosity hypothesis by Shakura and Sunyaev (1973). This SS model also assumes that disks are in stationary state; i.e., the same amount of gas is continuously put into the disk from the outer rim, and the exactly same amount of the gas is accreted onto a central star through the inner rim, so the disk luminosity L_d is kept constant in time. By relaxing some of the basic assumptions adopted in the standard SS model. we can derive various kinds of disk instabilities, which may be responsible for observable time variabilities on a wide rage of timescales.

In Table I, we summarize the basic assumptions of the SS disk and the relevant disk instabilities. It is very interesting to study global response of local instabilities, and to construct a more sophisticated time-dependent disk models, by which detailed comparison with observations is possible.

TABLE I
Disk Instabilities

Standard Model	Time-Dependent Model	Observational Appearance
HII (ionized)	HI→HII transition	DN, XN, AGN (?)
—	$P_{rad} \leftrightarrow P_{gas}$ transition	QPO (?), AGN (?)
1-T (one-temperature)	1-T→2-T transition	high-low transition (?)
no pairs	$\gamma\gamma \leftrightarrow e^+e^-$ transition	flickering (?)
no irradiation	irradiation-induced instability	AGN activity (?)
Kepler rotation	radial pulsation	QPO (?)
	nonradial pulsation	LFN (?)
axisymmetric	tidal instability	superhumps

Throughout this review, we neglect some other effects, which could be important; such as nonthermal emission processes and various kinds of magnetic activity. We also use the α model for disk viscosity. Unfortunately, the nature of the disk viscosity is poorly known, although fascinating models for creating large-magnitude viscosity in accretion disks are being proposed recently, which include shear instabilities (Drury 1977; Papaloizou and Pringle 1984), thermal convection (Lin and Papaloizou 1980), spiral shocks (Sawada et al. 1986), internal waves (Vishniac and Diamond 1989), magnetohydro-turbulence (Ichimaru 1976; Kato and Horiuchi 1986), and magnetohydro-dynamic (MHD) instabilities (Balbus and Hawley 1991). However at this moment, it is extremely difficult to develop alternative viscosity models because of complex behavior of hydrodynamic or magneto-hydrodynamic turbulence.

Instead of seeking possible seeds for disk viscosity, one may use the theory of disk instabilities as a probe to refine the model for the disk viscosity through comparison with the observations. The thermal limit cycle instability (explained in §3) is, in this respect, the most successful case; in fact, many groups of people claim that α should be larger in HII state than in HI state. From the fitting light curves of DN, roughly we find

$$\alpha \sim (0.05 - 0.2) \left[\left(\frac{T}{10^4 \text{ K}} \right) \left(\frac{r}{10^{10} \text{ cm}} \right) \right]^{0.4-0.6}, \tag{37}$$

for a temperature range $3000 \leq T \leq 10^5$ (K).

We thus finally wish to stress the importance of studying the time-dependent properties of unstable accretion disks. This is also a useful way to get information on disk viscosity.

Acknowledgements

The author would like to thank Eiko Kawazoe for drawing figure 2, and Itoh Science Foundation for support.

References

Abramowicz, M. A., Czerny, B., Lasota, J. P., and Szuszkiewicz, E.: 1988, *Astrophysical Journal* **332**, 646.

Alexander, D. R., Johnson, H. R., and Rypma, R. L.: 1983, *Astrophysical Journal* **272**, 773.

Balbus, S. A., and Hawley, J. F.: 1991, *Astrophysical Journal* **376**, 214.

Begelman, M. C., McKee, C. F., and Shields, G. A.: 1983, *Astrophysical Journal* **271**, 70.

Bell, K. R., Lin, D. N. C., and Ruden, S. P.: 1991, *Astrophysical Journal* **372**, 633.

Blumenthal, G. R., Yang, L. T., and Lin, D. N.: 1984, *Astrophysical Journal* **287**, 771.

Cannizzo, J. K., Ghosh, P., and Wheeler, J. C.: 1982a, *Astrophysical Journal* **260**, L83.

Cannizzo, J. K., Wheeler, J. C., and Ghosh, P.: 1982b, *Pulsations in Classical and Cataclysmic Variables*, eds. J. P. Cox and C. J. Hansen (Boulder, University of Colorado Press), p. 13.

Cannizzo, J. K., Wheeler, J. C., and Ghosh, P.: 1985, *Proc. Cambridge Workshop on Cataclysmic Variables and Low-Mass X-ray Binaries*, eds. D. Q. Lamb and J. Patterson (Dordrecht, Reidel), p. 307.

Cannizzo, J. K., Wheeler, J. C., and Polidan, R. S.: 1986, *Astrophysical Journal* **301**, 634.

Cox, A. N., and Stewart, J. N.: 1970, *Astrophysical Journal, Supplement Series* **19**, 243.

Cox, J. P., and Guili, R. T.: 1968, *Principle of Stellar Structure* (New York, Gordon and Breach).

Drury, L. O'C.: 1977, PhD. Thesis, Cambridge University.

Faulkner, J., Lin, D. N. C., and Papaloizou, J. C. B.: 1983, *Monthly Notices of the RAS* **205**, 359.

Hirose, M. and Osaki, Y.: 1990, *Publications of the ASJ*, **42**, 135.

Honma, F., Matsumoto, R., and Kato, S.: 1991, *Publications of the ASJ* **43**, 147.

Hōshi, R.: 1979, *Prog. Theor. Phys.* **61**, 1307.

Ichimaru, S: 1976, *Astrophysical Journal* **208**, 701.

Kato, S.: 1978, *Monthly Notices of the RAS* **185**, 629.

Kato, S.: 1983, *Publications of the ASJ* **35**, 249.

Kato, S.: 1989, *Publications of the ASJ* **41**, 745.

Kato, S., and Horiuchi, T.: 1986, *Publications of the ASJ* **38**, 313.

Kusunose, M., and Mineshige, S: 1992, *Astrophysical Journal* **390**, in press.

Lewin, W. H. G. et al.: 1976, *Astrophysical Journal* **207**, L95.

Lightman, A. P. and Eardley, D. M.: 1974, *Astrophysical Journal* **187**, L1.

Lin, D. N. C., and Papaloizou, J. C. B.: 1980, *Astrophysical Journal* **191**, 37.

Lin, D. N. C., and Papaloizou, J. C. B.: 1985, *Protostars and Planets II* (Tucson, University of Arizona Press).

Matsuoka, M.: 1992, this volume.

Meyer, F.: 1984, *Astronomy and Astrophysics* **131**, 303.

Meyer, F., and Meyer-Hofmeister, E.: 1981, *Astronomy and Astrophysics* **104**, L10.

Meyer, F., and Meyer-Hofmeister, E.: 1982, *Astronomy and Astrophysics* **106**, 34.

Meyer, F., and Meyer-Hofmeister, E.: 1984, *Astronomy and Astrophysics* **140**, L35.

Mineshige, S.: 1987, *Astron. Space Sci.* **130**, 331.

Mineshige, S., Hirose, M., and Osaki Y.: 1992, *Publications of the ASJ Letters* **44**, in press.

Mineshige, S., Kim, S.-W. and Wheeler, J. C.: 1990, *Astrophysical Journal* **358**, L5.

Mineshige, S., and Osaki, Y.: 1983, *Publications of the ASJ* **35**, 377.

Mineshige, S., and Osaki, Y.: 1985, *Publications of the ASJ* **37**, 1.

Mineshige, S., and Shield, G. A: 1990, *Astrophysical Journal* **351**, 47.

Mineshige, S., and Wheeler, J. C.: 1989, *Astrophysical Journal* **343**, 241.

Nicolis, G., and Prigogine, I.: 1977, *Self-Organization in Nonequilibrium Systems* (John Wiley and Sons, New York).

Novikov, I. D. and Thorne, K. P.: 1974, *Black Holes, Les Houches 1972*, eds. C. DeWitt and B. S. DeWitt (Gordon and Breach, New York), p. 343.

Okazaki, A. T.: 1992, this volume.

Okuda, T. and Mineshige, S.: 1991, *Monthly Notices of the RAS* **249**, 684.

Okuda, T., and Mineshige, S.: 1992, this volume.

Osaki, Y.: 1974, *Publications of the ASJ* **26**, 429.

Osaki, Y.: 1989, *Publications of the ASJ* **41**, 1005.

Osaki, Y., Hirose, M., and Ichikawa, S.: 1992, this volume.

Paczynski, B.: 1969, *Acta Astron.* **19**, 1.

Paczynski, B.: 1977, *Astrophysical Journal* **216**, 822.

Papaloizou, J., Faulkner, J., and Lin, D. N. C.: 1983, *Monthly Notices of the RAS* **205**, 487.

Papaloizou, J. C. B., and Pringle, J. E.: 1984, *Monthly Notices of the RAS* **208**, 721.

Pringle, J. E.: 1981, *Annual Review of Astronomy and Astrophysics* **19**, 137.

Pringle, J. E., and Rees, M. J.: 1972, *Astronomy and Astrophysics* **21**, 1.

Pringle, J. E., Rees, M. J., and Pacholczyk, A. G.: 1973, *Astronomy and Astrophysics* **29**, 179.

Saio, H.: 1992, this volume.

Sawada, K., Matsuda, T., Hachisu, I.: 1986, *Monthly Notices of the RAS* **219**, 75.

Schwarzenberg-Czerny, A.: 1989, *Astronomy and Astrophysics* **210**, 174.

Shakura, N. I., and Sunyaev, R. A.: 1973, *Astronomy and Astrophysics* **24**, 337.

Shakura, N. I., and Sunyaev, R. A.: 1976, *Monthly Notices of the RAS* **175**, 613.

Shapiro, S. L., Lightman, A. P., and Eardley, D. M.: 1976, *Astrophysical Journal* **204**, 187.

Shibazaki, N., and Hosi, R.: 1975, *Prog. Theor. Phys.* **54**, 706.

Shields, G. A., McKee, C. F., Lin, D. N. C., and Begelman, M. C.: 1986, *Astrophysical Journal* **306**, 90.

Smak, J.: 1982a, *Acta Astron.* **32**, 199.

Smak, J.: 1982b, *Acta Astron.* **32**, 213.

Smak, J.: 1984a, *Publications of the ASP* **96**, 5.

Smak, J.: 1984b, *Acta Astron.* **34**, 161.

Tainaka, K., Fukazawa, S., and Mineshige, S.: 1992, this volume.

Van der Klis, M.: 1989, *Annual Review of Astronomy and Astrophysics* **27**, 517.

Vishniac, E. T., and Diamond, P.: 1989, *Astrophysical Journal* **347**, 447.

Whelan, J. A. J., et al.: 1977, *Monthly Notices of the RAS* **180**, 657.
Whitehurst, R.: 1988, *Monthly Notices of the RAS* **232**, 55.

MOTION OF A CHARGED PARTICLE AROUND A BLACK HOLE PERMEATED BY MAGNETIC FIELD AND ITS CHAOTIC CHARACTERS

Y. NAKAMURA

Department of Mathematics, Faculty of Science, Ibaraki University, Mito, 310, Japan

and

T. ISHIZUKA

Department of Physics, Faculty of Science, Ibaraki University, Mito, 310, Japan

Abstract. Motion of a charged particle around a black hole immersed in magnetic field is calculated. It is shown that this motion has a chaotic property depending on initial parameters.

1. Introduction

Chaotic phenomena in conservative Hamiltonian system have been found in classical mechanics, for example, in galactic dynamics (Contopoulos 1987). Relativistic equations of the motion of a particle in the curved space-time due to a black hole are separable in variables for particular metrics of Schwarzschild and Kerr (Misner et al. 1973). In that case orbits are classified in the diagram of effective potential depending on the total energy and angular momentum of a particle. Compared with classical cases, in general relativistic one we expect trapping of the particle by the central black hole under some condition. We can estimate these conditions for the motion in the Schwarzschild and Kerr metrics.

On the other hand we have a similar Hamiltonian system for the motion of a charged particle in the curved space-time with magnetic field, but in this case separation of variable has not been found, and therefore this system may show nonintegrability (Prassana and Varma 1977). So in this case orbital motion becomes complicate depending on initial parameters. We may be able to expect some chaotic behaviours of motion as found out in the quantum states of a hydrogen-like atom in a magnetic field (Hasegawa and Takami 1991).

2. Equations

As one of the properties of black hole, it does not have a magnetic field, but we imagine a magnetic field around the black hole and it has been considered that energetic phenomena (such as the jet beams) in the active galactic nuclei relate with this field.

Astrophysics and Space Science **210**: 105–108, 1993.
© 1993 *Kluwer Academic Publishers.*

We calculate some particle's orbits in the off-equatorial plane near central hole immersed in a dipole type magnetic field, of which effect appears in the time component of vector potential **A**. The magnetic field around the black hole is assumed to be generated by a current flowing in the accretion disk situated at the distance R from the central black hole. Then the components of vector potential **A** are

$$A_t = \frac{-3a\mu}{2\gamma^2 \Sigma}[[r(r-M) + (a^2 - Mr)\cos^2\theta]$$
$$\times \frac{1}{2\gamma}\log\left(\frac{r-M-\gamma}{r-M-\gamma}\right) - (r - M\cos^2\theta)]$$

and

$$A_\phi = \frac{-3\mu\sin^2\theta}{4\gamma^2\Sigma}[(r-M)a^2\cos^2\theta + r(r^2 + Mr + 2a^2)$$
$$-[r(r^3 - 2Ma^2 + a^2 r) + \Delta a^2\cos^2\theta] \times \frac{1}{2\gamma}\log\left(\frac{r-M+\gamma}{r-M-\gamma}\right)$$

where μ is a dipole moment and γ is given by $\sqrt{(M^2 - a^2)}$ (Petterson 1975). Taking account of this contribution, we have the Lagrangian

$$L = \frac{1}{2}[-(1 - \frac{2Mr}{\Sigma})\dot{t}^2 - \frac{4Mar}{\Sigma}\sin^2\theta\dot{t}\dot{\phi}$$
$$+ \frac{A}{\Sigma}\sin^2\theta\dot{\phi}^2 + \frac{\Sigma}{\Delta}\dot{r}^2 + \Sigma\dot{\theta}^2 + qA_t\dot{t} + aA_\phi\dot{\phi}.$$

If the motion is limited within the $\theta - r$ plane, the θ- and r- components of the equation of motion become

$$\frac{d^2\theta}{ds^2} = \frac{-a^2\sin\theta\cos\theta}{\Sigma\Delta}\dot{r}^2 - \frac{2r}{\Sigma}\dot{r}\dot{t} + \frac{a^2\sin\theta\cos\theta}{\Sigma}\dot{\theta}^2$$
$$+ \frac{2Mra^2\sin\theta\cos\theta}{\Sigma^3}\dot{t}^2 - \frac{4Mra(r^2 + a^2)}{\Sigma^3}\sin\theta\cos\theta\dot{\phi}\dot{t} + \frac{\sin\theta\cos\theta}{\Sigma^3}$$
$$\times[(r^2+a^2)^3 - (r^2 + a^2 + \Sigma)\Delta a^2\sin^2\theta]\dot{\phi}^2 + \frac{q}{\Sigma}(A_{\phi,r}\dot{\phi} + A_{t,r}\dot{t}).$$

$$\frac{d^2r}{ds^2} = \frac{M(r^2 - a^2\cos^2\theta) - ra^2\sin\theta^2}{\Sigma\Delta}\dot{r}^2 + \frac{r\Delta}{\Sigma}\dot{\theta}^2 + \frac{2a^2\sin\theta\cos\theta}{\Sigma}\dot{r}\dot{\theta}$$

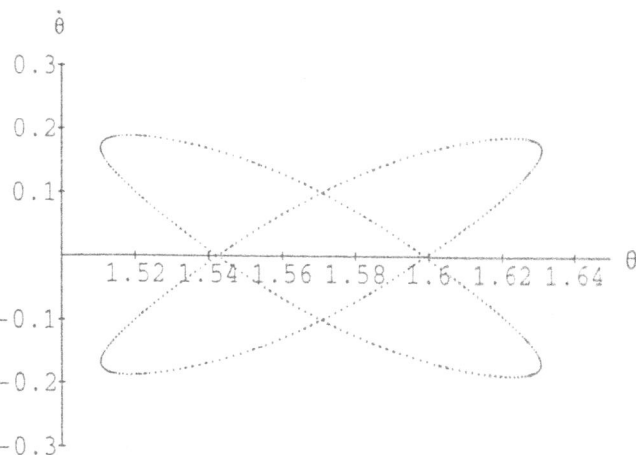

Fig. 1.   A regular pattern in motion with initial values of $\dot{\theta}_0 = 0.10$.

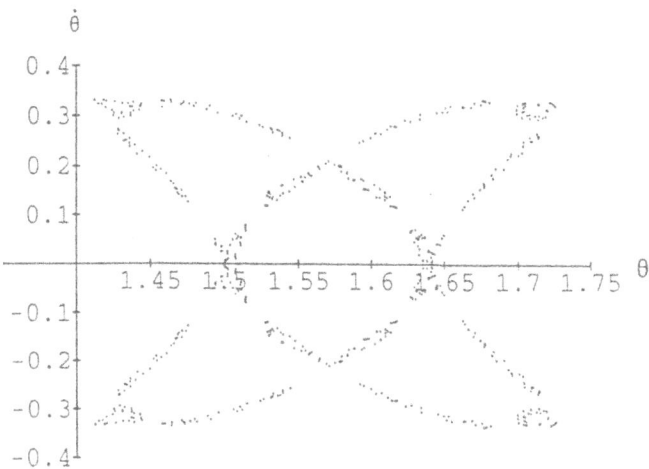

Fig. 2.   An intermediate case of $\dot{\theta}_0 = 0.21$.

$$-\frac{M\Delta}{\Sigma^3}(r^2 - a^2 \cos^2 \theta)\dot{\theta} + \frac{\Delta \sin^2 \theta}{\Sigma^3}[r^5 + 2r^3 a^2 \cos^2 \theta - Mr^2 a^2 \sin^2 \theta$$

$$+ra^4 \cos^2 \theta + (M - r)a^4 \sin^2 \theta \cos^2 \theta]\dot{\phi}^2 + \frac{2\Delta M a \sin^2 \theta}{\Sigma^3}$$

$$\times (r^2 - a^2 \cos^2 \theta)\dot{\phi}\dot{t} + \frac{q\Delta}{\Sigma}(A_{\phi,r}\dot{\phi} + A_{t,r}\dot{t}),$$

where q is the charge of the particle, and Σ, Δ, ϕ, t, r, and θ are usual variables in the Kerr metric.

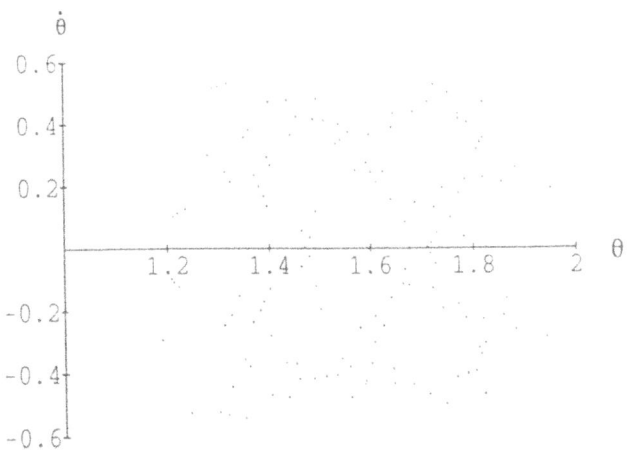

Fig. 3. An irregular case of $\dot{\theta}_0 = 0.25$.

3. Results

As parameters used in these equations we select 0.999, 5, 100, and 300 for a, E_0, L, and B in geometrical units. Furthermore, we consider the case where the particle starts from $r = 4.559$ to any direction with various initial values of $\dot{\theta}$, but with $\dot{r} = 0$ and $\dot{\phi} = 0$.

Figure 1 shows a case of regular pattern in which initial value of $\dot{\theta}$, $\dot{\theta}_0$, is equal to 0.1. The case of $\dot{\theta}_0 = 0.21$ near irregular one is shown in figure 2. In the figure 3 irregular case of $\dot{\theta}_0 = 0.25$ is given.

From these figures we find characteristic patterns which are similar to the Poincare's surface appearing in the Hamiltonian system with two degrees of freedom. Se we can conclude that motions of a particle around the black hole immersed in the magnetic field show chaotic characters.

References

Contopoulos, G. et al.: 1987, *Astronomy and Astrophysics* **172**, 55.
Hasegawa, H. and Takami, T.: 1991, *Butsuri* **46**, 750 (*in Japanese*).
Misner, C. W. et al.: 1973, *Gravitation*.
Petterson, J. A.: 1975, *Physical Review D:* **12**, 2218.
Prassana, A. R. and Varma, R. K.: 1977, *Pramana* **8**, 229.

NONLINEAR EXPANSION TRIGGERED BY MAGNETIC STRESS IN ACCRETION FLOW ONTO A BLACK HOLE

M. YOKOSAWA

Department of Physics, Ibaraki University, Mito 310

Abstract. The paper is an investigation of the magnetohydrodynamics(MHD) of the quasi-radial accretion onto a rotating black hole. It is demonstrated that the nonlinear effects in the accretion flow accelerate the rate of expansion of the Alfven mode. The accretion flow becomes unstable at the weak strength of magnetic field which is about one-thirtieth of the expected value by the linear theory.

1. Introduction

The stability pf MHD accretion towards a body was investigated by Williams (1975). It was derived from a perturbation method that, if the flow is super-Alfvénic at infinity but sub-Alfvénic further in, a smooth transition at the Alfvén surface would be unstable. The results were generalized to show that any ideal aligned-field flow where gas passes smoothly from a super-Alfvénic region into a sub-Alfvénic region will be unstable, without exception. It was concluded that super-Alfvénic flow towards a body most become sub-Alfvénic via a shock, followed by turbulent flow. The above conclusion was based on the linear analysis. We investigate here the nonlinear effects of the stability of the MHD accretion with full numerical MHD code.

The magnetic field lines near a rotating black hole are twisted by a frame-dragging effect and, thus, the rotational energy of the black hole is stored in the magnetosphere around the black hole. The twisted magnetic field exerts a torque on an ambient plasma. The particles acted on by magnetic stress either increase or decrease their angular momentum. The increased kinetic energy of particles is changed in some cases to thermal energy, wave energy or radiation energy. We first consider the possible magnetohydrodynamical structure formed around a black hole in a simple case in which initially a homogeneous field is changed by infalling gas with no angular momentum at infinity.

The two types of process were proposed regarding the extraction of rotational energy from a Kerr black hole by means of magnetic fields. One is an electromagnetical extraction (Blandford and Znajek 1977) and the other is a hydromagenetical-type extraction (Ruffini an Wilson 1975). The former type magnetic field is a force-free field generated by the surrounding matter outside the event horizon. The interaction of a magnetic field with the hole's rotation produces a "battery-like" behavior of the hole's horizon (Thorne et al. 1986). The latter extraction process is caused by the fact that two shells of different radii formed outside the event horizon are dragged due to the rotation of the black hole with a characteristic angular velocity. If connections between the two shells are made (ropes, springs or in the case of the magnetosphere magnetic lines of forces), rotational energy can be pumped out from the innermost shell to an external one, and the rotational energy can be extracted from a black hole (Ruffini 1977). Whereas electromagnetic extraction processes have been studied by many authors, the latter case has not. When the magnetic energy builds up to equipartition with the kinetic energy of infalling gas, the extraction energy could be comparable with the observed radiation energy from active galactic nuclei (AGN). Therefore, we consider the hydromagnetical extraction process. The maximal extraction rate is determined in this process by the maximum energy flux of electromagnetic field in a stabile accretion.

2. Model of Magnetohydrodynamical Accretion onto a Rotating Black Hole

We set the initial conditions such that the magnetic field and the gas are homogeneous, and the gas observed in locally non-rotating frame (LNRF) is rest. Two types of boundary conditions are adopted at the outer boundary. One is the free fall condition in which the gas density and the magnetic field strength are given by the analytic solutions of the free fall (Yokosawa, Ishizuka and Yabuki 1991). Other is the quasi-stationary condition in which the gas density is constant and the magnetic field strength slowly increases. The latter case is calculated in order to evaluate the energy and angular momentum transport rate of a black hole to the ambient matter in a stationary state. The initial gas temperature is selected either to be cool or to be hot. In the hot case, the gas pressure behind the shock front is comparable with the magnetic pressure. The calculating space is $r = 1.07 r_h \sim 10 r_h$ and $\theta = 0 \sim \pi/2$, where r_h is the horizon radius. The mesh size is 125 in the r-direction and 100 in the θ-direction. Computations were performed on the workstation, MIPS RS3230.

3. Dynamics and Stability of MHD Accretion

(a) Dynamics of accretion flows bounced by the magnetic stress has been presented. The first bounce occurs at the poles in the vicinity of the horizon (figure 1). The meridian motion is remarkable in the bounced region, which is a characteristic common to all strong magnetic stress either in the case of rapidly rotating black hole or in slow rotating black hole. When the gas pressure is comparable with the magnetic stress at the bounced region, no remarkable meridian motion appears. If the gas pressure at the shock front is higher than the magnetic pressure, the shock wave becomes quasi-stationary, that is, its wave is a standing wave formed near the event horizon. When

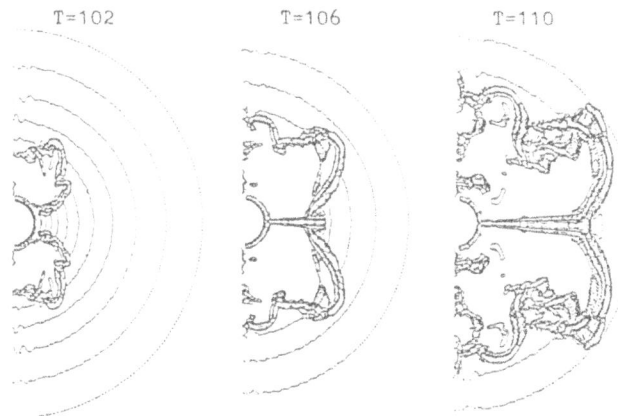

Fig. 1. The dynamical evolution of MHD accretion onto a rotating black hole. The density contours are presented by lines with the level n, $10^{n/5}$. The first instability occurs at the poles in the vicinity of the horizon.

the temperature of the falling gas is so low as the magnetic pressure predominates over the gas pressure, the magnetic energy stored at the shocked shell is released in a radially expanding wave. After a while, the expanding gas returns to falling onto the black hole with frozen magnetic field.

(b) The maximum energy density of the magnetic field obtained in quasi-stationary accretion flows is $B_{\mathrm{MAX}}^2 \approx 10^{-3}\rho c^2$. The more strong magnetic field produces the shock structure in the flow. The rotational velocity of the fluid reaches to $v_{\phi,\mathrm{MAX}} \approx 0.1c$. The stored energy in the magnetosphere around a black hole is about 0.1 percent of the rest mass energy of the accreting matter.

(c) The linear theory on the stability of trans-Alfvenic flow showed that magnetized accretion flow becomes unstable at the Alfvenic surface, i.e. $M_A = 1$. The numerical simulation shows that the quasi-pulsation is triggered by the magnetic stress at the Alfven Mach number $M_A \approx 30$.

References

Blandford, R. D., and Znajek, R. L.: 1977, *Monthly Notices of the RAS* **179**, 433.

Ruffini, R.: 1977. in *Proceedings of the First Marcel Grossmann Meeting on General Relativity*, ed. R. Ruffini (North-Holland Publishing Company, Amsterdam), p. 49.

Ruffini, R., and Wilson, J. R.: 1975, *Physical Review D:* **12**, 2959.

Thorne, K. S., Price, R. H., Macdonald, D. A., Suen, W. M., and Zhang, X. H.: 1986, in *Black Holes: The Membrane Paradigm* eds. Thorne, K. S., Price, R. H., and Macdonald, D. A (Yale University Press, New Haven London), p. 67.

Yokosawa, M. Ishizuka, T., and Yabuki, Y.: 1991, *Publications of the ASJ* **43**, 427.

Williams, D. J.: 1975, *Monthly Notices of the RAS* **171**, 537.

VARIABILITY OF X-RAY EMISSION FROM CEN X-3
OBSERVED WITH GINGA

R. KANETAKE

Astronomical Institute, Tôhoku University, Sendai 980, Japan

T. TAKESHIMA

Institute of Space and Astronautical Science, Sagamihara 229, Japan

K. MAKISHIMA

Department of Physics, University of Tokyo, Bunkyo-ku, Tokyo 113, Japan

and

M. TAKEUTI

Astronomical Institute, Tôhoku University, Sendai 980, Japan

Simplified stellar models like as one-zone models can be oscillated in chaotic state (Saitou *et al.* 1989). It is natural to consider that the irregular variations of Galactic X-ray sources might be caused from nonlinear oscillation of accretion disks or related flow. We studied the variability of Cen X-3 for applying the nonlinear theory to it.

The X-ray of Cen X-3 observed with Ginga from 22nd to 24th of March, 1989 are analyzed. Since Large Area Counters (LAC) of which the area of detector is 4,000 cm^2 are used. The data used here is obtained the range of 1,000 sec from 14 h 25 m on March 23rd with the time resolution of 0.25 sec. The energy bands are 1.7~4.6 keV, 4.6~9.3 keV, 9.3~19 keV, and 19~37 keV, respectively.

Takeshima *et al.* (1991) show that Cen X-3 is at the high state, when the X-ray from the source is not veiled by non-ionized matter (Schreier *et al.* 1976). So that we ignore the effect of circumstellar matter around the source.

First, we analyzed and obtained 4.817 sec by using the folding method. To search for deterministic chaos in the temporal behavior of Cen X-3, we used the method of VAS (Grassberger *et al.* 1983). We can obtain the second order dimension (Voges *et al.* 1987) by the following:

$$D_2 = \lim_{r \to 0} \frac{\log C(r)}{\log r}. \tag{1}$$

This is called the correlation dimension. Here $C(r)$ is defined as

$$C(r) = \lim_{N \to \infty} \frac{1}{N^2} \sum_{i,j=1}^{N} H(r - |X_i - X_j|), \tag{2}$$

where H: Heaviside function. If $C(r)$ can be described in the from,

Astrophysics and Space Science **210**: 113–115, 1993.
© 1993 *Kluwer Academic Publishers.*

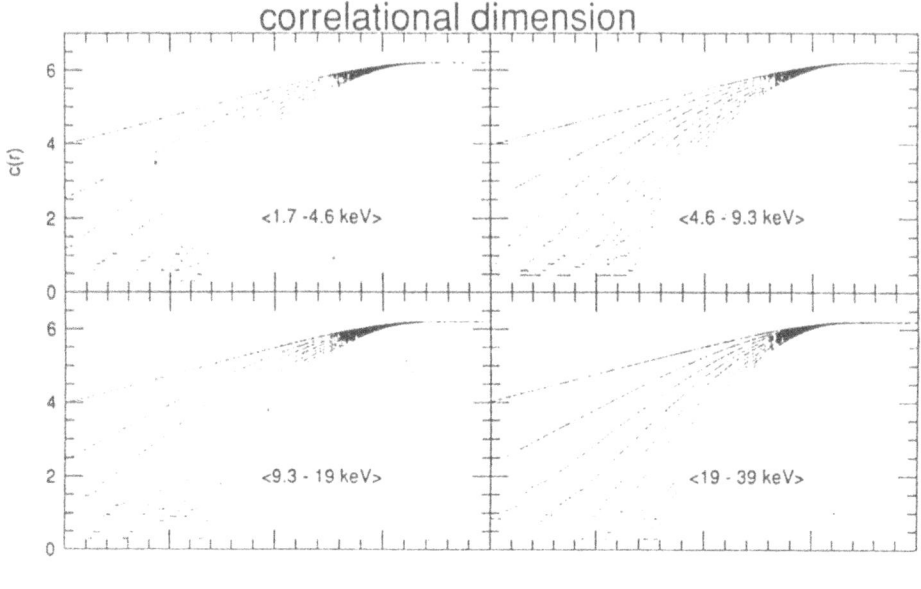

$$C(r) = r^{\nu}, \tag{3}$$

then ν is correlation dimension. Let X_i and X_j be vectors in a phase space of n dimensions. If each data set contains N, (N is the number of points), with a temporal resolution,

$$X_i = (x_i, x_{i+m}, \ldots x_{i+(d-1)m}) \tag{4}$$

where d is called the embedding dimension. The resulting data sets X_i define a d-dimensional phase space, and the attractor can be reconstructed in the space. For each embedding dimension d, $C(r)$ expresses the number of distances which are less than r, and could obtain a slope ν. If the density of the calculated points is high enough, ν can't increase any more, so we could determine the number of d by a comparison of the slopes between successive values of $C(r)$. For example, if d is smaller than 3, then it will explain that the dimension phase-space for the attractor is less than three. Deviations from the averaged 4.8 sec oscillation are calculated. The deviations change point by point. Even though we tried to make a running average over 11 points, it still remains the irregularity. We analyzed the running averages of the every 4 energy bands by using a practical method (Sato et al. 1987). The result is illustrated in Fig. 1~4. It is shown that the embedding dimension is not so small as 3~4.

Since the irregular variation of all components is not a low-dimensional phenomena, the variability may express the fluctuation of accretion flow from the inner edge of the disk or related flow.

References

Grassberger, P. and Procaccia, I.: 1983, *Physical Review Letters* **50**, 346.
Kolláth, Z.: 1990, *Monthly Notices of the RAS* **247**, 377.
Nagase, F.: 1989, *Publications of the ASJ* **41**, 1.
Saitou, M., Takeuti, M., and Tanaka, Y.: 1989, *Publications of the ASJ* **41**, 297.
Sato, S., Sano, M., and Sawada, Y.: 1987, *Prog. Theor. Phys. (Letters)* **77**, 1.
Schreier, E. *et al.*: 1972, *Astrophysical Journal* **172**, L79.
Takeshima, T. *et al.*: 1991, *Publications of the ASJ* **43**, L43.
Voges, W. *et al.*: 1987, *Astrophysical Journal* **320**, 794.

SPATIAL PATTERN FORMATION OF INTERSTELLAR MEDIUM

K. TAINAKA, S. FUKAZAWA, H. NISHIMORI, M. YOKOSAWA and
S. MINESHIGE
Department of Physics, Ibaraki University, Mito 310, Japan

Abstract. Population dynamics of multi-phased interstellar medium (ISM) is investigated by using the lattice model in position-fixed reaction. Interactions between three distinct phases of gas, cold clouds, warm gas, and hot gas give rise to cyclic phase changes in ISM. Such local phase changes are propagated in space, and stochastic steady-state spatial pattern is finally achieved. We obtain the following two characteristic patterns:

(1) When the sweeping rate of a warm gas into a cold component is relatively high, cold clouds associated with warm gas form small-scale clumps and are dispersively distributed, whereas hot gas covers large fraction of space.

(2) When the sweeping rate is relatively low, in contrast, warm gas and cold clouds are diffusively and equally distributed, while hot gas component is substantially localized.

1. Introduction

Interstellar medium (ISM) is thought to be composed of multi-phased gas (for a review, see Ikeuchi 1988). The understanding of ISM has made revolutionary development when McKee and Ostriker (1977) proposed the three-phase model for ISM. They inferred the coexistence of hot gas (HG) with temperature $T = 10^5 - 10^6$ K, intercloud gas (or warm gas, WG) with temperature $T \sim 10^4$ K, and HI clouds (or cold cloud, CC) with $T \sim 100$ K in pressure equilibrium.

These three components of gas cannot be in static equilibrium. Ikeuchi and Tomita (1983) investigated phase changes between distinct phases of ISM. However, spatial propagation of such phase changes are not described in their model, which treats systems in mean-field approximation. To obtain mass density dynamics, such as clumping or dispersive behavior of HG, we need to study spatial pattern formation of the fundamental phase changes in ISM. This is the main aim of this paper.

2. Physical Assumptions and Basic Equations

To study spatial pattern formation, we use the position-fixed reaction model (Tainaka 1988). The rate equations for each phase of gas are then:

$$\frac{dX_C}{dt} = AX_W - PP_{CH}, \tag{1}$$

$$\frac{dX_H}{dt} = PP_{CH} - P_{HW}, \tag{2}$$

Astrophysics and Space Science **210**: 117–120, 1993.

$$\frac{dX_W}{dt} = P_{HW} - AX_W, \tag{3}$$

where A and P are numerical constants, X_C, X_W, and X_H represent the mass densities of CC, WG, HG, respectively, and $X_C + X_W + X_H = 1$, and P_{ij} denotes the joint probability that a species j lies in the nearest neighbor of a species i [$i, j = $ C (cold cloud), W (warm gas), or H (hot gas)]. The relation $P_{ij} = P_{ji}$ thus holds. For details of the methods of simulations, see Tainaka et al. (1992).

3. Mean-Field Theory (MFT)

In the mean-field theory (MFT), we may write

$$P_{ij} = X_i X_j, \tag{4}$$

where i, j, k are one of (CC, WG, HG). By inserting equation (4) into basic equations (1)–(3) and by setting all the time derivatives to be zero, we obtain stationary solutions in MFT as,

$$X_{C,0} = \frac{1-A}{1+P}, \tag{5}$$

$$X_{H,0} = A, \tag{6}$$

$$X_{W,0} = \frac{P(1-A)}{1+P}. \tag{7}$$

It is rather easy to show that this is a stable solution as long as $0 \leq A \leq 1$.

4. Simulation Results

We fix $P = 1$ and change A (sweeping rate of a warm gas into a cold component) between 0 and 1. When A is relatively large, clustering behavior of CC is remarkable. Moreover, CC tends to surround WG, and HG is diffusively distributed. When A is relatively low, in contrast, HG begins to form one-dimensional cluster (chain structure), whereas mass densities of CC and WG increase.

Figure 1 depicts how mass fractions change as functions of A. The mean-field approximation requires that the mass densities of CC and WG should be the same when $P = 1$ (see equations 5–7). The simulation results, however, show a clear tendency that X_W also decreases together with the decrease in X_H as A decreases.

Global dynamics of phase changes in ISM leads to a paradox. It is obvious that the density of WG, X_W, decreases, as the sweeping rate (A) increases,

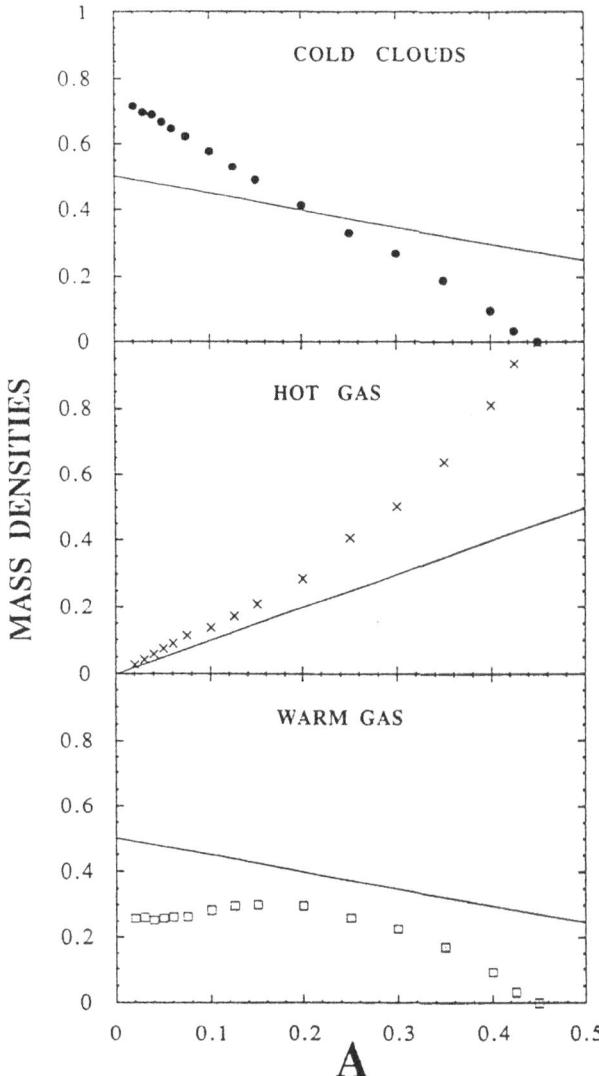

Fig. 1.   Mass fraction of each component of gas as a function of A. The solid lines are the result of mean-field theory (MFT).

because WG is transformed into CC. However, we found that even when A decreases towards zero, X_W again decreases. This can be understood in terms of clumping behavior of HG. Note that WG is created from HG via collision. When A is very small, HG generates clumping structure so that WG has less chance to collide with HG than the case that HG is dispersively distributed. This phenomena cannot be predicted nor explained by the mean-field theory.

5. Conclusion

We have developed the dynamical theory for the phase changes of the three-phased ISM. We find distinct features in its population dynamics which is not yet known in the mean-field theory. Especially, clumping behavior of HG is responsible for such qualitative discrepancy. The investigation of such global pattern formation is thus essential to understand the population dynamics.

The numerical computations were performed on Hitac M660H at Computer Center of Ibaraki University.

References

Ikeuchi, S.: 1988, *Fundamentals of Cosmic Physics* **12**, 255.
Ikeuchi, S., and Tomita, H.: 1983, *Publications of the ASJ* **35**, 77.
McKee, C. F., and Ostriker, J. P.: 1977, *Astrophysical Journal* **218**, 148.
Tainaka, K.: 1988, *J. Phys. Soc. Japan* **57**, 2588.
Tainaka, K., Fukazawa, S., and Mineshige, S.: 1992, *Publications of the ASJ*, submitted.

II. OBSERVATIONAL FACTS

SUPERGIANT VARIABLES: RECENT OBSERVATIONAL RESULTS

J. R. PERCY

Erindale Campus, University of Toronto, Mississauga, Ontario, Canada L5L 1C6

Abstract. Recent observations of several types of supergiant variable stars are reviewed: massive blue, yellow and red supergiants; classical and population II Cepheids; RV Tauri stars; yellow semi-regular (SRd) variables, including UU Herculis stars; and R Coronae Borealis stars. The emphasis is on non-linear aspects such as: amplitude and shape of the light and velocity curves; multiperiodicity, irregularity and chaos; long-term changes in period and amplitude; episodic and continuous mass loss.

1. Introduction

Supergiants are rare but interesting stars which can shed much light on stellar properties, processes and evolution. Almost all of the extreme supergiants are variable, and there are enhanced regions of instability among the B supergiants, the supergiants in the Cepheid instability strip, and among the M supergiants. Supergiants can be seen at great distances, and can be useful as distance indicators. Low-gravity supergiants undergo continuous and/or episodic mass loss, which has important implications for both stellar and galactic evolution. This mass loss may be related to the other forms of variability in these stars. Supergiants have relatively short life times, so changes in their properties and behaviour may occur on a short time scale. Non-linear effects of many kinds may occur in supergiants, ranging from dynamical instabilities to amplitude changes, multimode pulsations, irregularity and possibly chaos. This review deals with some recent observational results on supergiant variables. Other papers at this conference have dealt with theoretical results; see also the excellent reviews by Buchler, Kovacs, Takeuti and others in *The Numerical Modelling of Nonlinear Stellar Pulsations* (Buchler 1990).

2. Massive Supergiant Variables in General

Variability is almost universal among the most luminous stars. Maeder (1980) showed that the amplitude of photometric variability increased with increasing luminosity, and that the characteristic time scale P is related to the bolometric magnitude Mbol and the effective temperature Te by

$$\log P = -0.346\text{Mbol} - 3\log\text{Te} + 10.60$$

The mass loss $\dot{m}$ is also related to the luminosity L; Nieuwenhuijzen and de Jager (1990) derive

Astrophysics and Space Science **210**: 123–136, 1993.

$$\dot{m} \propto L^{1.42} m^{0.16} R^{0.81}$$

where m is the mass and R is the radius.

De Jager et al. (1991) have carried out a comprehensive study of the atmospheric motions in luminous stars, and conclude that the observed light and velocity variations are caused by high-mode internal gravity waves. Thus, the time scale P is much greater than the fundamental radial period, as noted for instance by Percy and Welch (1983). The observed irregularities in these stars may therefore be due to multimode pulsation. On the other hand, Buchler and Goupil (1988) have pointed out that stars which are close to dynamical instability may show chaotic behaviour as a result of interaction between the fundamental mode and an overtone. This effect may occur in some extreme supergiants.

2.1. MASSIVE BLUE SUPERGIANT VARIABLES

These stars, also known as Luminous Blue Variables (LBV's) or S Doradus variables, show low-amplitude photometric variations on time scales of weeks, with occasional large outbursts. P Cyg, for instance, varied between magnitudes 3 and 6 in the 1600's and now varies by about $\Delta V = 0.2^{\rm m}$ on a time scale of about a month (Percy et al. 1988). There are also abrupt increases and decreases of a few $\times 0.01^{\rm m}$ on time scales of two or three days. Spectroscopically, P Cyg shows a strong wind (indeed, it is the prototype star with "P Cygni line profiles"), along with evidence of shell ejections on a time scales of months (Markova and Kolka 1988). On the basis of polarimetric observations, Hayes (1985) and more recently Taylor et al. (1991) have developed a model in which discrete clouds of gas are ejected in the wind. Coordinated observations of this star would be extremely useful.

Much effort has gone into understanding the nature of the wind and its variability. Leitherer et al. (1989) and Blomme (1991) conclude that the magnitude of the mass loss can probably be explained as a radiative effect, with iron opacities playing a crucial role. Pauldrach and Puls (1990) and Blomme et al. (1991) have investigated the process whereby the stars cycle between high- and low-mass-loss states. They have identified a discontinuity ("bistability") in the mass loss rate and wind velocity at a specific value of the effective gravity. It is not clear, however, what drives the star between these two states.

Moskalik and Dziembowski (1991, preprint) have shown that, using the new Iglesias and Rogers (1991) opacities, the pulsation of the β Cephei stars can at last be explained; it is due to the classical κ - mechanism. Furthermore, the instability seems to extend to the B-type supergiants, thus explaining this region of enhanced instability in the H-R diagram (Maeder 1980).

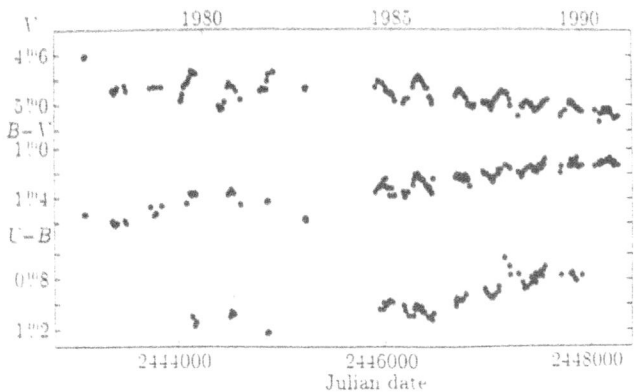

Fig. 1.   UBV photometric observations of V509 Cas (HR8752) since 1975 (Percy and Zsoldos 1992, preprint). In the last decade, the star has become fainter and bluer; the amplitude of the pulsation has decreased, and it has become more complex.

2.2. MASSIVE YELLOW SUPERGIANT VARIABLES

These stars lie in the upward extension of the Cepheid instability strip and of the period - luminosity relation. Similar stars are called "Leavitt Variables" in the Magellanic Clouds. They show irregular variations in brightness and radial velocity on time scales of about a year, presumably due to pulsation. Their light curves also show other interesting and unique features which may be related to the extreme properties of these stars.

The prototype stars are ρ Cas (F8Ia-0, variable) and V509 Cas (HR8752; G0Ia-0, variable), which turn out to be virtual twins in many respects. Rho Cas has a photometric period of 298.5 days; it underwent a deeper-than-average minimum in 1946/7 and a larger-than-average pulsation cycle in 1985/6 (Zsoldos and Percy 1991). Since 1985, it has decreased slowly in mean magnitude and in amplitude. The photometric behaviour of V509 Cas in the last decade (Fig. 1) has been very similar (Halbedel 1991; Percy and Zsoldos 1992, preprint). Between 1986 and 1991, this star became fainter and bluer. At the same time, it was pulsating with two (possibly three) periods with changing amplitudes. The unusual shape of the light curve may be connected with the ejection of one or more shells in the last decade or two. Takeuti (1988) has pointed out that episodes of enhanced mass loss may occur if the luminosity temporarily exceeds the Eddington luminosity.

Rho Cas and V509 Cas have been monitored spectroscopically by Sheffer and Lambert (1986, 1987). They find periods of 520 days in ρ Cas, and 421 and 315 days in V509 Cas. At this point, it is not clear if there is enough data to tell whether the photometric and spectroscopic observations are in agreement or in conflict. A decade or more *coordinated* observations are probably needed to address this question.

2.3. MASSIVE RED SUPERGIANT VARIABLES

These stars, typified by α Ori, α Sco and μ Cep, are designated as SRc or Lc types in the *General Catalogue of Variable Stars*. They are of spectral types M1-M4 Ia-Ib. See Johnson and Querci (1986) for a comprehensive review of M-type stars.

The light curves of SRc/Lc variables are irregular, but one can define two characteristic time scales, both of which depend on the temperature and luminosity of the star (Stothers and Leung 1971). The shorter time scale is in approximate agreement with the fundamental radial pulsation period, and can be represented by

$$\text{Mbol} = -7.20 \log P + 12.8$$

The longer time scale is about 10 times the shorter time scale; its nature and cause are unknown. Note, however, that long-term variations also occur in most small-amplitude red variables (Percy and Sen 1991), in the RVb variables discussed later in this review, and in a few other cool variables.

Red supergiants lose mass at rates of typically $10^{-6} m_\odot / a$. There are possibly discrete episodes of mass loss on time scales of 10^4 years, resulting in shell structures within the circumstellar envelope. Pulsation may be important in driving the mass loss, as it is in the M giants (e.g. Bowen 1988).

It is not clear whether the light curves of SRc/Lc variables are multiperiodic, chaotic or irregular. The long-term variability of μ Cep and RS Cnc often takes the form of occasional large-amplitude cycles of the short-term variations, as in the massive yellow supergiant variables described in the previous section.

3. Cepheid Variables

3.1. CEPHEID VARIABLES: FUNDAMENTAL PROPERTIES

Much progress has been made in determining the fundamental properties of Cepheids, and this makes it possible for theorists to construct hydro dynamical models with some confidence. The mass of SU Cyg has been found to be 5.9–$6.2 m_\odot$ from its binary orbit (Evans and Bolton 1990), and the mass of other binary Cepheids are within reach, thanks to the work of Evans, Szabados and others. Indirect determinations of Cepheid masses ("pulsation" mass, "beat" mass, "bump" mass) seem now to be in agreement with direct mass and evolutionary masses thanks to the new Iglesias and Rogers opacities (Petersen 1990; Moskalik et al. 1991, preprint).

Cepheid luminosities have been reviewed by Feast and Walker (1987) and by Madore and Freedman (1991). Fernie (1992, preprint) has recently refined the Cepheid and δ Scuti period - luminosity relations by combining these two groups of young population I stars. The resulting relations for Cepheids are:

$$< Mv > = -2.902 \log P - 1.203$$
$$\pm 0.030 \qquad \pm 0.029$$

$$\log(R/R_\odot) = 1.1116 + 0.7385 \log P$$
$$\pm 0.0060 \qquad \pm 0.0060$$

It will be interesting to compare these results with the forthcoming results from the HIPPARCOS mission. Cepheid reddenings and intrinsic colours have been investigated by Fernie (1990a) and mass loss rates by Deasy (1988) and by Welch and Duric (1988).

The evolution of intermediate-mass stars such as Cepheids has reviewed by Chiosi (1990). An important recent development has been the identification and study of many Cepheids in NGC 1866 and other Magellanic Cloud clusters (e.g. Welch et al. 1991).

3.2. CEPHEID VARIABLES: LIGHT AND VELOCITY CURVES

In the last decade, Fourier decomposition has proven to be an effective tool in the interpretation of the shapes of Cepheid light and velocity curves (Simon and Lee 1981). It has been used, for instance, to interpret the Hertzsprung progression - a systematic trend, with period, in the shape of these curves. A secondary maximum or bump in the curve is found to coincide with the primary maximum at a period of 9 to 10 days. This is found to be due to a resonance between the fundamental (P_0) and second overtone (P_2) radial periods: $P_2/P_0 \sim 0.5$ (Simon and Davis 1983; Buchler et al. 1990; Kovacs et al. 1990). Until recently, agreement between observations and predictions from hydrodynamic models did not occur at the correct period and/or mass, but this deficiency seems to be remedied by the new Iglesias and Rogers opacities.

Fernie (1990b) has investigated the structure of the Cepheid instability strip, and found that (a) light amplitude increases with increasing period. (b) there is no tendency for the amplitude to be greater at the red or blue edge of the instability strip, (c) the amplitude is generally greater in the middle of the instability strip, but large- and small- amplitude stars can co-exist at any place in the strip, (d) the strip seems to be wider in (B-V) than would be expected theoretically, and (e) there is a non-uniform distribution of stars in the strip, with relatively fewer stars in the upper left and lower right sections. Sasselov (this meeting) has suggested that some of these effects may be due to the problem of determining a valid mean (B-V) for a pulsating atmosphere. This may possibly explain the well-documented apparent presence of non-variable stars in the Cepheid instability strip.

Moskalik et al. (1991, 1992) have found that hydrodynamic models of Cepheids with periods of 25-40 days (or 22.5 days with Iglesias and Rogers opacities) show incipient RV Tauri behaviour, due to a 3/2 resonance between P_0 and P_1. There is no evidence of this phenomenon in classical

Cepheids (Fernie, private communication), but it may be present in population II Cepheids. Winzer (1973) suspected that the light curve of the 3.7-day Cepheid RT Aur might show fine structure at the 0.01^m level, but that is close to the noise level of the observations.

3.3. CEPHEID VARIABLES: OVERTONE PULSATORS

Most Cepheid variables pulsate in the fundamental radial mode. The double-mode or "beat" Cepheids pulsate in the fundamental and first overtone modes. Theoretical models have so far not succeeded in reproducing stable double-mode behaviour (Moskalik et al. 1992). The amplitude ratios (A_1/A_0) of the first overtone to fundamental mode do not show any obvious trend with period, which suggests that some other parameter may affect this ratio.

There are also Cepheids with small amplitudes, sinusoidal light curves, and periods less than five days which, on the basis of Fourier decomposition, seen to be separate from "normal" Cepheids. These are thought to be first overtone pulsators. Two examples are SU Cas and CO Vir. Welch et al. (1991) have found one such variable in NGC 1866 among fundamental mode pulsators. Antonello et al. (1990) have identified about 20 of them morphologically. They do not follow the Hertzsprung progression e.g. in the ϕ_{21}–P and ϕ_{31}–P diagrams. Diethelm (1990) has classified Cepheids with periods between one and three days according to their pulsation mode and their evolutionary state or metallicity.

3.4. CEPHEID VARIABLES: LONG-TERM CHANGES

The light and velocity curves of classical Cepheids change relatively slowly. The double-mode Cepheids are an obvious but well-understood exception. A less well-understood exception is HR 7308 (V473 Lyr), the classical Cepheid with the shortest known period (1.49^d). The photometric amplitude of this star varies from about 0.05^m to 0.40^m in a period of 1210 days (Percy and Ford 1981). The cause of this modulation is unknown. Another exception is Polaris, whose light and velocity amplitudes have been decreasing for at least two decades, any may reach zero by the year 2000 (Dinshaw et al 1989). The light amplitude of Y Oph may also be decreasing (Fernie 1990c). Fernie (1990c) points out that both Polaris and Y Oph appear to be situated in the middle of the instability strip.

The observed period changes in classical Cepheids are in good agreement with those predicted by evolutionary models (Saitou 1990). Fernie (1990c) has pointed out that the period change in Y Oph is not linear (the O–C diagram is a cubic), and that this is to be expected from evolutionary models. Szabados (1991) has detected apparent "phase shifts" in the O–C diagrams of some classical Cepheids, but the evidence for these is not compelling. This is fortunate, since no obvious explanation for such a phenomenon comes to mind!

4. Population II Cepheid Variables

These variables (also known as W Virginis stars) are low-mass stars of the disc or halo population which are situated in the Cepheid instability strip. The least luminous of these are the BL Her subgroup, which are evolving away from the horizontal branch, and which have periods of one to three days. Since these are classified as giants rather than supergiants, they will not be discussed further here, except to point out that Moskalik and Buchler (this meeting) have predicted from hydrodynamic models that some BL Her stars should show period doubling i.e. alternating larger and smaller amplitude. This phenomenon should be looked for in the observations.

More luminous population II Cepheids are believed to be executing "blue loops" from the asymptotic giant branch (AGB) due to shell flashes in the interior. These loops occur on time scales of a few thousand years, so period changes and perhaps even amplitude changes may be observable. RU Cam is a population II Cepheid whose amplitude decreased from over a magnitude to nearly zero in 1962–64 (Fernie and Demers 1966; Huth 1966). This may provide a "natural experiment" on the effect of pulsation amplitude on period and mean colour. All bright population II Cepheids should be systematically monitored - preferably photoelectrically but, if not, photographically or visually.

A few luminous population II Cepheids should be contracting rapidly from the AGB to the white dwarf stage. One or two such stars show some evidence for RV Tau behaviour (Arp 1955).

Population II Cepheids can also be sub-classified, according to their light curves, into "crested" and "flat top". Fadeyev and Muthsam (1990) have postulated, on the basis of hydrodynamic models, that this effect is due to a resonant coupling between the fundamental and first overtone modes. These same authors also comment on the origin of the shock waves which are prominent in the spectra of population II Cepheids at some phases.

5. RV Tauri Variables

RV Tauri stars are rare, luminous pulsating variables which tend to show alternating deep and shallow minima in their light curves. They are believed to be post-AGB stars on the basis of their location on the H-R diagram, their period changes (Percy et al. 1991), and their dust envelopes (e.g. Jura 1986). There is some disagreement as to whether they left the AGB about 10^2 years ago (Alcolea and Bujarrabal 1991) or about 10^3 years ago (Jura 1986; Percy et al. 1991); the latter number seems more reasonable.   In an important recent development, Buchler, Kovacs and their co-workers have shown that, in a sequence of models of low-mass yellow supergiants of decreasing effective temperature, there is a transition from periodic behaviour, through successive period doublings. to chaos (Buchler et al. 1987: Buchler and Kovacs 1987; Kovacs and Buchler 1988). The population II Cepheids, RV Tauri variables, and SRd variables may represent the observational manifestation of this sequence, though it might be more correct to think of the observational sequence as being a continuous one. Since the transition may be a function of luminosity, the "structure" of the yellow supergiant region of the H-R diagram could be quite complex.

RV Tauri-like behaviour has been found in a number of theoretical models, including Christy (1966), Stobie (1969), Deupree and Hodson (1976), and Fadeyev (1984). Deupree and Hodson attribute the behaviour to the effect of convection, Buchler and Kovacs to resonant interaction between modes. It is entirely possible that two or more physical mechanisms may contribute to the behaviour.

A long-standing question is whether the physical period of an RV Tauri star is the "single" period (between adjacent minima) or the "double" period (between adjacent deep minima). Recent spectroscopic studies have been divided on this issue (e.g. Lebre and Gillet 1991; Mozurkewich et al. 1987) but, if two modes are present in the atmosphere in 2:1 resonance, then it is possible that both answers to the question are correct.

5.1. RV TAURI STARS: HOW REGULAR?

Although RV Tauri stars are defined as showing alternating deep and shallow minima, the range of behaviour is actually very wide. A preliminary examination of the long-term light curves of RV Tauri stars complied by Payne-Gaposchkin et al. (1943) shows that the incidence of alternating deep and shallow minima ranges from 99% in AC Her and V Vul to 60% or less in U Mon, TT Oph, AR Pup and RV Tau (Fig. 2). This analysis is very crude, being based on sparse visual and photographic data. It should be repeated with better data. Five-day means of the dense visual observations in the AAVSO database (Mattei and Percy, this meeting) should be adequate for the purpose. Individual RV Tauri stars can go through exceptionally irregular phases (e.g. R Sct; Lebre and Gillet 1991).

Another approach to this question is to construct return maps for RV Tauri stars. This has been done by Veldhuizen and Percy (1989), Saitou et at. (1989) and Kollath (1990). In each case, the authors have used visual observations (since these are dense and extend over many years) but, in each case, the authors conclude that visual observations are only marginally suitable for the purpose at hand. Perhaps if the visual observations were processed in a more sophisticated way (such as by spline interpolation). they would yield a more definite result. It is also possible that RV Tauri stars are inherently irregular!

It is nevertheless interesting to look at the range of behaviour shown by Saitou et al.'s (1989) return maps of RV Tauri and SRd stars. These maps

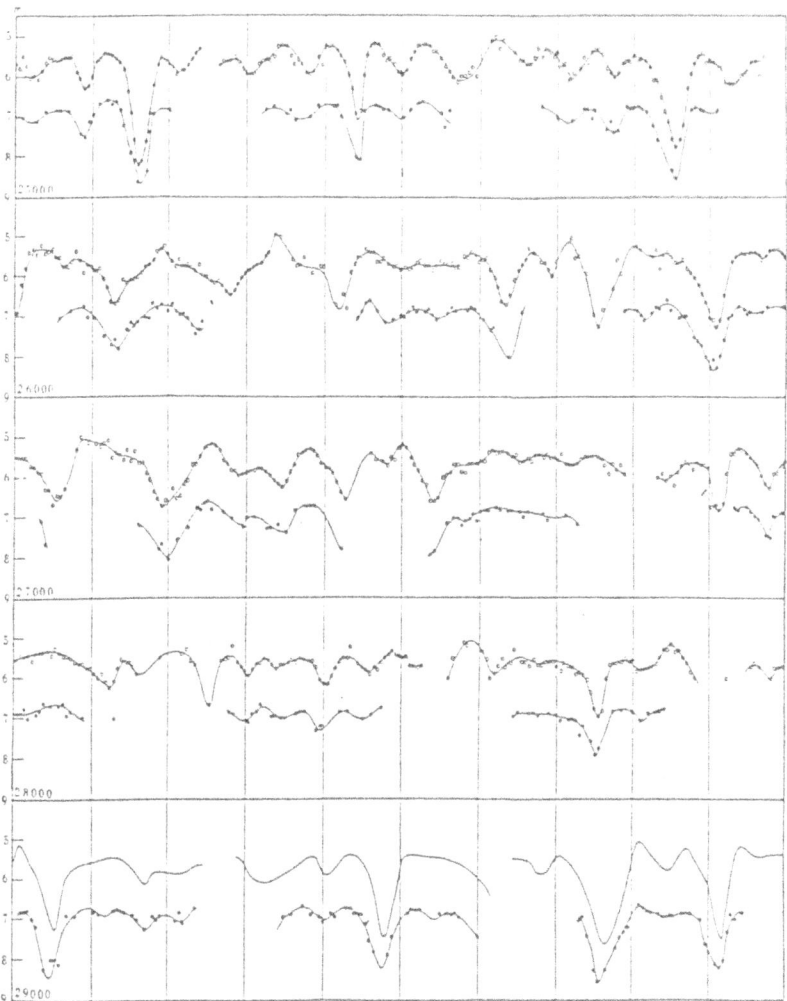

Fig. 2.   Long-term visual (open circles) and photographic (filled circles) observations of the RV Tauri star R Sct (Payne-Gaposchkin et al. 1943). Note that alternating deep and shallow minima do not always occur; the pulsation can be quite irregular.

support the idea that RV Tauri stars may exhibit a smooth progression from period-2 behaviour to irregularity or chaos. The return map of S Vul is particularly interesting, and supports the suggestion that this 68.8-day "Cepheid" shows RV Tauri characteristics.

5.2. RV Tauri Stars: Long Term Changes

The periods of RV Tauri stars exhibit a wide variety of long-term changes (see Percy et al. 1991 and references therein). There are *abrupt* changes whose magnitude, when averaged over the duration of the observations, is in rough agreement with evolutionary predictions. There are *cyclic* changes (especially prominent in AC Her; see Zsoldos 1988) which are not due to, for instance, light travel time in a binary system. There are *irregular* changes which can be explained by random changes in period from one cycle to the next (Percy, Csatary and Zorgdrager, unpublished results), which are similar to the period fluctuations in Mira stars. All of these period changes are useful in constraining evolutionary models of RV Tauri stars.

About half of all RV Tauri stars (the RVb stars) show long-term changes in mean brightness, on time scales of up to 2500 days. The long period is roughly 10 times the shorter one. In the case of U Mon, the long-term light curve is similar to that of a totally-eclipsing binary (!). The amplitude of pulsation becomes very small when the star is at its long-term minimum. Long-term radial velocity variations have been observed in U Mon, but the cause of the RVb phenomenon has yet to be identified.

6. Semi-Regular (SRd) Yellow Supergiants

This is a heterogeneous group of seldom-studied, poorly-understood stars. In the *General Catalogue of Variable Stars*, there are 122 RV Tauri stars and 78 SRd variables. In many cases, there may be insufficient observations to classify the star reliably. Zsoldos and his collaborators have published a number of studies of individual RV Tauri and SRd stars, utilizing new UBV photometry as well as archival data. See Eggen (1986) for a recent review of SRd variables.

As mentioned in Section 5, theoretical results suggest that the sequence W Vir → RV Tau → SRd may show a continuous spectrum of behaviour from periodicity through successive period-doubling to chaos. Mantegazza (1984) used factor analysis to separate partially the RV Tauri and SRd variables on the basis of DDO colours; the SRd variables are cooler, in agreement with the theoretical results.

The problems of identifying period-4, period-5, and chaotic behaviour in SRd variables are in most cases similar to those in RV Tauri stars, or worse: too little data and/or too little precision. There is also the very real problem of determining the effective temperature and luminosity.

6.1. METAL DEFICIENT YELLOW SUPERGIANTS

Several authors (most recently Waelkens et al. (1991); see also references therein) have called attention to this small but interesting group. HR4049 and HD 52961 have [Fe/H] = −4.8, low photometric amplitudes, and periods of several weeks. They are presumably post-AGB stars, but how long do they spend in this stage, and where are their progenitors?

6.2. UU Herculis Stars

These are defined as high galactic latitude F and G supergiants with low photometric amplitudes and periods of several weeks. They were thought by some to be massive, young stars (in which case, how did they reach high galactic latitude?) but are now generally believed to be low-mass stars of the old disc population, in the post-AGB stage of evolution (Arellano Ferro et al. 1989; Luck et al. 1990).

The best-studied examples are UU Her itself, 89 Her (Fernie 1991a) and HD161796 (Fernie 1991a). 89 Her is fairly regular with a period of 65.0±1.2 days; HD161796 is more irregular. The latter star is cooler, and its greater irregularity conforms to the theoretical predictions of Buchler, Kovacs and their collaborators.

Fernie and Sasselov (1989) have shown that the lack of observable period and colour variations in UU Her place strong constraints on its evolutionary status.

7. R Coronae Borealis Variables

R CrB stars are rare carbon-rich hydrogen-deficient yellow supergiants which undergo unpredictable rapid decreases in brightness, followed by slower recovery to normal brightness. The fadings are due to obscuration by dust clouds. Most R CrB stars (and cool hydrogen-deficient carbon stars in general) also undergo small-amplitude brightness variations on time scales of 40 to 200 days or more, probably due to pulsation (Lawson et al. 1990).

The pulsations are most obvious in RY Sgr. In this star, they are sufficiently regular that the observed period changes can be compared with an evolutionary model (of a 0.7 solar mass star in the post-AGB contraction phase). The agreement is good, at least to first approximation. In R CrB itself, Fernie (1991b, and references therein) has found that the pulsations are somewhat irregular, though there is a characteristic time scale of 40 to 50 days. A subsequent analysis by Lawson (1991) of 1985–90 photometry reveals two or more periods in the 1986–89 data, and a different period in the 1985 and 1990 data. Thus, it is not clear whether the pulsations are multiperiodic, chaotic or irregular.

Feast (1986) has proposed a model in which dust clouds are ejected every 40 days in a cone angle of 40°, in random directions. The average rate of mass loss is $10^{-6}m_\odot/a$. These ejections are adequate to explain the observed frequency of fadings.

The similarity between the time scale of the ejections in this model, and the time scale of the photometric variations, is one of several indications that the pulsations may be connected with the ejections. But why are discrete clouds of dust produced? Is it because they are somehow triggered by non-radial pulsation? G. Clayton, B. Whitney and their collaborators (1991, preprints) have proposed that the dust forms within one or two stellar radii of the photosphere, and that individual dust clouds form as a result of density enhancements caused by the passage of shocks through the atmosphere.

8. Discussion

The study of non-linear processes in supergiant variables requires long series of observations of good and consistent quality. Long-term projects are often not encouraged or supported in astronomy. This is unfortunate, because they provide types of information not otherwise available.

Long-term visual and photographic observations exist for many supergiants. In "raw" form, these observations may be inadequate for some purposes, such as to identify multiperiodicity or chaos. It may be possible to process these observations so as to extract the maximum possible information. This requires that the observations be preserved and archived in electronic form - and continued, of course.

Long-term photoelectric observations of some supergiants are now becoming available, both from Automatic Photometric Telescopes (APT's) and networks of skilled amateur astronomers such as those in the AAVSO photoelectric program. The papers by Fernie (1991a,b) and by Zsoldos and Percy (1991) illustrate the type of research which can be done. The most interesting results, however, often take a decade or more of data.

Some degree of coordination is essential. APT's and human observers should share the monitoring of selected stars. Long-term coordinated photometric, spectroscopic and polarimetric observations should be made of selected stars; objects such as P Cyg, ρ Cas, α Ori, R Sct, 89 Her and R CrB come immediately to mind. The planning of such observations should be an important part of future meeting of this kind.

Acknowledgements

I am grateful to the Natural Sciences and Engineering Research Council of Canada for research and travel support, and to the Local Organizing Committee of this meeting for their support and hospitality, and for organizing such a pleasant and productive meeting.

References

Alcolea, J. and Bujarrabal, V.: 1991, *Astronomy and Astrophysics* **245**, 499.
Antonello, E., Poretti, E. and Reduzzi, L.: 1990, *Astronomy and Astrophysics* **236**, 138.
Arellano Ferro, A., Giridhar, S., Chavez, M. and Parrao, L.: 1989, *Astronomy and Astrophysics* **214**, 123.
Arp, H. C.: 1955, *Astronomical Journal* **60**, 1.
Blomme, R.: 1991, *Astronomy and Astrophysics* **246**, 199.

Blomme, R., Vanbeveren, D. and Van Rensbergen, W.: 1991, *Astronomy and Astrophysics* **241**, 479.

Bowen, G. H.: 1988, *Astrophysical Journal* **329**, 299.

Buchler, J. R. (editor) 1990, *The Numerical Modelling of Nonlinear Stellar Pulsations* (Kluwer Academic Publishers).

Buchler, J. R. and Goupil, M. -J.: 1988, *Astronomy and Astrophysics* **190**, 137.

Buchler, J. R., Goupil, M.-J. and Kovacs, G.: 1987, *Phys. Lett.* **A126**, 177.

Buchler, J. R. and Kovacs, G.: 1987, *Astrophysical Journal* **320**, L57.

Buchler, J. R., Moskalik, P. and Kovacs, G.: 1990, *Astrophysical Journal* **351**, 617.

Chiosi, C.: 1990, *Publications of the ASP* **102**, 412.

Christy, R. F.: 1966, *Astrophysical Journal* **145**, 337.

Deasy, H.: 1988, *Monthly Notices of the RAS* **231**, 673.

De Jager, C., de Koter, A., Carpay, J. and Nieuwenhuijzen, H.: 1991, *Astronomy and Astrophysics* **244**, 131.

Demers, S. and Fernie, J. D.: 1966, *Astrophysical Journal* **144**, 437.

Deupree, R. G. and Hodson, S. W.: 1976, *Astrophysical Journal* **208**, 426.

Diethelm, R.: 1990, *Astronomy and Astrophysics* **239**, 186.

Dinshaw, N., Matthews, J. M., Walker, G. A. H. and Hill, G. M.: 1989, *Astronomical Journal* **98**, 2249.

Eggen, O. J.: 1986, *Astronomical Journal* **91**, 890.

Evans, N. R. and Bolton, C. T.: 1990, *Astrophysical Journal* **356**, 630.

Fadeyev, Yu. and Muthsam, H.: 1990, *Astronomy and Astrophysics* **234**, 188.

Feast, M. W.: 1986, in *Hydrogen Deficient Stars and Related Objects*, eds. Hunger, Schönberner and Rao (D. Reidel Publ. Co.), p. 151.

Feast, M. W. and Walker, A. R.: 1987, *Annual Review of Astronomy and Astrophysics* **25**, 345.

Fernie, J. D.: 1990a, *Astrophysical Journal, Supplement Series* **72**, 153.

Fernie, J. D.: 1990b, *Astrophysical Journal* **354**, 295.

Fernie, J. D.: 1990c, *Publications of the ASP* **102**, 905.

Fernie, J. D.: 1991a, *Publications of the ASP* **103**, 1087.

Fernie, J. D.: 1991b, *Publications of the ASP* **103**, 1091.

Fernie, J. D.: 1992, *Astrophysical Journal*, in press.

Fernie, J. D. and Sasselov, D. D.: 1989, *Publications of the ASP* **101**, 513.

Halbedel, E.: 1991, *IAU IBVS* **#3600**.

Hayes, D. P.: 1985, *Astrophysical Journal* **289**, 726.

Huth, H.: 1966, *Die Sterne* **42**, 129.

Iglesias, C. A. and Rogers, F. J.: 1991, *Astrophysical Journal* **371**, L73.

Johnson, H. R. and Querci, F.: 1986, *The M-Type Stars*, NASA SP-492.

Jura, M.: 1986, *Astrophysical Journal* **309**, 732.

Kollath, Z.: 1990, *Monthly Notices of the RAS* **247**, 377.

Kovacs, G. and Buchler, J. R.: 1988, *Astrophysical Journal* **334**, 971.

Kovacs, G. and Buchler, J. R.: 1989, *Astrophysical Journal* **346**, 898.

Kovacs, G., Kisvarsanyi, E. G. and Buchler, J. R.: 1990, *Astrophysical Journal* **351**, 606.

Lawson, W. A.: 1991, *Monthly Notices of the RAS* **253**, 625.

Lawson, W. A., Cottrell, P. L., Kilmartin, P. M. and Gilmore, A. C.: 1990, *Monthly Notices of the RAS* **247**, 91.

Lebre, A. and Gillet, D.: 1991, *Astronomy and Astrophysics* **246**, 490.

Leitherer, C., Schmutz, W. and Abbott, D. C.: 1989, *Astrophysical Journal* **346**, 919.

Luck, R. E., Bond, H. E. and Lambert, D. L.: 1990, *Astrophysical Journal* **357**, 188.

Madore, B. F. and Freedman, W. L.: 1991, *Publications of the ASP* **103**, 933.

Maeder, A.: 1980, *Astronomy and Astrophysics* **90**, 311.

Mantegazza, L.: 1984, *Astronomy and Astrophysics* **135**, 300.

Markova, N. and Kolka, I.: 1988, *Astrophysics and Space Science* **141**, 45.

Moskalik, P., Buchler, J. R.: 1991, *Astrophysical Journal* **366**, 300.

Moskalik, P. and Buchler, J. R. and Marom, A.: 1992, *Astrophysical Journal* **385**, 685.

Mozurkewich. D., Gehrz, R. D., Hinkle, K. H. and Lambert, D. L.: 1987. *Astrophysical Journal* **314**, 242.

Nieuwenhuijzen. H. and de Jager, C.: 1990, *Astronomy and Astrophysics* **231**, 134.

Pauldrach, A. W. A. and Puls, J.: 1990, *Astronomy and Astrophysics* **237**, 409.

Payne-Gaposchkin, C., Brenton, V. K. and Gaposchkin. S.: 1943, *Ann. Harvard Coll. Obs.* **113**, #1.

Percy, J. R. and Ford, R. P.: 1981, *JAAVSO* **10**, 53.

Percy, J. R. and Sen, L. V.: 1991, paper presented at the Fall 1991 Meeting of the AAVSO.

Percy, J. R. and Welch, D. L.: 1983, *Publications of the ASP* **95**, 491.

Percy, J. R. et al.: 1988, *Astronomy and Astrophysics* **191**, 248.

Percy, J. R., Sasselov, D. D., Alfred, A. and Scott, G.: 1991, *Astrophysical Journal* **375**, 691.

Petersen, J. O.: 1990, *Astronomy and Astrophysics* **238**, 160.

Saitou, M.: 1990, *Publications of the ASJ* **42**, 341 .

Saitou, M., Takeuti, M. and Tanaka, Y.: 1989. *Publications of the ASJ* **41**, 297.

Sheffer, Y. and Lambert, D. L.: 1986, *Publications of the ASP* **98**, 914.

Sheffer, Y. and Lambert, D. L.: 1987, *Publications of the ASP* **99**, 1277.

Simon, N. R. and Davis, C. G.: 1983, *Astrophysical Journal* **266**, 787.

Simon, N. R. and Lee, A. S.: 1981, *Astrophysical Journal* **248**, 291.

Stobie, R. S.: 1969, *Monthly Notices of the RAS* **144**, 485.

Stothers, R. Leung, K.-C.: 1971, *Astronomy and Astrophysics* **10**, 290.

Szabados, L.: 1991, *Konkoly Obs. Comm.* **11**, 125.

Takeuti, M.: 1988, *Astrophysics and Space Science* **140**, 431.

Taylor, M., Nordsieck, K. H., Schulte-Ladbeck, R. E. and Bjorkman, K. S.: 1991, *Astronomical Journal* **102**, 1197.

Veldhuizen, T. and Percy, J. R.: 1989, *JAAVSO* **18**, 97.

Waelkens, C., Van Winckel, H., Bogaert, E. and Trams, N. R.: 1991, *Astronomy and Astrophysics* **251**, 495.

Welch, D. L. and Duric, N.: 1988, *Astronomical Journal* **95**, 1794.

Welch, D. L., Mateo, M., Coté. P., and Madore, B. F.: 1991, *Astronomical Journal* **101**, 490.

Winzer, J. E.: 1973, *Astronomical Journal* **78**, 618.

Zsoldos, E.: 1988, IAU IBVS #3192.

Zsoldos, E. and Percy, J. R.: 1991, *Astronomy and Astrophysics* **246**, 441.

THE AAVSO DATABASE OF VARIABLE
STAR OBSERVATIONS

J. R. PERCY

Erindale Campus, University of Toronto, Mississauga, Ontario, Canada L5L 1C6

and

J. A. MATTEI

Director: AAVSO 25 Birch Street, Cambridge, MA 02138-1205, USA

The American Association of Variable Star Observers (AAVSO) is the largest organization of variable star observers in the world, with members in 42 countries. The purpose of the AAVSO is to coordinate variable star observing, done primarily by amateur astronomers, evaluate the accuracy of these observations, compile, process and publish them, and make them available to researchers and educators around the world. Over 6.5 million observations of variable stars have been complied since the AAVSO was founded in 1911. Those since about 1960 are in computerized form, and it is intended to have all observations in this form within the next year. About 250,000 observations are submitted to and archived by the AAVSO each year, over half of them from outside the USA. Most of the observations are visual, but there is also an active photoelectric program which concentrates on semi-regular and irregular stars such as small-amplitude red variables. In 1990, over 200 requests for AAVSO data and services were received from researchers and educators; this number has increased by a factor of 10 in the last two decades.

AAVSO services are sought by astronomers (a) to receive real-time, up-to-date information on unusual stellar behaviour, (b) to assist scheduling and execution of variable star observing programs using Earth-based large telescopes, and instruments aboard satellites; (c) to request simultaneous optical coverage of stars being studied during Earth-based or satellite observing programs, and immediate notification of their activity; (d) to correlate optical data with data obtained using other techniques (photometry, spectroscopy, polarimetry), and at other wavelengths from gamma-rays to radio waves; (e) to carry out collaborative long-term data analyses of the behaviour of (particularly) large-amplitude variable stars; and (f) to assist educators in setting up observing projects, and students in carrying them out.

For many hundreds of stars, the AAVSO has almost continuous visual observations, using consistent sets of comparison stars, and extending over many decades. These can be used to study long-term changes in period, amplitude, mean magnitude, and light curve shape in large-amplitude stars

Astrophysics and Space Science **210**: 137–138, 1993.
© 1993 *Kluwer Academic Publishers.*

such as Mira variables (Percy et al. 1990). Although the precision of a single
visual observation is relatively low (typically, 0.2 to 0.3 magnitude), useful
studies of smaller-amplitude variables can be done by Fourier analyzing the
observations, grouping them into means, or otherwise processing them so as
to extract the maximum amount of information (e.g. the study of Rho Cas
by Percy et al. 1985). AAVSO photoelectric observations, which for some
stars now extend over almost a decade, are well-suited for studying long-
term changes in smaller-amplitude variables, and for distinguishing between
regular and irregular behaviour (e.g. the study of EU Del by Percy et al.
1989).

AAVSO visual observations have also been used to search for chaotic
behaviour in Mira stars (Cannizzo et at. 1990), the cataclysmic variable SS
Cyg (Cannizzo and Goodings 1988; Hempelmann and Kurths 1990), and
the RV Tauri star R Sct (Veldhuizen and Percy 1989). There are several
dozen RV Tauri and SRd variables in the AAVSO visual and photoelectric
observing programs. These observations, if suitably processed and analyzed,
may be extremely useful for studying the nature of the irregularity in these
stars.

To request AAVSO data and services, please contact the Director at the
address above, or by e-mail at AAVSO@CFA8.BITNET.

References

Cannizzo, J. K. and Goodings, D. A.: 1988, *Astrophysical Journal* **334**, L31.
Cannizzo, J. K., Goodings, D. A. and Mattei, J. A.: 1990, *Astrophysical Journal* **357**, 235.
Hempelmann, A. and Kurths, J.: 1990, *Astronomy and Astrophysics* **232**, 356.
Percy, J. R., Colivas, T., Sloan, W. B. and Mattei, J. A.: 1990, in *Confrontation between Stellar Pulsation and Evolution*, eds. C. Cacciari and G. Clementini (ASP Conference Series #11).
Percy, J. R., Fabro, V. A. and Keith, D. W.: 1985, *JAAVSO* **14**, 1.
Percy, J. R., Landis, H. J. and Milton, R. E.: 1989, *Publications of the ASP* **101**, 893.
Veldhuizen, T. and Percy, J. R.: 1989, *JAAVSO* **18**, 97.

THEORETICAL MODES FITTING ON MIRAS LIGHT CURVES

D. BARTHÉS

GRAAL, Université Montpellier II/ CNRS, F-34905 Montpellier Cedex 05, France

Y. TUCHMAN

Racah Institute. Hebrew University of Jerusalem, Israel

M. O. MENNESSIER

GRAAL, Université Montpellier II/ CNRS, F-34905 Montpellier Cedex 05, France

and

J. A. MATTEI

AAVSO. 25 Birch street, Cambridge, MA 02138-1205, USA

1. Observational Data Analysis

Visual observations of long period variable stars over 20 years were provided by the American Association of Variable Stars Observers, and were analysed as part of the preparation of the HIPPARCOS mission.

A set of frequencies is extracted from the light curve by using Fourier transform, preliminary Van Cittert deconvolution and comparison of the results obtained through different kinds of spectral windows. The same procedure is applied to the residual obtained after nonlinear fit of the main frequency. After final comparison of both sets, a nonlinear fit of the common frequencies gives the 'clean' power spectrum.

2. Theoretical Model

Different equilibrium stellar model (i) give theoretical linear nonadiabatic pulsation modes $(\nu_j)_i$ with their growth rates $(\eta_j)_i$ (Tuchman 1978). The metallicity is taken between 0.005 and 0.02; the mixing length is $\lambda = 1 \pm 0.2$; the upper bound is $r = 0.7$. Assuming two peaks of the power spectrum to be the fundamental (ν_0) and first overtone (ν_1) modes, one looks for the corresponding models. The best one is selected by checking the other theoretical overtones they give. So are obtained the mass. the luminosity, the effective temperature and the effective radius of each star.

3. Results

With power spectra of S CMi, R LMi and S UMi one can find a model in excellent agreement not only for fundamental and first overtone modes but also for the next ones. For example. in the case of S CMi (Table I), ν_0, ν_1, ν_2, ν_3 and ν_5 appear in the observed spectrum. while ν_4 (with negative growth

Astrophysics and Space Science **210**: 139–140. 1993.

rate) does not. This model gives $L = 7200 L_\odot$, $M = 1.9 M_\odot$, $T_{\rm eff} = 2155$ K. $R_{\rm eff} = 515 R_\odot$.

TABLE I
S CMi

Observed power spectrum		Assumption	Model			
$\nu(d^{-1})$	A (mag)		ν	η	P (days)	
0.00059	0.16	ν_0	0.00062	+	1610	ν_0
0.00132	0.17					$\nu_3 - \nu_2, \nu_2 - \nu_1, 2\nu_0$
0.00241	0.13					$\nu_1 - \nu_0$
0.00299	2.18	ν_1	0.00298	+	335	ν_1
0.00315	0.23					$\nu^{\bullet}$
0.00422	0.15		0.00413	+	242	ν_2
0.00560	0.20		0.00555	+	180	ν_3
0.00601	0.21					$2\nu_1$
			0.00680	−	147	
0.00838	0.10		0.00833	+	120	$\nu_5, 3\nu_1 - \nu_0$
0.00896	0.23					$3\nu_1$
			0.00971	−	103	
			0.01075	+	93	
0.01186	0.10		0.01190	+	84	$4\nu_1, \nu_8$
			0.01266	−	79	

In each explained remain a frequency ν^* and/or its combination with the main one. This can be explained by using a rough atmospheric model: pure H_2; $\rho \propto r^{-2}$; $g \propto r^{-2}$; $\gamma = cte$; lower bound at $r = R_{\rm eff}$; $P_{\rm rad}/P = 0.05$. Then ν^* corresponds to the fundamental radial adiabatic mode of the atmosphere (0.00309 d^{-1} for S CMi).

4. Conclusion

- These three miras are pulsating on the *first overtone*;
- Within the physics (convection + surface conditions) used in the theoretical models, one can predict the *masses* and the *luminosities* of Mira variables by using very clear and precise observational parameters (periods of the main modes);
- Only a combination of a stellar model with an atmospheric one can provide a complete interpretation of the power spectrum of the light curve of a Mira star.

References

Tuchman, Y., Sack, N.. and Barkat, Z.: 1978, *Astrophysical Journal* **219** 183.

ON THE OBSERVED COMPLEXITY OF CHAOTIC STELLAR PULSATION

Z. KOLLÁTH

Konkoly Observatory, P.O. Box 67, H-1525 Budapest, Hungary

1. Introduction

The existence of some variable stars producing very complicated light curves is well known. Theoretical calculations suggest that the irregular behaviour of the pulsation models in the RV Tauri and W Virginis regime is low dimensional is the result of period-doubling or tangent bifurcations (see e.g. Buchler and Kovács 1987, Kovács and Buchler 1988, Tanaka and Takeuti 1988).

The light variation of the RV Tauri star R Scuti covering 150 years was analyzed by Kolláth (1990). A striking similarity was found between the reconstructed attractor of the Rössler model and that of the light variation of R Scuti. This result confirms the theoretical prediction of the existence of chaos, but a discrepancy still exists between the theory and observation. We could not find evidence for low dimension ($D = 2 - 3$). The analyses of other stars also show rather erratic behaviour (e.g. Cannizzo and Goodings 1988, Cannizzo et al. 1990). A possible answer for this discrepancy is the treatment of stochastic perturbations by convection (Perdang 1991).

It was shown (see e.g. Mitschke et al. 1988) that a relatively simple system (e.g. low pass filter) can increase the correlation and information dimension of a signal. The inner pulsation - light variation transfer mechanism can be represented by a smooth function for many kind of variable stars, but it is more complicated for red giant and supergiant stars. In the latter case the shock waves play an important role and the light variation can be modeled only with the combined dynamics of the pulsation and the transfer mechanism. We discuss the effect of a simple model to the complexity of the observable light curves: even a linear transfer mechanism can increase the dimension of the attractor.

2. Characterization of the Complexity of the Light Variation

In order to get information about the complexity of a time series one can calculate the Fourier transform of the data. Multiperiodic signals can be easily detected by this method, but the spectra cannot distinguish between stochastic and low dimensional chaotic processes.

Astrophysics and Space Science **210**: 141–143, 1993.

One can usually derive qualitative information from the observational data describing the view of the reconstructed phase trajectories. For extracting qualitative dynamics from the one-dimensional observable. i.e. the light curve, the Takens-theorem (method of delays) was applied (Takens 1980). The vectors of the reconstructed phase space are defined as $(v_k, v_{k+L}, v_{k+2L}, ..., v_{k+(N-1)L})$, where v_k means the original signal and N is the embedding dimension.

The simplest method to get quantitative information is the calculation of the correlation dimension (see e.g. Grassberger and Procaccia 1983). If the phase space has more complex structure, its dimension is usually higher. The correlation integral $(C(r))$ counts the number of point pairs whose distance is less than r. If the $\log(r) - \log C(r)$ curve has a linear component, the correlation dimension is well estimated by its slope.

While the correlation dimension misses the information of the temporal running the nonlinear prediction (see e.g. Casdagli 1989) is based on the dynamics. The future of a deterministic system can be predicted to some degree from the surrounding phase trajectories representing the history of the system in the past. A very important measure of the prediction is the run of the prediction error (σ) versus the number of predictee (M). It has the scaling law: $\log \sigma \sim \log M/D$, where D is the information dimension of the attractor (see e.g. Casdagli 1989).

A similar measure on the observed complexity is the Lyapunov spectrum. The determination of the largest Lyapunov-exponent from a time series is relatively easy (see e.g. Wolf et al. 1985).

3. Effect of Inner Pulsation - Light Variation Transfer

For yellow and red supergiant stars the shock waves play an important role in the visible light variation. High resolution spectroscopic and spectrovelocimetric observations of R Sct and AC Her were reported by Gillet et al. (1990). These observations clearly describe the shock phenomena in the photosphere of RV Tau stars. The acceleration-curves calculated from the observed radial velocities have sharp peaks indicating the driving of shocks. The kinematics of the shocks is well estimated by ballistic motion except for the shock producing phase.

We used this fact to construct a simple transfer mechanism. The input is an arbitrary time series and the output is exactly determined by the variation of the input function at its maxima. Shocks were accelerated at the maxima of this time series and the model light variation of the stars was calculated from the temporary place and the intensity of the shocks. We used the most simple approximation. we did not want to describe the real light-curves only to predict the possible effect of the transfer mechanism.

This mechanism can neither create nor destroy information, but it redistributes the information on the time axis. Since the additional Lyapunov exponent due to the transfer mechanism is negative, it is expected, that the largest Lyapunov exponent remains unchanged. The dimension of the combined system, however, may be higher than that of the original dynamics (e.g. Mitschke et al. 1988). Even an added linear part can increase the dimension of the attractor if the added Lyapunov exponent is larger than the largest negative Lyapunov exponent.

As an input, we used the simplest chaotic models, i.e. the solution of the Lorenz (Lorenz 1963) and Rössler equations (Rössler 1976) with parameters according to the well developed chaos. The reconstructed trajectories were calculate from the input and output signals for both models. The simple view of the attractors indicates the differences in the complexity. The correlation integrals and their derivatives were also investigated. Although we can find well defined scaling regions in the case of the input function, this behaviour disappears for the output, and the correlation dimension is increased.

The increased complexity is also confirmed by the predictability of the data. The prediction errors were increased and the slope of the number of predictee $-\sigma$ curves were decreased indicating the higher informational dimension. We found, that the observed light curve may be more complicated than the underlying physical oscillations. As it is expected, the largest Lyapunov exponent was unchanged.

It is very important to consider the effect of a more realistic inner pulsation - light variation transfer and to develop methods for an inverse approach: how can we describe the complexity of the underlying dynamics from the observable light variation.

References

Buchler, J. R. and Kovács, G.: 1987, *Astrophysical Journal, Letters to the Editor* **320**, L270.

Canizzo, J. K. and Goodings, D. A.: 1988, *Astrophysical Journal* **334**, L31.

Canizzo, J. K. Goodings, D. A. and Mattei, J. A.: 1990, *Astrophysical Journal* **357**.

Casdagli, M.: 1989, *Physica* **D35**, 335.

Gillet, D., Burki, G. and Duquennoy, A.: 1990, *Astronomy and Astrophysics* **237**, 159.

Grassberger, P. and Procaccia, I.: 1983, *Physical Review Letters* **50**, 346.

Kolláth, Z.: 1990, *Monthly Notices of the RAS* **247**, 377.

Kovács, G. and Buchler, J. R.: 1988, *Astrophysical Journal* **334**, 971.

Lorenz. E. N.: 1963, *J. Atmos. Sci.* **20**, 130.

Mitschke, F., Möller, M. and Lange, W.: 1988, *Physical Review A: General Physics* **37**, 4518.

Perdang, J.: 1991, in *ESO Workshop on Rapid Variability of OB-Stars: Nature and Diagnostic Value*, ed. D. Baade (ESO), p. 349.

Rössler, O. E.: 1976, *Phys. Lett.* **57A**, 397.

Takens, F.: 1980, *Dynamic Systems and Turbulence* (Warwick, Lecture Notes Math.) **898**, 366.

Tanaka, Y. and Takeuti, M.: 1988, *Astrophysics and Space Science* **148**, 229.

Wolf, A., Swift, J. B., Swinney, L. and Vastano, J. A.: 1985, *Physica D* **16**, 285.

PREDICTING VARIABLE STAR LIGHT CURVES

T. SERRE

DASGAL, Observatory of Paris-Meudon, URA CNRS 335, F-92195 Meudon, France

Abstract. I present a new method to test determinism, in particular the nonlinear behavior, in observed time series of pulsating stars, based on a recent prediction method which exploits the dynamical system theory. A method for filling gaps in data has thereby been constructed. Estimated bounds to the necessary embedding dimension can be obtained and chaotic divergences can be estimated.

1. Principles of the Prediction Method

This method uses a reconstructed representation of the signal $s(t)$, with vectors $\mathbf{X}(i\delta t) = (s(i\delta t), s(i\delta t + \tau), \ldots, s(i\delta t + (m-1)\tau))$ in a m-dimensional state space with $\tau = d\delta t$, the time delay and δt, the time step of the signal, by using the time delay method Takens (1980). The reconstructed attractor is equivalent to an attractor constructed with the true variables describing the dynamics of the underlying physical process. In order to faithfully reproduce the attractor, this reconstruction requires a sufficiently large number of observed cycles and low noise level, that depend on dynamics complexity.

The true local law, $\mathbf{F}$, in the neighbourhood of a point of an attractor, links a past vector $\mathbf{X}(i\delta t)$, to its future vectors $\mathbf{X}(i\delta t + T)$, with a time step T, i.e. $\mathbf{X}(i\delta t + T) = \mathbf{F}(\mathbf{X}(i\delta t))$. In this neighbourhood, the closest vectors (this learning set is found by fast sorting and searching methods), are used to perform a least square approximation $\mathbf{F}_a$ of $\mathbf{F}$, once the order of Taylor expansion for $\mathbf{F}_a$ is fixed. This process is then iterated by estimating F at each predicted point (Farmer and Sidorowich, 1988, see the review of Casdagli et al. 1991). The basic assumption of this method is that a deterministic behavior must be present to assure the existence of a stationary structure, i.e. an attractor, in the state space.

2. Astrophysical Interests of This Method

The above method enables to fill short gaps, which are seen as a missing trajectory in the attractor, in an observed signal. It can be of interest for signals resulting from multisite campaigns, by identifying aliases. It differs from the MEM (Maximum Entropy) Method (Fahlman and Ulrych 1982) mainly in the fact that it uses local fits: the predictions are made by interpolations on the reconstructed attractor and local extrapolations where the goodness is controled by using neighbor vectors, instead that MEM uses an ARMA predictor with global fits, which gives an extrapolation in the gaps, with statistical controls, based in an interpolation on the available signal.

Astrophysics and Space Science **210**: 145–147, 1993.

The method used here does not create information if a successful dynamics approximation on the known parts is performed in the limit of the prediction time. Random or unknown dynamics are not predictable if not contained in the available recorded signal. Applications on artificial signals are successful (Serre et al. 1992).

Applications in identifying aliases in RV Tauri star are successful, but are more difficult when applied to light curve of a δ Scuti star GX Peg. Successful aliases reduction on a model of this star (5 artificial frequencies) shows that the noise level is the principal limitation and further studies are needed.

This prediction method is also used for testing divergences in the chaotic attractors. The tests on Rössler chaotic attractor are successful. An application on a hydrodynamical model of type II cepheids (from Kovács and Buchler 1987) confirms his chaotic nature. A method for estimating the correct embedding dimension or a lower bound (the efficient embedding dimension), by testing the best predictions, is adapted (Serre and Buchler, 1992) from a method of Casdagli et al. (1991). A test on the RV Tauri star, R Scuti, has thereby given only a lower bound, 4 or 5, for the embedding dimension, because of the stochastic perturbations.

3. Conclusions

This approach provides a new way of analyzing signals based on state space representation. The dynamics can be characterized by local or global modellings of the attractor (Casdagli et al., 1991). These modellings have a great potential for a deeper understanding of chaotic hydrodynamical models. Presently, the noise is the main limitation to the full efficiency of this method. Stochastic perturbations due to instrumental uncertainties and atmospheric noise must be reduced to low level by data processing methods preserving the dynamics information. A method, based on prediction, developed by Kostelich and Yorke (1988) is an interesting approach on noise reduction problems.

Acknowledgements

I thank Dr. D. Kurtz and Dr. W. Dziembowski for useful remarks during the colloquium, and , Dr. M. Auvergne, M. J. Goupil and J. R. Buchler for useful discussions about this work.

References

Casdagli, M., Des Jardins, D., Eubank S., et al.: 1991, *Tech. Rep.* LA-UR-91-1637 (Los Alamos Nat. Lab.)
Fahlman G. G. and Ulrych T. J.: 1982, *Monthly Notices of the RAS* **199**, 53.

Farmer J. D. and Sidorowich J. J.: 1988, *Tech. Rep.* LA-UR-88-901 (Los Alamos Nat. Lab.)

Kostelich E. J. and Yorke J. A.: 1990. *Physica D* 41, 183.

Kovács G. and Buchler J. R.: 1988. *Astrophysical Journal* **334**, 971.

Serre, T., Auvergne, M., and Goupil, M. J.: 1992, *Astronomy and Astrophysics*. in press.

Serre T. and Buchler J. R.: 1992, in preparation.

Takens F.: 1981, in *Dynamical System and Turbulence*, eds. Rand D. and Young L. S.(Springer Lecture Notes in Mathematics **898**, Berlin), p. 366.

CHAOTIC BEHAVIOR AND STATISTICAL ANALYSIS OF SOME MIRA AND SR STARS

T. YANAGITA

The Institute of Statistical Mathematics, 4-6-7 Minami-Azabu, Minato-ku. Tokyo 106, Japan

H. SATOH

Institute of Astronomy. The University of Tokyo, Mitaka, Tokyo 181. Japan

and

K. SAIJO

Department of Science and Engineering, National Science Museum, Ueno Park, Taitou-ku, Tokyo 110, Japan

Abstract. Visual light curves of some Mira and semiregular variable stars are analyzed to search for their chaotic and statistical behaviour. Discussion are also given.

1. Data

We analyze light curves of some Mira and semiregular (SR) variable stars, which are listed in table I, to find out chaotic and statistical behavior. Light curves of these stars are obtained from the database of Variable Star Observers League in Japan (Saijo and Kiyota 1991). Periods of long-term observations are about 20000 days for these stars except for W Cyg, whose observation period is about 4000 days.

Before analysis, we average the same day's observations and smooth out the fluctuation with high frequency by averaged over 20 day's ones. After these arrangement, we get the time series with its unit time, 20 days.

TABLE I
Characteristics of Stars

Star	Type	Period	Magnitude	Sp.
o Cet	M	332	2.0–10.1	M5e
χ Cyg	M	408	3.3–14.2	S6e
R Leo	M	310	4.4–11.3	M6
V Boo	SRA	258	7.0–12.0	M6e
W Cyg	SRB	131	5.8–8.9	M4
RS Cnc	SRC	120	7.1–8.6	M6e

From GCVS 4th edition (1985)

Astrophysics and Space Science **210**: 149–151, 1993.

TABLE II
Period Obtained by AR Model

o Cet		λ Cyg		R Leo	
Period	Power	Period	Power	Period	Power
333.3	2.34	412.4	2.56	312.5	2.64
171.7	1.61	205.1	2.17	168.8	1.35
116.3	0.55	136.5	1.03	2105.3	1.21
		769.2	1.01		

V Boo		W Cyg		RS Cnc	
Period	Power	Period	Power	Period	Power
256.4	1.59	128.6	− 0.34	217.4	− 0.49
135.6	0.25	264.9	− 0.38	130.7	− 0.56
888.9	0.17			327.9	− 0.61

2. Statistical Analysis

To understand the statistical property of the time series, periodgram is a basic statistic and defined by Fourier transformation of time series. However, the periodgram analysis is difficult to determine the peak of spectrum. Therefore, we use Auto Regressive (AR) model to estimate the power spectrum (e.g. Akaike and Nakamura 1988). AR model expressed the value $x(t)$ as a linear combination of M past values, $x(t) = \sum_{m=1}^{M} a(m)x(t-m) + \varepsilon(t)$ ($t=1,2,\ldots$), here $\varepsilon(t)$ is Gaussian noise. The parameters, $a(m)$ and $< \varepsilon\varepsilon >$, are estimated by Yule-Walker method. After estimating the parameter of AR model, we determine the power spectrum by

$$p(f) = \frac{\sigma^2}{|1 - \sum_{m=1}^{M} a(m)\exp(-i2\pi fm)|^2}$$

The main periods of the stars are easily determined because the power spectrum has a rational function form. The periods obtained from AR model are shown in table II.

3. Dynamical System Approach

The time series of Mira and SR stars behaves like a low dimensional chaotic dynamical system such as Lorenz system. Lorentz plot (return map of time interval between one local maximum and next local maximum) is useful to determine the low dimensional chaotic dynamics. However, in this case. this return map is difficult to plot because of missing observation. So we plot a stroboscopic return map, $x(t)$ vs $x(t+T)$. Stroboscopic return map of Mira stars has circular structure because of the periodic behavior, while that of SR stars shows more irregular map like a Brownian motion.

Correlation integral is one of a tool to determine the dimension of chaotic attractor (Grassberger and Procaccia 1983).

$$C_p(d) = \lim_{m \to \infty} \frac{1}{m^2} \sum_{i,j=1}^{m} H(d - |\mathbf{x}_i - \mathbf{x}_j|)$$

$$\mathbf{x}_i = (x(i), x(i+\tau), \ldots, x(i+(p-1)\tau))$$

where H is Heaviside function defined by $H(x) = 1$ for positive x, 0 otherwise. It is difficult to determine the dimension of Mira and SR stars, because of the limitation of observation number m and observational error. However, we conclude that some Mira stars' dimension is slightly above one and SR stars have more higher dimensions than those of Mira stars.

4. Discussions

Power spectrum of each star is obtained by AR analysis. Resultant periods and powers of each star are shown in table II, in which higher frequencies of main period of each stars are presented. Besides higher frequencies, we find another periods for R Leo and SR stars. Power spectrum of R Leo indicates probable second period of 2105 days, which is about 7 times of main period (312 days). For V Boo, we cannot find secondary period of 8–9 times main period, which Wood (1975) defined from the spectrum analysis. For W Cyg, we can find secondary period of 265 days, which is remarked in fourth edition of GCVS (1985). For RS Cnc, we find periods of 217 and 328 days besides main period of 131 days. But we cannot find period of 1700 days in GCVS. Period of 769 days for χ Cyg, which is about 1.9 times main period, cannot distinguish definitely from noise level.

From a dynamical system point of view, we cannot find evidence of chaos in the light curves of these stars. As mentioned above, however, SR stars show more irregular behavior and have more higher dimension than those of Mira stars. We cannot know the spatial structure of stars from light curve only. This lack of information play a role of noise. This effect induce the difficulty to understand a low dimensional dynamical system. We should recognize the variable star as a "spatio temporal chaos".

References

Akaike, H. and Nakagawa, T.: 1988, *Statistical Analysis and Control of Dynamic Systems* (Kluwer Academic Pub).

Grassberger, P. and Procaccia, I.: 1983, *Physical Review Letters* **50**, 346.

Kholopov, P. N. (ed.): 1985, *General Catalogue of Variable Stars, 4th edition* (Astronomical Council of USSR Academy of Science, Moscow).

Saijo, K. and Kiyota, S.: 1991, *Bull. Natn. Sci. Mus., Ser. E.* **14**, 11.

Wood, P. R.: 1975, in *Multiple Periodic Variable Stars*, ed. W. S. Fitch, IAU Colloq., **29**, 69.

UNDERSTANDING CHANGES IN PERIOD RATIOS

J. O. PETERSEN

Copenhagen University Observatory, Copenhagen, Denmark

The discrepancies between observed and theoretically calculated period ratios of the double mode, classical Cepheids and other groups of Cepheid type variables have played a prominent role in investigations of the important physical characteristics of these stars during the last twenty years. Today, there is growing consensus that the drastic increases in astrophysical opacities proposed by Simon (1982) provide the correct solution of all the period ratio problems (e.g. Petersen, 1989; Iglesias and Rogers, 1992; Moskalik et al., 1991).

It seems at first strange that the *optimal* opacity change (see Andreasen, 1988) can produce *both* the decrease in period ratio for double mode classical Cepheids and the increased period ratio in models of δ Scuti stars. Here we use *κ-effect-functions* (Refsdal and Stabell, 1972) to provide both a detailed description of the effects of opacity changes and an improved and more direct understanding of these effects.

Denoting the period ratio $\Pi_{10} = \Pi_1/\Pi_0$, the *κ-effect-function* for the period ratio is defined by

$$\varphi(\log T) = \frac{\Delta \Pi_{10}}{\Delta \log \kappa \, \Delta \log T} \tag{1}$$

Thus φ is the increase in period ratio per unit increase in $\log \kappa$ and unit temperature interval $\Delta \log T$ (see Refsdal and Stabell for details). Fig. 1 compares the *κ-effect-functions* for models of a typical double mode Cepheid of period 2.82 days and a δ Scuti star of period 0.127 days. It is seen that φ behaves very differently in the two cases in the temperature interval $\log T = 5.2 - 5.8$, which is relevant for realistic (and the *optimal*) opacity changes. In the case of the double mode Cepheid φ is negative in this temperature interval. This means that an opacity increase results in a decrease in the period ratio. For the δ Scuti model φ is seen to be positive except in a small region. Therefore, the same opacity increase now results in an increased period ratio.

Very recently, Iglesias and Rogers (1992, Fig. 24) gave detailed corrections to the Cox-Tabor opacities. Using these correction factors and the *κ-effect-functions* shown in Fig. 1, we can calculate the resulting period ratio changes. We consider the model as consisting of (many) layers of thickness $\Delta \log T$. Then Eq. (1) tells us that each layer produces a correction to the period ratio of the model of $\Delta \Pi_{10} = \varphi \Delta \log \kappa \, \Delta \log T$, where φ is given in Fig. 1 and $\Delta \log \kappa$ can be determined from the information given in Iglesias and

Astrophysics and Space Science **210**: 153–155, 1993.

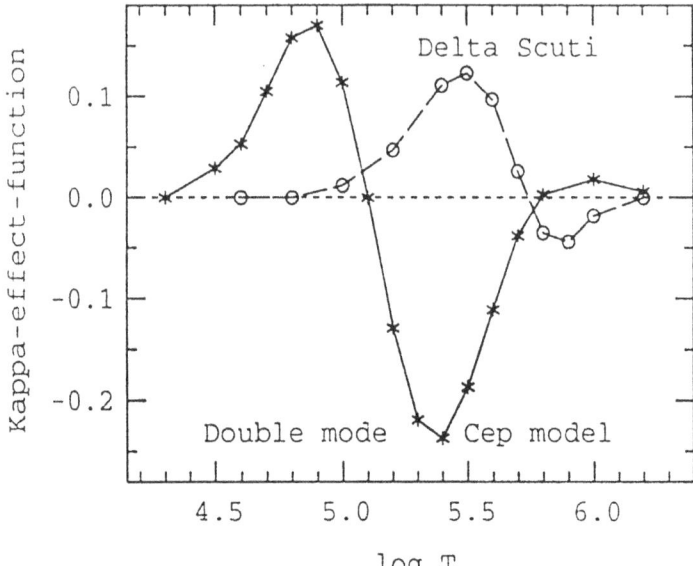

Fig. 1. *κ-effect-functions* for models of a double mode Cepheid of period 2.82 days and a δ Scuti star of period of 0.127 days.

Rogers. For the Cepheid case we obtain a total correction to the period ratio of $\Delta\Pi_{10} = -0.037$. This compares favourably with the difference between the observed values close to 0.70 and the theoretical values based upon the Los Alamos standard opacities of about 0.74. In the δ Scuti case we find $\Delta\Pi_{10} = +0.016$. This is in good agreement with the difference between observed period ratios close to 0.773 and the values of about 0.76 based upon the Cox-Tabor opacities.

Using the *κ-effect-functions* given in Fig. 1, it is very easy to calculate the effects resulting from arbitrary opacity changes. And besides, these curves give a natural feeling for the complicated relations between opacities and period ratios. Consider e.g. an opacity increase in the temperature interval $\log T = 4.5 - 5.1$. Fig. 1 immediately tells us that in the Cepheid case this increase will result in an increase in the resulting period ratio. In the δ Scuti case an opacity increase in $\log T = 4.5 - 5.1$ is seen to have a very small effect. We conclude that *κ-effect-functions* for period ratios give an improved understanding of the relations between opacities and period ratios, and that they are very useful for discussions of improved opacities.

This work has been supported by the Danish Natural Science Research Council through grant 11-9024.

References

Andreasen, G. K.: 1988, *Astronomy and Astrophysics* **201**, 72.
Iglesias, C. A. and Rogers, F. J.: 1992, preprint.

Moskalik, P., Buchler, J. R. and Marom, A.: 1991, preprint.
Petersen, J. O.: 1989, *Astronomy and Astrophysics* **226**, 151.
Refsdal, S. and Stabell, R.: 1972, *Astronomy and Astrophysics* **20**, 19.
Simon, N. R.: 1982, *Astrophysical Journal, Letters to the Editor* **260**, L87.

ON THE APPLICATION OF FOURIER DECOMPOSITION PARAMETERS

PARAMETERS

J. O. PETERSEN

Copenhagen University Observatory, Copenhagen, Denmark

1. Introduction

The application of Fourier decomposition parameters has revolutionized important areas of investigations of Cepheid type variables since the introduction of Fourier analysis in its modern form by Simon and Lee (1981).

In the literature several different representations of the results of Fourier analysis have been utilized. In view of the growing interest for applications of Fourier decomposition it is important to use and publish Fourier data in an optimal way. Most studies until now have used amplitude ratios and phase differences derived from traditional light curves giving the light variation in magnitudes, following the original recipe of Simon and Lee (1981). However, Stellingwerf and Donohoe (1986) advocated the use of phases rather than phase differences. Recently, Buchler et al. (1990) argued that the standard Simon & Lee form contains all relevant physics, and suggested analysis of flux-values rather than of magnitudes, because this removes the distorting effects of constant, false light. Thus there are many choices to be made in practical applications of Fourier analysis, and there is at present no convincing argument for preferring one specific representation.

2. Phase Definitions

For the precise definition of phases of the type proposed by Stellingwerf and Donohoe (1986; SD in the following) there are several possibilities. We here use the mean light curves of RRab variables in ω Centauri published by Martin (1938) to compare three natural choices.

Let us evaluate three different possibilities for the standard point defining the phases: (1) the point at which the rising branch crosses the mean magnitude, m_o, (2) the point at which the rising branch crosses the median magnitude, $m_{0.5} = 0.5(m_{\min} + m_{\max})$, and (3) light maximum. In order to make comparisons with the remarkable pattern in the phases of the simple pulsation models discussed by SD, we choose in cases (1) and (2) 0.5 for the phase of the standard point. For convenience we denote in the following the phases of cases (1) and (2) C_k and D_k, respectively. In order to compare with Payne-Gaposchkin's (1947) analysis we choose in case (3) as standard phase 0.0 (these phases are G_k in the following). Comparing C_k, D_k, and G_k

Astrophysics and Space Science **210**: 157–158, 1993.
© 1993 *Kluwer Academic Publishers.*

we find the most interesting results for the phases D_k. In nearly all stars D_k with (1-2) σ follow the prediction of the simple models of alternating phases $D_k = \pi/2$ or $3\pi/2$. This is in beautiful agreement with the "sawtooth" case, which should also correspond to light curves of high skewness. The phases C'_k show significant differences between their average values and the prediction from the simple models, and G_k plots resemble traditional phase difference plots (as in *e.g.* Petersen, 1984) with no indication of a simple explanation of their basic properties. We conclude, therefore, that both the phases C_k and G_k are less interesting than D_k, and that D_k should be preferred in future work.

In order to discuss bump progression sequences it is important to compare data for RRab variables with similar data for Cepheids. For periods 3–7 days classical Cepheids show a pattern very similar to that of the RRab stars in ω Cen: a linear decrease of D_1 with period and $D_2 \approx \pi/2$, $D_3 \approx 3\pi/2$, and $D_4 \approx \pi/2$. And the classical Cepheids of period 7–20 days show the Hertzsprung progression in D_k similar to the less well defined progression sequence in Type II variables of period 1–4 days (Petersen and Andreasen, 1987). We conclude that the D_k phases agree very well with the simple SD models outside resonances, and that they seem to provide improved possibilities for future comparisons of pulsation properties of RR Lyrae stars, Type II Cepheids, and classical Cepheids.

In view of the meager information on advantages and disadvantages of different representations of Fourier parameters, we recommend: (i) Published tables of observational results should give amplitudes and phases directly together with their standard errors and the SD standard phase; (ii) The original observational data should be easily available for calculation of new types of Fourier decomposition parameters.

This work has been supported by the Danish Natural Science Research Council through grant 11-9024.

References

Buchler, J. R. Moskalik, P., and Kovacs. G.: 1990, *Astrophysical Journal* **351**, 617.
Martin. W. C. 1938, *Leiden Annalen*, **XVII**, Part 2.
Payne-Gaposchkin, C.: 1947, *Astronomical Journal* **52**, 218.
Petersen, J. O.: 1984, *Astronomy and Astrophysics* **139**, 496.
Petersen, J. O. and Andreasen, G. K.: 1987, *Astronomy and Astrophysics* **176**, 183.
Simon, N. R. and Lee, A. S.: 1981, *Astrophysical Journal* **248**, 291.
Stellingwerf, R. F. and Donohoe, M.: 1986, *Astrophysical Journal* **306**, 183.

LINE-PROFILES OF F SUPERGIANT STARS AS CANDIDATES OF PROTO-PLANETARY NEBULAE

S. TAMURA, M. TAKEUTI and J. ZALEWSKI

Astronomical Institute, Tohoku University, Aobayama, Sendai 989, Japan

In this report the results of spectroscopic observations on selected F supergiant stars as the candidates of proto-planetary nebulae (PPNs) are presented. Volk and Kwok(1989) selected nearly one hundred new candidates of PPNs based upon the scenario on the evolution of them. They summarized four classes of stars which should be the PPNs and are described in various papers. These are (i) High galactic latitude supergiants, (ii) Nonvariable OH/IR stars, (iii) Low color temperature infrared objects, and (iv) R CrB. In order to examine the extended envelope of the PPNs we have started our project to obtain the Hα profile which should give us some clue to the structure of the envelope. We are mainly concerned with F supergiant stars. Our sample objects are mostly classified as pulsating or semiregular variables.

In August 1990 and February 1991 we observed 7 selected F supergiant stars suggested in the literatures as the candidates for PPNs including UU Her as a typical sample of F-type supergiants. Stellar samples are listed in Table I and were mostly chosen from Volk and Kwok (1989, Table 1) and

TABLE I
The List of Observed Stars

Star Name	Other desig.	RA	DEC	l	b
HD 46703	SAO 25845	6 33 49.3	+53 33 36	162	+20
	BD+53 1040				
HD 56126	SAO 96709	7 13 25.3	+10 05 09	207	+10
UU Her	SAO 65424	16 34 12.2	+38 04 05	64	+41
	BD+38 2803				
HD 161796	SAO 30548	17 43 41.3	50 03 48	77	+31
	BD+50 2457				
HD 163506	SAO 85545	17 53 24.0	+26 03 23	51	+23
	89 Her				
IRAS 18095+2704	—	18 09 31.0	+27 04 30	54	+20
HD 179821	SAO 124414	19 11 25.0	+00 02 18	36	−05
	AFGL 2343				
HD 187885	SAO 163075	19 50 00.7	−17 09 38	24	−21

from Luck and Bond (1984). We have obtained various types of Hα profiles.

Astrophysics and Space Science **210**: 159–161, 1993.

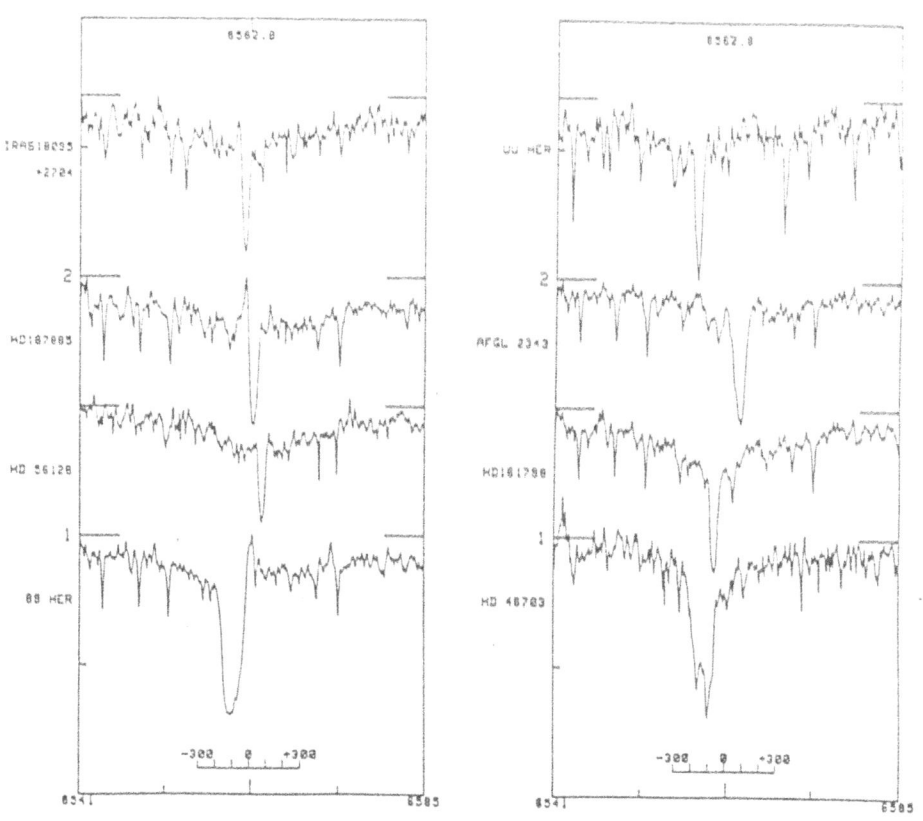

Fig. 1.   Obtained Hα profiles of eight F supergiant stars. The abscissa is the wavelength referred to the heliocentric system together with the velocity scale. The ordinate is an arbitrary intensity.

They are displayed in Fig. 1. The express report and a short discussion about these stars was already presented by Tamura and Takeuti(1991) and more detailed description on observed data are given in another place (Tamura et al. 1992) together with related data previously obtained.

   a. *Types of Hα profiles:* As shown in Fig. 1, the Hα profiles consist of absorption and emission components in complicated way. They show clear evidence of activity of extended atmosphere and they are considered to be candidates of PPNs.

   b. *Time variations:* In addition to a wide variety of the Hα profiles we have noticed their time variations ranging over different time spans.

c. Radial velocities: We can identify a couple of absorption lines like the FeIλ6569, FeIIλ6516, and CIλ6587 as well as the Hα in the wavelength range of our spectra. We have estimated the radial velocities for our sample stars.

References

Luck, R. E. and Bond, H. E.: 1984, *Astrophysical Journal* **279**, 729.
Tamura, S. and Takeuti, M.: 1991, *IAU Inform. Bull. Variable Stars*, No. 3561.
Tamura, S., Takeuti, M., and Zalewski, J.: 1992, *Sci. Rep. Tôhoku Univ.* Ser. 8, **13**.
Volk, K. M. and Kwok, S.: 1989, *Astrophysical Journal* **342**, 345.

LOW AMPLITUDE NEW TYPE VARIABLE STARS IN GLOBULAR CLUSTERS

YAO BAO-AN
Shanghai Observatory

ZHANG CHUNG-SHENG and QIN DAO
Purple Mountain Observatory

and

TONG JIAN-HUA
Beijing Observatory

Rapid progress in the stellar pulsation research has presented many new challenges to traditional pulsation theory. The following progress made by us should belong to one of the new challenges:

1) There are many variable stars located at the Horizontal Branch (HB) but outside the instability strip with amplitude larger than 0.02 mag.

Maybe some astronomers have already doubted of the 1950's conclusion-the RR Lyrae stars are confined to the narrow instability strip in the C-M diagram, the boundaries of the gap are extremely sharp and definite, beyond the edges no light variations occur with ranges greater than 0.02 mag. We are changing the conclusion by observations. Here we do not mean the microvariability. We still raise the question from the classical viewpoint. We want to show that there are variable stars outside the strip with peak to peak amplitudes larger than 0.02 mag.

Since 1975 we use M4 as the calibration cluster to measure our newly discovered flare stars and variable stars in the Oph dark cloud region. Because we use the double astrograph and the Schmidt telescope together, three plates were got simultaneously. As the by-product a large number of unusual suspected variable stars were found (1979; Yao et al. 1981). There are 12 stars located at the HB but blueward of the instability strip and 7 stars at HB but redward of the gap. All of them are the members of M4 (Cudworth 1990). We have checked 3 of them with the RCA CCD attached to the Zeiss 1-meter telescope of Yunnan Observatory and published the preliminary results (Yao 1987). (G206. G140. G481). Another star G327, was confirmed recently with the 60-cm reflector of MSSSO (Yao 1991). We believe that most of the other stars in our list are variable too because we chose the above 4 stars randomly and all of them appeared to be variables. Some of them have been observed with the CCD camera and the reduction is progressing. In fact other astronomers also pointed out the light variation of some stars common in our list. e.g. G537. 539, 258, 246 and 172 (Evans 1977; Alcaino 1975). According to our photographic observation none of them be-

Astrophysics and Space Science 210: 163–165, 1993.
© 1993 *Kluwer Academic Publishers.*

longs to the normal RR Lyrae stars. We hope that we will determine their periods.

2) There are variable stars at the tip of the RGB but leftward of the region occupied by long period variables.

R. E. White et al. (1975) gave evidence for variability of some stars located at the tip of RGB but leftward of the region occupied by Mira or long period variable stars (further down the giant and asymptotic-giant branches) in M15 and other globular clusters. But in an attempt to confirm and extend White's results D.E. Welty (1985) obtained photographic photometry of giant and asymptotic-giant branch stars in six globular clusters (M3, M5, M13, M15, M71, M92) by measuring several hundred plates of these clusters on the PDS. He did not confirm variability for any of these stars found by White et al. with amplitude > 0.2 mg. Obviously, further investigation of these stars is significant for understanding the stellar evolution on the giant and asymptotic-giant branches. However, we have found that the red giant star K1040 in M15 located at the same region has a period of $\sim$ 4.3 hour with amplitude $\sim$ 0.04 mag (Yao, 1990). The similar star G512 in M4 was published by us before (Yao et al. 1981). We share White's opinion and believe this is the common phenomena among these kind of stars. We are searching for the similar stars in globular clusters too.

3) There are variables located at the lower middle part of the RGB.

We have shown that there are variable stars located at the lower middle part of the red giant branch in the globular cluster M4(1979). G265 and G543 have shown certain periodicity (Yao, 1987). There are 7 similar stars in our list in M4 to be confirmed by CCD photometry. No results were reported in the past so the theorist did not pay attention to it. We will continue to observe them in the future.

4) Are the constant stars located within the gap of HB really constant?

People have already found a few stars located at the middle part of the instability strip of the horizontal branch in several globular clusters which do not show evidence of light variation. G392 in M4 is such a star (Cudworth 1990). We suggest that people should reobserve these stars with high accuracy because we have already found in 1979 that the G392 is really a variable star. Due to the low precision of the photographic photometry we have not got its period. It is not a normal RR Lyrae star (with amplitude < 0.2 mag and maybe multiperiodic). We are observing this star with CCD camera and try to get its periods soon.

5) The UV bright star K1082 in M15 is a short period variables indeed.

Using the photographic plates obtained with the 60-cm Zeiss reflector of P.M.O. by Kukarkin et al., Chu (1977) has reported that the UV Bright star K1082 in M15 is a new type variables with a period $P = 0.087004$ day and amplitude of 0.2 mag. However, the later observations (Hesser et al., 1979; Liller et al., 1980; Smith et al., 1979) show no evidence of variation with the reported period and amplitude. Due to the potential importance of the star, we have scanned the original plates again with the PDS of P.M.O. Our results have shown that it can be said with 90% probability that K1082 is a new type variable star. But the reported period is wrong. It may be a multi-period variables with the main period $P_1 = 0.1683$ day. It is possible that the second period $P_2 = 0.0985$ or 0.0437 day.

We have also observed the star with CCD photometer. The amplitude of the light variation is less than 0.01 mag in 1987 and ~ 0.04 mag in 1991 and the period is changing. The reduction is still in progress and we will continue to monitor K1082.

STELLAR PULSATION: (I) ANALYSIS OF
STABILITY OF ENVELOPE

YU ZHI-YAO

Shanghai Observatory, 80 Nandan Road, Shanghai 200030, China

Abstract. The life-time of the star on AGB is approximately 6×10^4 yr. We divide it into front half and back half of AGB (including to optical Mira variable and OH/IR star) according to their evolution character. The observations show that the star has non-pulsation, but constant mass loss rate ($\sim 5 \times 10^{-7}$ $M_\odot$ yr^{-1}) on front half of AGB. Its circumstellar envelope is formed. When the star has pulsation on back half of AGB, its mass loss rate is relative with time, and increases gradually. In this time the star is on the stage of optical Mira variable. When the mass loss rate reaches the value of $\sim 3 \times 10^{-6}$ $M_\odot$ yr^{-1}, the star evoluted from the stage of optical variable into the stage of radio bright OH/IR star. On the end of AGB the mass loss rate reaches $\sim 10^{-4}$ $M_\odot$ yr^{-1}. (Band and Habing 1983, Hermen and Habing 1985).

The contribution of the star to its circumstellar envelope are its radiation pressure F_R and gravitation. Here we assume its radiation pressure is first, and the gravitation is secondary. So we neglect the disturb of gravitation. We give the characters of the disturb as following.

1. FRONT HALF STAGE ON AGB

The disturb of F_R are not relative with time.

2. BACK HALF STAGE ON AGB

(1) The Stage of Optical Mira Variable

The disturb of F_R are periodicity.

(2) The Stage of OH/IR Star

The disturb of F_R raises with time gradually.

Assuming the circumstellar envelope is ideal gas and it is radial sphere-symmetrical adiabatic out-flowing, we have obtained and studied the equations of the disturbance of the stable state for the envelope, and the following conclusions have been obtained.

(1) The star has non-pulsation on front half of AGB, but constant mass loss rate ($\sim 5 \times 10^{-7}$ $M_\odot$ yr^{-1}). The solutions of stable state are Lyapunov stability.

(2) When the star is on the stage of optical Mira variable it has the pulsation, its mass loss rate is approximately between 5×10^{-7} and 3×10^{-6} $M_\odot$ yr^{-1}, and the all disturbance solutions of velocity, density, and pressure of gas in the envelope, pure heat got by its gas per unit volume, and its mass loss rate have the character of spheric-wave. If we fix on time t, the all disturbance solutions are functions, which values are attenuate with increase of radius r. If we fix on r, the all disturbance solutions are the harmonic functions of $\omega_* t$ (ω_* is $2\pi/T_*$, T_* is the optical period of the star).

Astrophysics and Space Science **210**: 167–168, 1993.
© 1993 *Kluwer Academic Publishers.*

(3) When the star is on the stage of OH/IR star it has the pulsation. its mass loss rate is approximately from 3×10^{-6} $M_\odot$ yr^{-1} to 1×10^{-4} $M_\odot$ yr^{-1}. the all disturbance solutions increase with increase of time. and solutions of stable state are non-stability. Our computational results indicate that the mass loss rate raises gradually. On the end of AGB the mass loss rate is 1.2×10^{-4} $M_\odot$ yr^{-1}. The circumstellar envelope is non-stability.

References

Band, B. and Habing, H. J.: 1983, *Astronomy and Astrophysics* **127**, 73.
Hermen, J. and Habing, H. J.: 1985, *Phys. Rep.* **124**, 175.

STELLAR PULSATION: (II) MULTIPLE DISTINCT SHELLS

YU ZHI-YAO

Shanghai Observatory, 80 Nandan Road, Shanghai 200030, China

Abstract. We can find the conclusion with our analysis for VLA observation of OH maser and CO (2-1) emission line that they are distributed on some different distinct shells in the circumstellar envelope, respectively.

For an expanding spherical envelope the angular radius R_0^A is given by

$$R_0^A = R^A(V) \left[1 - \frac{(V - V_0)^2}{V_e^2} \right]^{-\frac{1}{2}} \tag{1}$$

where $R^A(V)$ is the angular radius at LSR velocity V, V_0 is systemic radial velocity, V_e is circumstellar expanding velocity (Bowers et al. 1983). Formula (1) is changed as following

$$[R^A(V)]^2 = [R_0^A]^2 - \left[\frac{R_0^A}{V_e} \right]^2 (V - V_0)^2 \tag{2}$$

We may see from formula (2), a shell has determinate R_0^A and V_e for each shells, $[R^A(V)]^2$ is linear monotonous decrease by degrees function of $(V - V_0)^2$. Thus we may divide the envelope into shells according to these regions where the distribution of points, which coordinates are $[R^A(V)]^2$. $(V - V_0)^2$, is ranged by the orientation of linear monotonous decrease by degrees in $[R^A(V)]^2 - (V - V_0)^2$ map, then do statistic fitting. The error of statistic fitting is least.

The data of VY CMa selected by the paper are from Bowers et al. (1983). We have obtained that VY CMa OH 1612 MHz masers occur on three distinct expanding shells (i.e. shell A, B, and C), their corresponding angular radii $[R_0^A]$ and expanding velocities $[V_e]$ are 2.43, 3.08, 3.21 arcsec and 24.0, 30.4, 34.9 km s^{-1}, respectively. Given R_0^A, the linear radius R_0 can be determined if the distance is known. The distributions of VY CMa OH 1612 MHz points are seen from Fig. 1, 2, and 3.

We have also obtained the multiple expanding shells at the analysis and explanation of the CO (2-1) emission line. For an expanding spherical envelope parabola is adaptable (Knapp 1982). Therefore we may obtain

$$T_A^*(V) = T_A^*(\text{peak}) \left[1 - \frac{(V - V_0)^2}{V_e^2} \right] \tag{3}$$

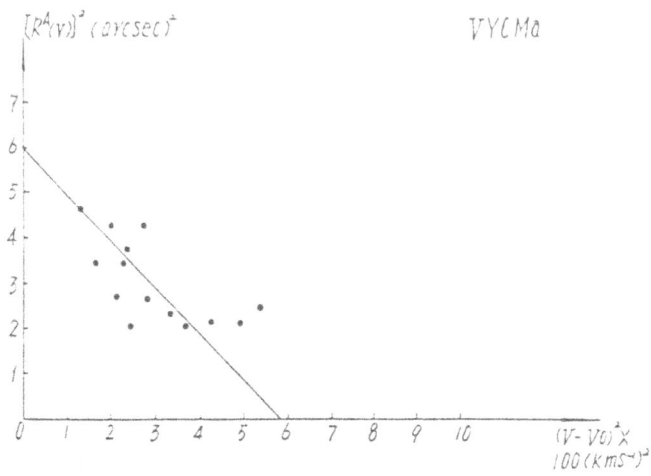

Fig. 1. The Distribution on Shell A

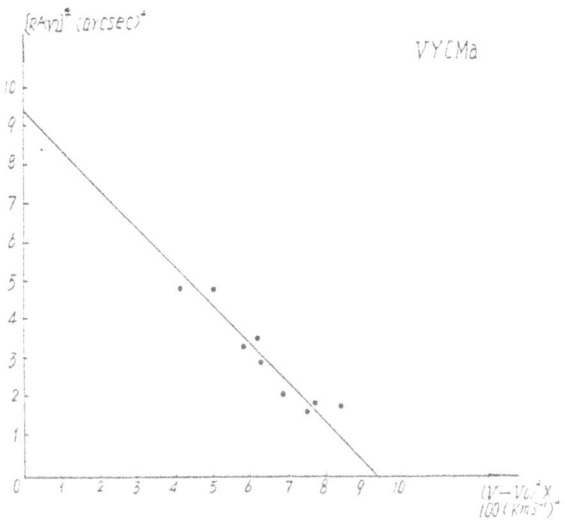

Fig. 2. The Distribution on Shell B

where T_A^* (peak) is the antenna temperature of peak. $T_A^*(V)$ is the antenna temperature at V. We take some points at the CO (2-1) emission line (27 points for CIT6). We measure a group of $T_A^*(V)$, V for each point, farther some groups of $(V - V_0)^2$, $T_A^*(V)$. We draw $T_A^*(V) - (V - V_0)^2$ map. We find that these points, which coordinates are $T_A^*(V)$ and $(V - V_0)^2$, are distributed at several regions from the $T_A^*(V) - (V - V_0)^2$ map (three regions for CIT6). At each region the distribution of the points is ranged by the orientation of linear monotonous decrease by degrees. According to formula (3) we may obtain

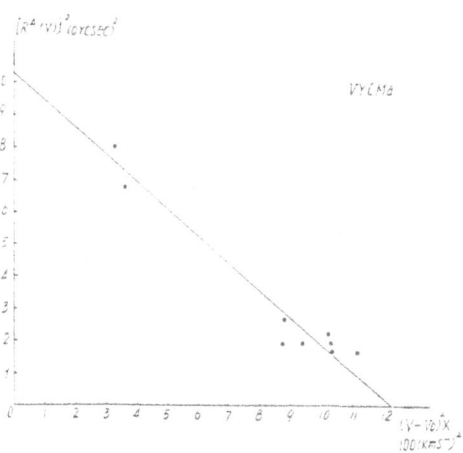

Fig. 3.   The Distribution on Shell C

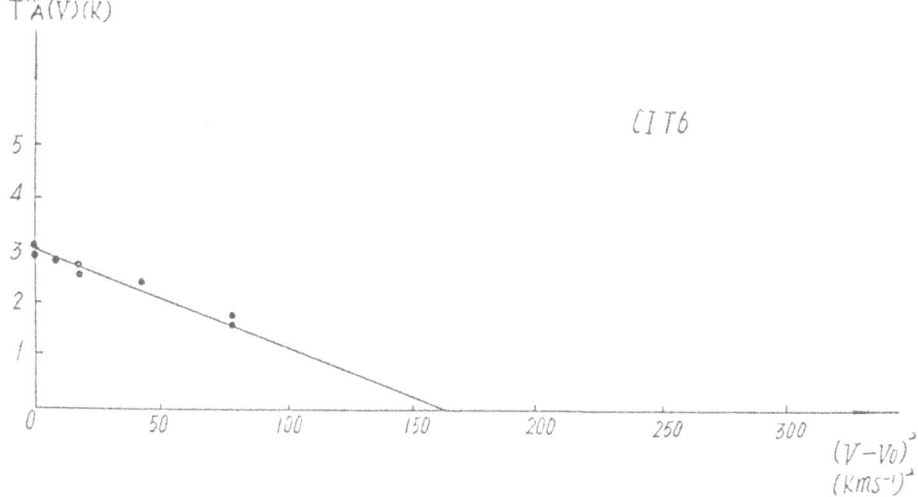

Fig. 4.   The Distribution on Shell D

$$T_A^*(V) = T_A^*(\text{peak}) - \frac{T_A^*(\text{peak})}{V_e^2}(V - V_0)^2. \tag{4}$$

For a shell which radial distance from stellar center is r, and expanding velocity is V_e, T_A^* (peak) and V_e are determinate. $T_A^*(V)$ is linear monotonous decrease by degrees function of $(V - V_0)^2$. Using $T_A^*(V) - (V - V_0)^2$ map and formula (4) we divide the envelope into shells according to the regions in $T_A^*(V) - (V - V_0)^2$ map where the distribution of points is ranged by the orientation of linear monotonous decrease by degrees.

The observational data are selected from Knapp (1982). We have obtained that CIT6 CO (2-1) molecules also occur on three distinct expanding shells

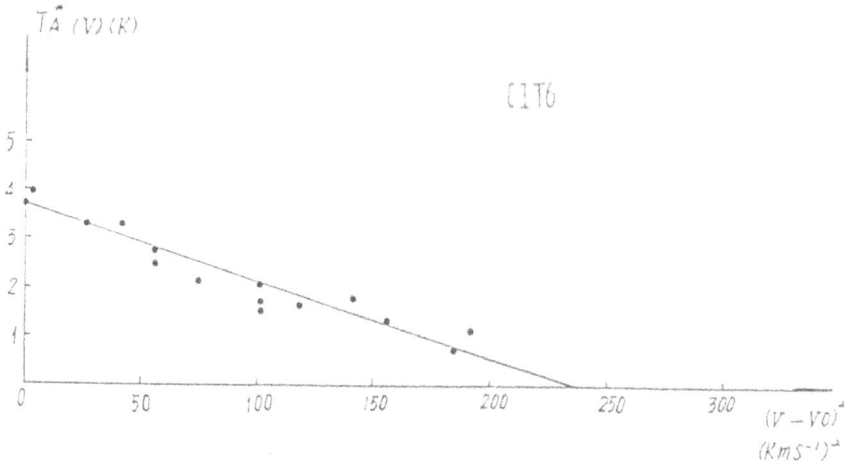

Fig. 5.   The Distribution on Shell E

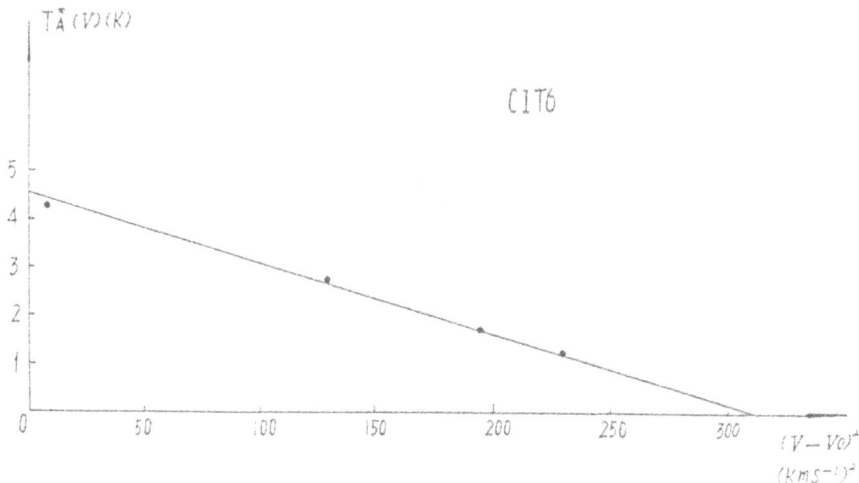

Fig. 6.   The Distribution on Shell F

i.e. shell D, E, and F), the corresponding peak antenna temperature $[T_A^*$ (peak)] and expanding velocity $[V_e]$ are 3.1, 3.8, 4.5 K and 12.7, 15.2, 17.4 km s^{-1}, respectively. The distributions of the points of shell D, E, and F are seen from Fig. 4, 5, and 6, respectively.

References

Bowers, P. F. et al.: 1983, *Astrophysical Journal* **274**, 733.
Knapp, G. R. et al.: 1982, *Astrophysical Journal* **252**, 616.

LONG-TERM AMPLITUDE AND PERIOD VARIATIONS OF

δ SCUTI STARS: A SIGN OF CHAOS?

M. BREGER
Institut für Astronomie, University of Vienna,
Türkenschanzstr. 17, A-1180, Wien, Austria

Abstract. On short time-scales of under a year, the vast majority of δ Scuti stars studied in detail show completely regular multiperiodic pulsation. Nonradial pulsation is characterized by the excitation of a large number of modes with small amplitudes. Reports of short-term irregularity or nonperiodicity in the literature need to be examined carefully, since insufficient observational data can lead to an incorrect impression of irregularity. Some interesting cases of reported irregularities are examined.

A few δ Scuti stars, such as 21 Mon, have shown stable variations with sudden mode switching to a new frequency spectrum. This situation might be an indication of deterministic chaos. However, the observational evidence for mode switching is still weak.

One the other hand, the case for the existence of long-term amplitude and period changes is becoming quite convincing. Recently found examples of nonradial pulsators with long-term changes are 4 CVn, 44 Tau, τ Peg and HD 2724. (We note that other δ Scuti pulsators such as X Cae and θ^2 Tau, have shown no evidence for amplitude variations over the years.) Neither the amplitude nor the period changes are periodic, although irregular cycles with time scales between a few and twenty years can be seen. While the amplitude changes can be very large, the period changes are quite small. This property is common in nonlinear systems which lead to chaotic behavior. There exists observational evidence for relatively sudden period jumps changing the period by about 10^{-5} and/or slow period changes near $dP/dt \leq 10^{-9}$. These period changes are an order of magnitude larger than those expected from stellar evolution.

The nonperiodic long-term changes are interpreted in terms of resonances between different nonradial modes. It is shown that a large number of the nonradial acoustic modes can be in resonance with other modes once the mode interaction terms, different radial orders and rotational m-mode splitting are considered. These resonances are illustrated numerically by the use of pulsation model. Observational evidence is presented that these interaction modes exist in the low-frequency domain.

1. Are the Variations of δ Scuti Stars Really Periodic?

The majority of the small-amplitude variables show nonrepetitive light curves and the periods reported by early investigators were cycle-count periods. The important question arises whether this seeming irregularity is really caused by irregular (possibly chaotic) variability or simply the combination of the light curves of several strictly periodic multiple pulsation modes.

The difficulty of distinguishing between the two hypotheses for each star is caused by insufficient observational data in both quantity and quality. Insufficient data may favor either of the two hypotheses in an unpredictable manner. Irregular variability might be deduced if limited data does not permit the detection of the multiple periods present. One the other hand, even for variability which is not strictly periodic, short data sets might be fit by a set of periods regarded as real, constant pulsation periods. The multiple periods seen in this example would fit only this data set and never be seen again. This observational dilemma can be avoid not only by the collection of long data sets (possibly from multiple site to avoid aliasing), but also by carefully examining power spectra of independent data sets.

For most δ Scuti stars discovered so far too little is known in order to discuss the regularity or stability of the periods of variation. For about thirty small-amplitude variables some conclusions are possible and the case for irregular pulsation looks very weak. This can already be seen in the power spectra, where the widths of the individual peaks due to the different pulsation modes generally correspond to the widths expected from the length and spacing of the observational data. This is strong evidence in favor of stable and nondrifting periods, at least over time scales of several months, possibly even years. A good example of such a nonradial pulsator is the star θ^2 Tau (Breger et al. 1989).

In a series of papers (e.g. Morguleff, Rutily and Terzan 1976a) deduced irregular variability for some δ Scuti stars. For 14 Aur they state 'no periodicities are present in the pulsation'. In our view, they should have considered multiple modes as a more serious alternative for all their stars. Fitch and Wisniewski (1979) presented new photometry and an independent analysis of 14 Aur. They could describe the pulsation in this close binary system by nonradial modes. We have analyzed seven nights of unpublished photometry of 14 Aur obtained at McDonald Observatory and confirm the regularity reported by Fitch and Wisniewski.

Another star studied by Morguleff, Rutily and Terzan (1976b) was 44 Tau for which they also found no periodicities. Again, with this star we are dealing with a complex nonradial pulsator. Poretti, Mantegazza and Riboni (1991) present 25 new nights of photometry and present seven frequencies of pulsation. Since their set of nonradial pulsations can also fit the previous data, this star also provides no evidence for the existence of irregular pulsation in δ Scuti stars.

2. Mode Switching

One could also imagine a situation, in which the pulsation appears regular for time spans of several years before it becomes unstable and period (or mode) switching occurs. More than ten years ago, Kurtz (1980) examined the stability of the observed periods in δ Scuti stars. Except for the star 21 Mon, he found the previously reported period switching to be unconvincing.

Another example for possible mode switching is the star HN CMa (Breger, Balona and Grothues 1991). Between 1981 and 1990 the star became almost constant with a 'new' dominant frequency. In their paper, the authors emphasize that a conclusion of actual mode switching is not necessary, if one considers the long-term variability of nonradial pulsation amplitudes (see next section) in a star with small amplitudes near the limit of detectability.

A further reported example of possible variable frequency spectra is FM Com. While Antonello et al. (1985) could fit their observed light variations with three frequencies, on the basis of new photometry Paparo and Kovacs (1984) argue in favor of variable frequency spectra. This star might also be explained by variable amplitudes leading to an appearance of mode switching and we refer to another paper in this journal.

The previous discussions dealt with stars for which 'interesting' irregular variations had been reported. The discussion should not create the impression that suspected irregular behavior is the rule. One the contrary, the majority of the δ Scuti stars studied in detail show multiperiodic variations which are remarkably constant from cycle to cycle, or even year to year. The small observed variations in the periods $(dP/dt \sim 10^{-9})$ are within a one or more orders of magnitudes of the expected evolutionary effects.

3. Long-Term Amplitude and Period Variations

The recent years have seen a large improvement in the observational data of δ Scuti stars. The change was motivated by the realization of the complexity of the nonradially pulsating stars, which led to a concentration on selected stars and long observations (often multisite in order to decrease the aliasing problem) over a single observing season. Among the nonradial pulsators the frequency spectra of about two dozen stars can be regarded as understood.

The determination of the long-term constancy or variability of the individual amplitudes associated with individual pulsation modes requires another quantum jump in the quantity of data required: sufficient data for a multiple frequency solution has to be available for at least a second observing season. Over 100 nights of observation would not be an unrealistic requirement. The verification of constant amplitudes in a multiperiodic pulsator is probably one of the most difficult tasks in this kind of work. We note here that the appearance of variable amplitudes may be a sign of insufficient data as well as true amplitude variability. Table I lists δ Scuti stars whose reported amplitude variability we regard as reasonably reliable; the selection is probably neither complete nor perfect.

We note that the typical time scales of amplitude and period variations are years. However, even a single star can show a wide variety in behavior for its different pulsation modes. This is demonstrated by the star 4 CVn (see Breger 1990), but other stars have shown this variety as well:

(i) the amplitudes of some pulsation modes appear essentially constant for ten or even twenty years,

(ii) some amplitude variations, such as decreases in size, are steady over twenty years,

(iii) a steady increase in amplitude can be followed by a very rapid col-

TABLE I

Some δ Scuti star whose pulsation modes have variable amplitudes with long time scales

Star	Frequency (cycles per day)	Amplitude range (mag)	Time scale	Ref.
GN And	single	0.007 to 0.022	~15 years	1
τ Peg	single	≤0.005 to 0.012 irregular	decade(s)	2
4 CVn	f_1, 8.60	0.011 to 0.023	decade(s)	3
	f_2, 5.85	0.08 to 0.018	decade(s)	
	f_3, 5.05	0.004 to 0.025	decade(s)	
	f_5, 7.38	0 to 0.014	a few years	
44 Tau	f_2, 7.01	0.003 to 0.021	decade(s)	4
	f_7, 9.56	0.006 to 0.021	decade(s)	
BK Cet	f_1, 11.1	0.046 to 0.01 steady decline	decade(s)	5,6,7
HN CMa	f_1, 4.5	≤0.001 to 0.005		8,9
HD 2724	7.38	0.012 to 0.036	a few years	10

References:
1: Garrido et al. (1985), 2: Breger (1991), 3: Breger (1990), 4: Poretti et al. (1991), 5: Lampens and Rufener (1990), 6: Poretti (1989), 7: Kurtz (1990). 8: Breger et al. (1991), 9: Baade and Stahl (1982), 10: Lampens (1991)

lapse and decrease to near zero in less than two years. However, the oscillation does not disappear and slowly starts up again. As an example, for 4 CVn the amplitude of the 7.37 cycles per day oscillation (with a possible identification as P_2, $l = 2$, $m = -1$) decreased from a reliably determined 0.014 mag in 1974 to near zero during 1976 and 1977 with a subsequent increase to an intermediate value of 0.007 mag.

This variety is typical of chaotic behavior. Nevertheless, the question also arises whether or not the observed amplitude and period variability could be the result of beating between two close frequencies. We regard this explanation as unlikely for the following reason: the observed variations can be modelled by two close frequencies only for stars where little data from different years are available. For the stars with extensive data, three or more close frequencies beating with each other would be required to provide a reasonable fit to the observations. This suggests 'true' amplitude and period variability.

4. The Resonance Hypothesis as an Explanation
for the Irregularities

An explanation for the amplitude variability might be found in the resonance condition between different nonradial pulsation modes. On the theoretical side, it was shown by Däppen (1985) that nonradial modes can contribute to irregular behavior more than radial modes. Moskalik (1985) extended the calculations of Dziembowski (1980, 1982) to resonant mode coupling. Takeuti (1990) showed that nonlinearly coupled oscillations can show apparent period switching and amplitude variations, while a radial resonance condition ($P_o/P_3 = 2.0065$) was examined by Takeuti and Zalewski (1991).

There exists a variety of different resonance conditions which can lead to amplitude and period variations. We note that resonances can occur from interaction modes originating from the different pulsations. Nonradial pulsation naturally produces a large number of possible resonances through the interaction terms ($\nu_k - \nu_{k-1}$). Let us demonstrate this with an example using a model of a typical δ Scuti star of 2 solar masses, 7340 K, 3.53 solar radii, a rotation period of 5 days and $l = 2$ modes. For the pulsational frequencies we use the models of Fitch (1981), while the rotational effects including the second order were taken from Saio (1981). Despite criticism of the Fitch modes (e.g. see Dziembowski and Krolikowska 1990) we use these models because the $m = 0$ frequencies agree with the observed values for the star 4 CVn, and because the demonstration of the interaction terms is insensitive to the model used. In fact, for radial and nonradial pulsation, at high radial orders the frequency spacing, ($\nu_k - \nu_{k-1}$), becomes equidistant.

Figure 1 shows the computed frequency spectrum in both the nonrotating frame (observer) and the rotating stellar frame. We note that in the rotating frame the rotational frequency splitting is, of course, quite small and the second-order effect becomes relatively large.

Now consider the interaction modes, ($\nu_k - \nu_{k-1}$). Even for different values of m, the frequencies of these interaction terms lie in a narrow range. Due to the large number of possible interaction terms (consider the range in k and m values even at a constant l), a number of different resonances can occur. These may fulfill the conditions of deterministic chaos. We note in this regard that the variations in amplitude can be very large, while the effects on the periods are relatively small (less than 10^{-4}).

The present explanation predicts the existence of interaction terms with frequencies ($\nu_k - \nu_{k-1}$). We would expect that even if the amplitude of the individual interaction terms were too small to be observed, the resonances would kick some interaction term towards detectability.

These interaction terms may already have been observed. We refer to the star 1 Mon (Shobbrook and Stobie 1974), the large-amplitude pulsator Al Vel (Walraven, Walraven and Balona 1991), and 4 CVn. For the star X

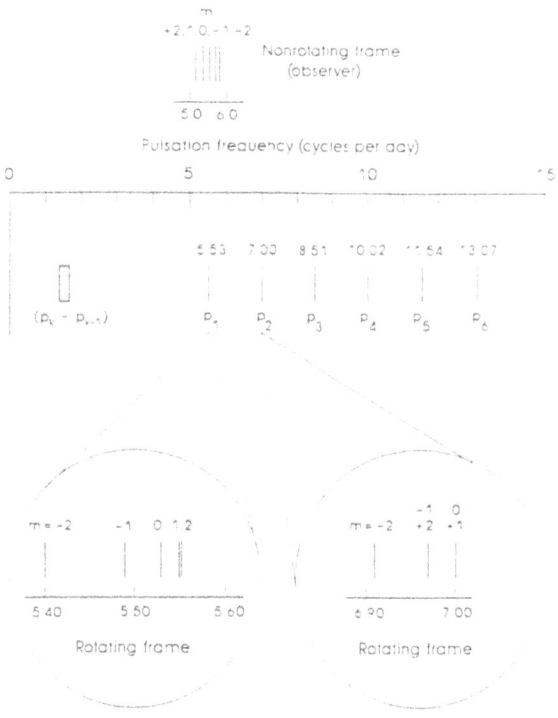

Fig. 1.   Frequency spectrum for the nonradial acoustic $l = 2$ pulsation modes computed from the Fitch (1981) model 2.0M48 with first and second-order rotational effects by Saio (1981). A rotational frequency of $\Omega = 0.2$ revs per day was assumed. The results are shown for the frames of reference of both the star and the observer.

Cae, Mantegazza and Poretti (1991) present evidence in favor of coupled oscillations and excitation by resonance (though no difference terms and no amplitude variations). Furthermore, there are a number of papers in which small frequencies outside the p-mode range have been reported. Whether these frequencies are caused by observational errors, g-modes, rational effects, or interaction terms is still an open question. Examples are HR 8210 (0.64 and 0.34 cycles per day, Kurtz 1979); HD 93044 (2.146 cycles per day, Li Zhi-ping et at. 1991). Finally, we would like to present some evidence for a mode interaction term in 4 CVn for the year 1974, for which 27 nights of unpublished photometry by Fitch is available. If the seven-frequency solution in the 5 to 9 cycles per day range applicable for 1966 to 1984 (Breger 1990) is subtracted, very strong power near 1.4 cycles per day remains (Figure 2). This value (or one of its aliases) can be fit by several differences of observed p-modes. For the other years the amplitude of the interaction mode is weak.

Further work should therefore concentrate more on reliable observations in the 0 to 3 cycles per day frequency domain in order to establish the

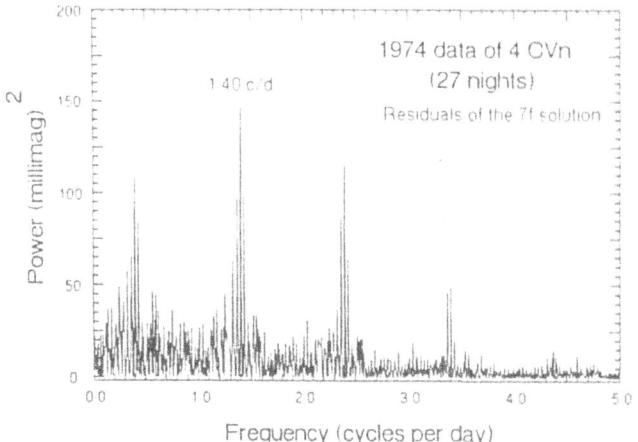

Fig. 2.   Observational evidence for low-frequency power in unpublished data of 4 CVn by Fitch. The overall solution has already been subtracted. We note that the power near 1.40 cycles per day is also present in the raw data. This frequency (or one of its aliases) is interpreted here as an interaction mode with a large amplitude temporarily pushed up by resonance.

link between amplitude variability and interaction-term resonances. It might even be possible to observe one of the 'kicks' which lead to transfer of power to different pulsation mode.

Acknowledgements

It is a pleasure to thank W. Däppen and R. Dvorak for many helpful discussions.

References

Antonello, E., Guerrero, G., Mantegazza, L., and Scardia, M. 1985, *Astronomy and Astrophysics* **146**, 11.

Breger, M.: 1990, *Astronomy and Astrophysics* **240**, 308.

Breger, M.: 1991, *Astronomy and Astrophysics* **250**. 107.

Breger, M., Balona, L. A., and Grothues, H. -G.: 1991, *Astronomy and Astrophysics* **243**, 164.

Breger, M., Garrido, R., Huang, L., Jiang, S.-y., Guo, Z.-h., Frueh, M., and Paparo, M.: 1989, *Astronomy and Astrophysics* **214**, 209.

Däppen, W.: 1985, in *Chaos in Astrophysics*, eds. Buchler, J. R., Perdang, J. M., and Spiegel, E. A. (D. Reidel. Dordrecht), p. 273.

Dziembowski, W. A.: 1980, *Lecture Notes Phys.* **125**, 22.

Dziembowski, W. A.: 1982, *Acta Astron.* **32**, 147.

Fitch, W. S. and Wisniewski, W. Z.: 1979, *Astrophysical Journal* **231**, 808.

Garrido, R., Gonzalez, S. F., Rolland, A., Hobart, M. A., Lopez de Coca, P., and Peña, J. H.: 1985, *Astronomy and Astrophysics* **144**, 211.

Kurtz, D. W.: 1979, *Monthly Notices of the RAS* **186**, 567.

Kurtz, D. W.: 1980, *Monthly Notices of the RAS* **193**, 61.

Kurtz, D. W.: 1990, *Delta Scuti Star Newsletter (Vienna)* **2**, 17.

Lampens, P.: 1991, *Delta Scuti Star Newsletter (Vienna)* **3**, 20.

Lampens, P. and Rufener, F.: 1990, *Astronomy and Astrophysics, Supplement Series* **83**, 145.

Li Zhi-ping, Jiang Shiyang, and Cao Ming: 1991, *Astronomy and Astrophysics*, preprint.

Mantegazza, L. and Poretti, E.: 1991, *Astronomy and Astrophysics*, preprint.

Morguleff, N., Rutily, B., and Terzan A.: 1976a, *Astronomy and Astrophysics* **52**, 129.

Morguleff, N., Rutily, B., and Terzan A.: 1976b, *Astronomy and Astrophysics, Supplement Series* **23**, 429.

Moskalik, P.: 1985, *Acta Astron.* **35**, 229.

Paparo, M. and Kovacs, G.: 1984, *Astrophysics and Space Science* **105**, 357.

Poretti, E.: 1989, *Astronomy and Astrophysics* **220**, 144.

Poretti, E., Mantegazza, L., and Riboni, E.: 1991, *Astronomy and Astrophysics*, in press.

Takeuti, M.: 1990, *Delta Scuti Star Newsletter (Vienna)* **2**, 8.

Takeuti, M. and Zalewski, J.: 1991, *Delta Scuti Star Newsletter (Vienna)* **4**, 14.

Walraven, Th., Walraven, J., and Balona, L. A.: 1991, preprint.

THE DELTA SCUTI STAR GX PEGASI: A THEORETICAL INVESTIGATION OF ITS POWER SPECTRUM

E. MICHEL, M. J. GOUPIL, Y. LEBRETON and A. BAGLIN
Observatorie de Paris, DASGAL, URA CNRS 335, France

1. Introduction

Target of a STEPHI multisite campaign, the Delta Scuti star GX Pegasi has been found to oscillate with at least five simultaneous, close frequencies (table I).

TABLE I
GX Peg's frequencies ($\pm 0.25\ \mu$Hz) (from Michel et al., 1992a)

	ν_1	ν_2	ν_3	ν_4	ν_5
frequency (μHz)	187.2	227.6	230.4	237.5	240.5

Mode identification together with informations about the star that such an identification can provide are outlined below (see also Michel et al, 1992b).

The mode identification is carried out by means of a comparison between the observed frequencies and the adiabatic frequencies of models appropriate to this star. Models that match GX Peg's position in a Hertzsprung-Russell diagram have masses in the range $1.9 - 2M_\odot$. When included, convective core overshoot is handled as in Maeder and Meynet (1989). According to these models, GX Peg is a rather evolved, main sequence star.

Adiabatic frequencies, ν_{nl} (degree l, radial order n), and eigenfunctions are calculated with a programme kindly supplied by Christensen-Dalsgaard (1982).

2. Mode Identification

On the basis of geometrical visibility effects (Christensen-Dalsgaard et al., 1982) as well as energy and mode trapping arguments (Lee 1985, Dziembowski and Krolikowska, 1990), we consider modes $l = 0, l = 1$ as most probably detected.

Only the first and second radial overtones belong to the observed frequency interval, in the range of effective temperatures and masses of interest. The frequency ν_1 is identified with the first radial overtone. Comparison

Astrophysics and Space Science **210**: 181–183, 1993.
© 1993 *Kluwer Academic Publishers.*

between theoretical and observed period ratios shows that ν_3 corresponds to the second radial overtone. when Livermore opacities (Rogers and Iglesias, 1991) are used.

Models without overshoot offer the possible choice of $l = 1$ modes with radial order $n = 2$ or 3 for the three other observed frequencies. As a consequence of their different evolutionary status, models with convective core overshoot (which has to exist to some extent. see also Dziembowski et al., 1991) only allow the mode $l = 1$, $n = 3$, whose frequency then falls very close to the observed frequency, ν_4.

We therefore retain the following identification as the most probable one: ν_1, ν_2 as the radial first and second overtones modes and (ν_3, ν_4, ν_5) as the triplet $l = 1$, $n = 3$ with azimuthal orders $m = 1, 0, -1$ respectively.

3. First Step Towards Seismology of GX Pegasi

The $l = 1$ modes in the frequency range $\nu_1 - \nu_5$ all have amplitude in the inner gravity propagative region except the ($n = 3$) mode which is trapped in the envelope. It is precisely this last mode which is detected in GX Peg's pulsation. This supports the idea that mode trapping plays a role mode selection as suggested by Dziembowski et al. (1991).

Assuming that the star (which belongs to a spectroscopic binary system) is synchronized ($\Omega_{rot} = 5 \mu Hz$), the value of the splitting $(\nu_5 - \nu_2)/2$ shows that GX Peg's rotation cannot be solid. If the surface rotation is Ω_{rot} and a linear variation of the rotation with radius is assumed, the core rotation then amounts up to $\sim 2\Omega_{rot}$.

Departure from equidistance for the triplet $l = 1$ (expectantly due to rapid rotation and binarity) is found very asymmetric and large but in agreement with theoretical calculations (Saio, 1981; Dziembowski et al., 1990).

The quantity $C_{nl} = (\nu_{nl} - \nu_{n0})/(\nu_{n0} - \nu_{n-1,0})$ sharply varies with $1/\nu_{nl}$ when the mode (n, l) develops amplitude in the inner gravity propagative region. This variation is only slightly sensitive to mass and to external regions but very sensitive to core overshoot. C_{nl} then tests the deep regions of convective core models. Its variation is found very close. indeed, to the asymptotic arctangent behavior expected for mixed modes. which depends on the extension of the inner evanescent and gravity propagative regions.

The observed value C_{31} for GX Peg falls slightly below the theoretical C_{31} curves. The discrepancy decreases when overshoot is included. Small changes in metallicity and amount of overshoot must be tested before turning to a more sophisticated treatment of the very inner convective regions.

Acknowledgements

We gratefully thank J. Christensen-Dalsgaard for providing his code and helpful discussions, and W. Dziembowski for fruitful comments during this conference.

References

Christensen-Dalsgaard, J: 1982, *Monthly Notices of the RAS* **199**, 735.
Christensen-Dalsgaard, J. and Gough D. O.: 1982, *Monthly Notices of the RAS* **198**, 141.
Dziembowski, W. A. and Goode, P. R.: 1992, preprint.
Dziembowski, W. A. and Krolikowska, M.: 1990, *Acta Astron.* **40**, 19.
Dziembowski, W. A. and Pamyatnykh, A. A.: 1991, *Astronomy and Astrophysics* **248**, L11.
Lee, U.: 1985. *Publications of the ASJ*, **37**, 279.
Maeder, A. and Meynet, G.: 1989, *Astronomy and Astrophysics* **210**, 155.
Michel, E. et al.: 1992a, *Astronomy and Astrophysics*, in press.
Michel, E., Goupil, M. J., Lebreton, Y.. and Baglin, A.: 1992b, *Astronomy and Astrophysics*, in prep.
Rogers, F. J. and Iglesias, C. A.: 1991, preprint.
Saio, H.: 1981, *Astrophysical Journal* **244**, 299.

SHORT-TERM AMPLITUDE VARIATION OF
FM COM (=HR 4684)

M. PARARÓ
Konkoly Observatory, Budapest, Hungary

J. PEÑA and R. PEÑICHE
Instituto de Astronomia, UNAM, Mexico

and

C. IBANOGLU, Z. TUNCA and S. EVREN
EGE University Observatory, Izmir, Turkey

Abstract. Using unpublished data obtained in 1982 at Mexico and Turkey combined with Hungarian observations the process of amplitude variation of the frequency at 18.5 c/d is shown.. The time scale of the amplitude increase is less than 100 days.

For a long time a basic problem is in the study of Delta Scuti stars what is their characteristic behaviour. Three possibilities were suggested: i) they pulsate with constant periods and amplitudes, ii) the periods are constant but they show amplitude modulation, iii) mode switching may exist.

Most of Delta Scuti stars with large amplitude and the well-analysed low amplitude $78\theta^2$ Tau (Breger et al. 1989, Kovács and Paparó 1989) belong to case i). The periods and amplitudes are constant over years.

The case iii) may represented by only one star, 21 Mon (Stobie et al. 1977) where completely different frequency sets were found with two years difference. The reanalysis done by Kurtz (1980) could not change this result.

Concerning to case ii), lately a definite amplitude variation was found by Breger et al. (1990) for 4 CVn where the time scale covers years or decades. Short time scale amplitude variation was reported in θ Tuc by Stobie and Shobbrook (1976), however, the reanalysis done by Kurtz (1980) raised up the beating of frequencies as an alternative solution.

HR 4684 was reported by Paparó and Kovács (1984) to have a very complicated spectrum which seems to change on a very short time scale, however, this finding based on six nights of observation from one site was not convincingly proved. Fortunately HR 4684 was also observed in Mexico and Turkey in 1982. The unpublished four Turkish and four Mexican observing runs excellently match in time to the Hungarian observations. 14 nights obtained at three different site over 90 days (coverage is 2.8%) gave the possibility to investigate the amplitude variation on short time scale. (The detailed analysis will be published elsewhere). The most compact dataset (60 days, coverage is 3.3%) was separated into four slightly overlapping tracks. The amplitude spectrum of the separated tracks could be seen in Figure 1.

Astrophysics and Space Science **210**: 185–187, 1993.
© 1993 *Kluwer Academic Publishers.*

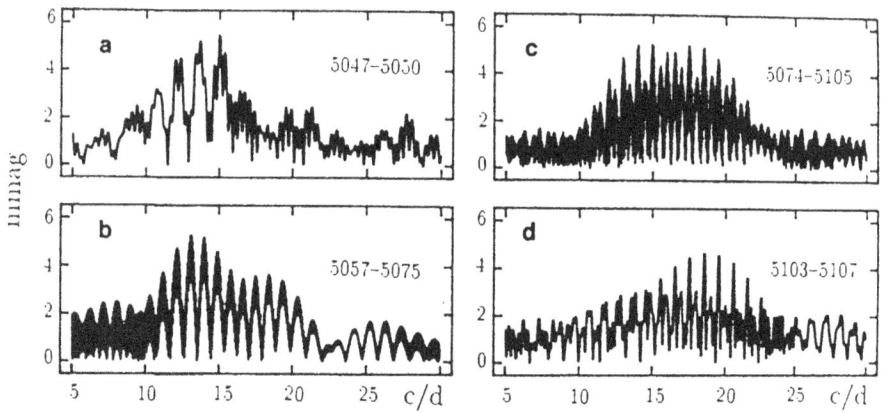

Fig. 1.   Amplitude spectrums of HR 4684

In the panels from a) to d) a continuous increase could be noticed in the amplitude of the frequency at 18.5 c/d. The first panel (a) based altogether on four nights of observation (coverage is 24%) from Turkey, Mexico and Hungry shows the dominant frequency at 15.08 c/d. Practically there is no sign of the frequency at 18.5 c/d. The second panel (b) based on Turkish and Mexican observations (three nights, coverage is 3%) shows how the amplitude of 18.5 c/d frequency is increasing but the amplitude is still less than the amplitude of 15.08 c/d. In the third panel (c) using two Mexican and two Hungarian observations (coverage is 2.6%) the amplitude of both frequencies are the same. In the last panel (d) using four Hungarian observations (coverage is 16%) the dominant frequency is at 18.5 c/d. The amplitude of the frequency at 15.08 c/d decreased to the half of the previous value.

Regarded the rather good coverage of the first and last panels and the advantages coming from multisite observations the short time scale amplitude variation in HR 4684 during 1982 seems to be well-established. The amplitude variation of a single frequency as an explanation, instead of beating of closely spaced frequencies, is base on the detailed analysis of different years.

Looking at the long-term amplitude variation of 4 CVn and the short-term amplitude variation in HR 4684 the question is raised: what is the theoretical time scale of amplitude variation in Delta Scuti stars? As it was kindly checked by W. Dziembowski during the meeting the time scale may be less than a year or may cover years or decades. It means, both the long-term and short-term amplitude variations could be explained by the present state of the theory.

References

Breger, M., Garrido, R., Huang Lin, Jiang Shi-yang, Guo Zi-he, Frueh, M. and Pararó, M.: 1989, *Astronomy and Astrophysics* **214**, 209.

Breger, M., McNamara, B. J., Kerschbaum, F., Huang Lin, Jiang Shi-yang, Guo Zi-he and Poretti, E.: 1990, *Astronomy and Astrophysics* **231**, 56.

Kovács, G. and Pararó, M.: 1989, *Monthly Notices of the RAS* **237**, 201.

Kurtz, D. W.: 1980, *Monthly Notices of the RAS* **193**, 61.

Paparó, M. and Kovács, G.: 1984, *Astrophysics and Space Science* **105**, 357.

Stobie, R. S. and Shobbrook, R. R.: 1976, *Monthly Notices of the RAS* **174**, 401.

Stobie, R. S., Pickup, D. A. and Shobbrook, R. R.: 1977, *Monthly Notices of the RAS* **179**, 389.

PERIOD VARIATIONS AND EVOLUTION OF
DELTA SCUTI VARIABLES

JIANG SHI-YANG

Beijing Astronomical Observatory, Chinese Academy of Sciences

The Short periods of Delta Scuti Stars allow the observational detection of the period changes expected from the stellar evolution within several tens years. For about 30 years we are keeping this topic as a small systematic observational program with our 60 cm telescope. Here we publish the period variation of 18 stars in table I. The data of 28 And are taken from R. Garrido et al. in *AAp* **144** (1983), 211; the period variation rate of 4 CVn is given by M. Breger. Both of them are low amplitude variables.

From these data we find out that both the population I and II variables can have period increasing and decreasing. The rate is between 2×10^{-6} to 8×10^{-8} days/year. Due to the period range of about 0.2 days, the time scale of period variation in one direction is limited within about 2 million to 200 million years. Usually we take the mass of this A to F type variables as 1.5 to 2.0 solar mass, so the main sequence life time is about several billion years. If all these A to F type main sequence stars will be variable in some period of its life time the possibility for a star to show Delta Scuti type variation is about hundredth, especially for the high amplitude type. Up to now we have found some 50 high amplitude variables with $V < 10.5$, and the total number of A3 to F5 with $V < 10.5$ is about 1.2×10^5, so the incidence of high amplitude variables is about 0.04%.

It seems that these period variations are mainly caused by stellar evolution. The period variation is smooth and continuous in general from the diagram of $(O-C)$ of the time of light maxima. Both for the population I and II, there are always quite clear random fluctuations of the time of light maxima within an observational season. Normally this kind of fluctuation can be as large as 1 to 2 percentages of the pulsation period itself, so if in some season only having few times of light maxima, you can not have any true idea about the period variations. Some people often suggest abrupt variation for some stars may just caused by this reason.

For BL Cam there are only four groups of time of light maxima. From the first two groups D. H. McNamara get a period of 0.0390976 days. The error should be less than ± 0.0000001, but from the observations of McNamara to the next observations of E. Rodriguez et al. in *IBVS* No. 3428, 1990, the time span is too large so that we can not easy to find correct cycle number for them. If we use the formula given by McNamara, we have 110335.5423. So the cycle number may be 110335 or 110336. The first makes period increasing and the second makes the period decreasing. Now we have some

Astrophysics and Space Science **210**: 189–191, 1993.

TABLE I

star	P_0	P and P/P_0	ΔV	Tp	Sp
28 And	0.069304115 11	$(1.2\pm0.5)\times10^{-8}$ 1.7×10^{-7}	0.03	2×10^7	F0IV
BS Aqr	0.197822854 56	$-(9.2\pm2.0)\times10^{-9}$ 4.7×10^{-8}	0.52	2×10^7	A8-F3
GY Aqr	0.0610384097 21	$-(4.6\pm0.0)\times10^{-9}$ 2.5×10^{-8}	0.54	4×10^7	A2-A8
YZ Boo	0.104091579 2	$(3.0\pm0.4)\times10^{-9}$ 2.9×10^{-8}	0.50	6×10^7	A6-F1
BL Cam	0.03909783 2	$(1.5\pm0.1)\times10^{-7}$ $3.9\times10{-6}$	0.33	1×10^6	PEC
AD CMi	0.12297422 1	$(1.89\pm0.09)\times10^{-8}$ 1.5×10^{-7}	0.30	1×10^7	F0III-F3III
VZ Cnc	0.178364047 5	$-(3.4\pm0.8)\times10^{-8}$ 1.9×10^{-7}	0.45-0.80	4×10^6	A7III-F2III
XX Cyg	0.13486509 2	$(1.9\pm0.1)\times10^{-9}$ 1.4×10^{-8}	0.85	8×10^7	A5
RS Gru	0.14701153 9	$-(2.2\pm0.7)\times10^{-8}$ 1.5×10^{-7}	0.59	8×10^6	A6-A9IV-F0
DY Her	0.14863130 .1	$-(5.2\pm0.6)\times10^{-9}$ 3.5×10^{-8}	0.51	3×10^7	A7III-F4III
KZ Hya	0.05951104 2	$(1.8\pm0.1)\times10^{-8}$ 3.0×10^{-7}	0.80	8×10^6	A0
EH Lib	0.088413258 62	$-(7.6\pm2.9)\times10^{-10}$ 8.6×10^{-9}	0.60	2×10^8	A5-F3
SZ Lyn	0.12053482 2	$(1.9\pm0.2)\times10^{-8}$ 1.6×10^{-7}	0.64	8×10^6	A7-F2III-IV
HD79889	0.09586955 6	$-(8.0\pm1.9)\times10^{-8}$ 9.3×10^{-7}	0.40	3×10^6	A3
ZZ Mic	0.0671796 2	$-(5.2\pm2.0)\times10^{-8}$ 7.7×10^{-7}	0.42	4×10^6	A3-A8IV
DY Peg	0.072926362 1	$-(2.3\pm0.07)\times10^{-9}$ 3.2×10^{-8}	0.70	7×10^7	A3-F1
SX Phe	0.054964509 2	$-(3.2\pm0.2)\times10^{-9}$ 5.8×10^{-8}	0.40-0.77	5×10^7	A2-F4
4 CVn	0.11635	$-(1.3\pm?)\times10^{-7}$	<0.026	2×10^6	F3III-IV

observations with a focal reducer CCD camera on the 1 metre telescope of Yunnan observatory in February of 1991, which makes a cycle number differences of 132547.6091. So it may be 132547 or 132548 from the beginning. But after we consider about the period variations we choose the 110333 for the first of Rodriguez and the 132544 for the first of our observation. That means the first cycles differences makes a phase differences of 2.5423 cycles. the second one makes a phase differences of 3.6091 cycles which is quite good match with the cycles differences.

From all these we think all these stars are very close to the main sequence, and the evolution are not directly but spiral like to leave the main sequence.

PHYSICAL CHARACTERS OF HD 93044 *

LI ZHIPING

Beijing Astronomical Observatory, Chinese Academy of Sciences, Beijing 100080, China

Abstract. The measurements of $uvbyH_\beta$ of HD 93044 were obtained in April 1991. and the observational results that the star locates nearly in the middle of δ Scuti instability strip with somewhat deviation to red edge. According to Crawford (1979) and Philip's (1979) calibrations, the effective temperature, absolute visual magnitude and surface gravity are obtained to be $T_{eff} = 7300 \pm 200$ K, $M_v = 1.^m 33 \pm 0.39$ and $\log g = 3.7 \pm 0.15$, respectively. The observational results of $\Delta m_1 = 0.01$ give an estimate of $[Fe/H] = -0.003 \pm 0.18$, so the opinion of metallic deficient is not supported obviously. The observations show the reddening index $E(b-y)$ to be 0.014 which is 1.4 times as large as the standard deviation of Crawford's (1979) statistics.

1. Introduction

HD 93044 was discovered to be a δ Scuti variable (Heynderickx 1990, Li et al. 1990.) with a dominant frequency 11.90808 cycle per day (Heynderickx 1990). There is some argument about if it has multiperiods. By means of period analysis method of PDM-technique, Heynderickx (1990) analyzed 214 measurements obtained between December 1984 and February 1987 and pointed out, "one frequency is clearly present and after prewhitening no other significant frequency could be found." But because the fitting is rather poor during some nights, he then explained that it might be that the pulsation amplitude of HD 93044 is variable in some erratic way. On the other hand, by using MEMPOW method Li et al. (1991) analyzed 177 V band measurements obtained from April 9 to May 6, 1990 and presented two pulsating frequencies of HD 93044, the ratio between the frequencies is about 0.51 which is in agreement with the ratio of radial pulsation fundamental frequency and third overtone one.

Other interesting questions of HD 93044 originate from the metallic abundance and luminosity. From the Geneva photometry measurements, Heynderikx (1990) concluded HD 93044 to be a near zero main sequence and somewhat metallic deficient λ Bootis star. On the other hand, by using 72 inch telescope of Lowell Observatory and a 40 Å/mm spectrograph, Slettbak (1968) obtained a spectral observation and concluded that the MK classification of HD 93044 is A7 III. By checking the $uvby$ measurements of Slettbak (1968) and the character of pulsating frequencies of HD 93044, Li et al. (1991) thought a significant reddening index possibly existent. All those opinions above are not in agreement completely. In order to solve these problems, we made a $uvbyH_\beta$ observations, here we give some results.

* Project supported by Young Foundation of Beijing Astronomical Observatory, China

2. Observations and Analysis

On the night of April 4. 1991. by using the 60 cm reflector working on DC mode (Shi et al. 1987) and controlled by a microcomputer at Xinglong Station of Beijing Observatory, we observed HD 93044 in $uvbyH_\beta$ filters for about two hours. The standard stars were HR 3974 and HR 3951 which were selected from the Catalog of bright $uvbyH_\beta$ standard stars (Perry et al. 1987). The observed results are listed in Table I. where n represents the number of observations and σ is the standard deviation of observations.

TABLE I
The $uvby\beta$ observations of HD 93044

Star	V	$b-y$	m_1	c_1	β	n
HD 93044	7.11	0.174	0.172	0.841	2.740	5
HR 3974	4.486	0.106	0.201	0.876	2.837	5
HR 3931	5.35	0.416	0.234	0.388	2.599	5
σ	0.004	0.011	0.010	0.028	0.0095	

According to Crawford's (1979) empirical calibrations of the $uvbyH_\beta$ systems, we obtain $(b-y)_0 = 0.188$, $\delta m_1 = 0.01$. $\delta c_1 = 0.181$ and $M_v = 1^m.33 \pm 0.39$, the error of absolute visual magnitude is caused by observations and the uncertainty of the calibrations. Then we obtain the reddening index $E(b-y) = 0.014$ and which is 1.4 times as large as the standard deviation of the statistics of the calibrations of Crawford (1979).

Using the empirical formula of Hack (1978) $[Fe/H] = -11.3\Delta m_1(b-y) \pm 0.18$, we get $[Fe/H] = -0.003 \pm 0.18$, and using the grid for the calculation of $\log g$ and T_{eff} from Strömgren four colour indices (Philip et al. 1979), we obtain $\log g$ and T_{eff} to be 3.7 ± 0.15 and 7300 ± 200 K respectively.

3. Discussion

Heyderickx (1990) thought, HD 93044 is rather close to the ZAMS with visual absolute magnitude $1^m.96$. This magnitude is about $0^m.5$ fainter than that we obtained. And according to the MK spectral classification of Slettbak (1968) HD 93044 is about $1^m.0$ above the ZAMS. So we tend to think that the magnitude value of Heynderickx (1990) is larger than the real value and it was caused by neglecting the reddening. Meylan (1980) give the intrinsic temperature parameter $(B_2 - V_1)$ of different MK classification, corresponding to A7 III, the value of $(B_2 - V_1)$ is -0.001. this is far smaller than the observed value 0.078 of HD 93044 (Heynderickx 1990), this difference implied a significant reddening possibly existent also.

The λ bootis type metallic deficient of HD 93044 is not confirmed in our observations of $uvbyH_\beta$. Is the metallic abundance index m_1 not sensitive to the λ bootis type star? Or is another cause existent? In order to solve this question, more photoelectric photometry and spectroscopic observations are more useful.

Acknowledgements

The financial support for author to attend this conference was given by the L.O.C. of I.A.U. Colloquium No. 134. I would like to thank Prof. Kodaira. K., Takeuti, M., Hamada, T., Tanaka, Y.. Ando. H.. and Kogure. T. et al. for their warm reception during my visit of Japan.

References

Crawford, D. L.: 1979, *Astronomical Journal* **84**, 1858.
Hack, B.: 1978, *Astronomy and Astrophysics* **63**, 273.
Heynderickx. D.: 1990, *Astronomy and Astrophysics* **232**, 79.
Li Zhiping, Tang Qingquan. Cao Ming: 1990, *IBVS*, 3467.
Li Zhiping. Jiang Shiyang, Liu Yanying. Cao Ming: 1991, *Astronomy and Astrophysics* **245**, 485.
Meylan, G., Python, M.. Hauck, B.: 1989, *Astronomy and Astrophysics* **90**, 83.
Perry, C. L., Olsen, E. H.. Crawford, D. L.: 1987. *Publications of the ASP* **99**, 1184.
Philip, G. A., Relyea, L. J.: 1979, *Astronomical Journal* **84**, 1743.
Shi, C. M., Du, B. T.. et al.: 1987, *Acta Astrophys. Sin.* **7**, 230.
Slettebak A., Wright R. R., Graham J. A.: 1968, *Astronomical Journal* **73**, 152.

PERIOD ANALYSIS OF THE δ SCUTI STAR HD93044

LIU ZONG-LI

Beijing Astronomical Observatory, Chinese Academy of Sciences, Beijing, China

Abstract. HD93044 was observed electrophotometricaly on April 21, May 1, 2 and 4, 1991 at Xinglong Station of Beijing Astronomical Observatory. Combining the data with Li Zhi-ping's data together and a period analysis was completed using a program which consists of a combination of Fourier transforms of prewhitened data and the multifrequency least squares of brightness residuals (LSR). Three pulsation frequencies (11.90809, 16.79553, 22.44827 cycles per day) with visual amplitudes between 0.0056 and 0.0203 mag were found. The solution fits the observations to +0.0071 mag which is equal to the mean square deviation of observations. The first frequency (11.90809 cycles per day) must be the right value of the fundamental frequency of HD93044.

Key words: δ Scuti – period analysis

1. Introduction

Slettebak et al. (1968) gave some parameters of HD93044 and classified it as an A7 giant which is conformable with the mean colours of the star in the Geneva photometric system (Rufener 1981, Golay 1980). Heynderickx (1990) observed this star (from December 1984 to March 1985, and from January 1986 to February 1987) and obtained its one period (11.90808 cycles per day) by means of the PDM-technique described by Stellingwerf (1978). Li Zhi-ping et al.(1991) observed it in 1990 and obtained two frequencies (11.38927 and 22.49097 cycles per day).

As stated above, Heynderickx obtained only one frequency. He noted that the agreement of the fitted light curve with the data was rather poor during some nights. However, he was unable to detect a second oscillation period in the residuals of the light curves. So he thought that the pulsation amplitude of HD93044 is variable in some erratic way. Li Zhi-ping et al. obtained two frequencies, but the fundamental frequency was different from Heynderickx's. In order to understand the nature of this star and to obtain the correct values of the frequencies it was decided to observe this star again.

2. Observations and Period Analysis

HD93044 was observed from April 21 to May 4, 1991 with the 60-cm telescope at the Xinglong Station of Beijing Astronomical Observatory, China. The single-channel integration photometer equipped with an EMI6256B photomultiplier was used in the DC mode with the output digitized with a V/F converter. The Johnson V filter was used. Two comparison stars were used: HD93457 and HD93664. No evidence for any variability of them was found.

Astrophysics and Space Science **210**: 197–200, 1993.
© 1993 *Kluwer Academic Publishers.*

The frequencies of HD93044 were determined using a program (Hao Jin-xin, 1991) which consists of a combination of Fourier transforms of prewhitened data and the multifrequency least squares of brightness residuals (LSR). And the program PERIOD (Breger. 1990) was used also. The calculated four frequencies are shown in Table I.

TABLE I
Four-frequency solution for HD93044

Frequency	Amplitude	Phase	Zeropoint	Residual
11.907±0.001	0.0209±0.0008	0.94±0.01	0.00015	0.0072
16.800±0.004	0.0074±0.0008	0.36±0.04		
21.756±0.005	0.0060±0.0008	0.58±0.05		
24.629±0.005	0.0060±0.0008	0.05±0.05		

The first frequency (11.907 cycles per day) is very closed to the value (11.90808 cycles per day) which was obtained by Heynderickx.

In order to obtain more accurate result the data obtained by Liu Zong-li were combined together with the data obtained by Li Zhi-ping et al. A period analysis of the new combined data was completed using the same method. A three-frequency solution was obtained. It is shown in Table II.

TABLE II
Three-frequency solution of the combined data

Frequency	Amplitude	Phase	Zeropoint	Residual
11.90809±0.00002	0.0203±0.0006	0.609±0.006	0.0003	0.0071
16.79553±0.00007	0.0064±0.0006	0.48±0.02		
22.44827±0.00008	0.0056±0.0006	0.50±0.02		

The fit of three-frequency solution of the combined data to the measurements is shown in Figure 1.

3. Conclusion and Discussion

The first frequency(11.90809 cycles per day) of three-frequency solution of the combined data is completely identical with the value (11.90808 cycles per day) obtained by Heynderickx. So it must be the right value of the fundamental frequency of HD93044. And it means that the fundamental frequency of this star has kept constant for at least 7 years from 1984 to 1991. The second frequency(16.79553 cycles per day) of three-frequency solution of the combined data is very closed to the value (16.780 cycles per day) of four-frequency solution of the data (from April to May 4. 1991). So it might

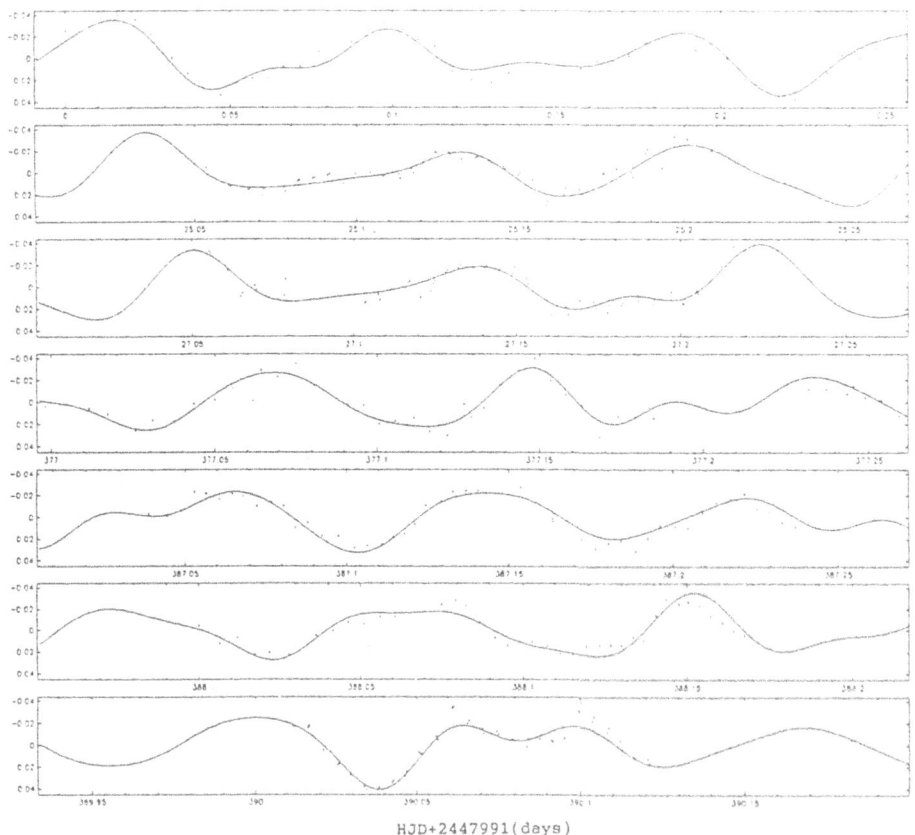

Fig. 1. Fit of three-frequency solution of the combined data to the measurements.

be the right value of the second frequency of this star. Furthermore it is obvious that the three-frequency fit is better than the one-frequency fit and two-frequency fit.

However, the value of the third frequency of the two sets of data are different each other. I think that the small data would be responsible for the disagreement between the different sets of data. In order to obtain the more accurate values of other frequencies, to give the more exact physical parameters to understand the nature of this star more observations and new period analysis are needed.

References

Breger, M: 1990, *Comm. Asteroseismology* **20**.
Golay, M.: 1980, *Vistas in Astronomy* **24**, 141.
Hao, J. -X.: 1991, *Publications of Beijing Astronomical Observatory* No. **18**.

Heynderickx. D.: 1990. *Astronomy and Astrophysics* **232**, 79.

Li, A. -P., Jiang, S. -Y.. Liu, Y. -Y., Cao. M.: 1991, *Astronomy and Astrophysics* **245**, 485.

Rufener, F.: 1981. *Astronomy and Astrophysics. Supplement Series* **45**, 207.

Slettebak, A.. Wright. R. R.. Graham, J. A.: 1968. *Astronomical Journal* **73**. 152.

Stellingwerf, R. F.: 1978, *Astrophysical Journal* **224**, 953.

OBSERVATION OF A VARIABLE, ZZ CETI
WHITE DWARF: GD154

B. PFEIFFER, G. VAUCLAIR and N. DOLEZ
Observatoire Midi-Pyrénées, Toulouse, France

M. CHEVRETON, J. R. FREMY and G. HERPE
Observatoire de Paris-Meudon, France

M. BARSTOW
Physics Department, University of Leicester, U. K.

S. J. KLEINMAN and T. K. WATSON
Astronomy Department, University of Texas at Austin, U.S.A.

J. A. BELMONTE
Instituto de Astrofísica de Canarias, Tenerife, Spain

S. O KEPLER, A. KANAAN and O. GIOVANNINI
Instituto de Fisica, Universidade Federal do Rio Grande do Sul, Bresil

R. E. NATHER, D. E. WINGET, J. PROVENCAL, J. C. CLEMENS,
P. BRADLEY and J. DIXSON
Astronomy Department, University of Texas at Austin, U.S.A.

A. D. GRAUER
*Department of Physics and Astronomy, University of Arkansas,
Little Rock, Arkansas, U. S. A.*

G. FONTAINE, P. BERGERON and F. WESEMAEL
Département de Physique, Université de Montréal, Québec, Canada

C. F. CLAVER
Astronomy Department, University of Texas at Austin, U.S.A.

T. MATZEH and E. LEIBOWITZ
Department of Physics and Astronomy, University of Tel-Aviv, Israel

and

P. MOSKALIK
Copernicus Astronomical Center, Warsaw, Poland

The ZZ Ceti stars form a class of variable white dwarfs: the hydrogen dominated atmosphere ones, which do pulsate in an instability strip in the effective temperature range 13000K–11500K. We know 22 such ZZ Ceti white dwarfs. Their variations are caused by nonradial g-mode pulsations with periods are in the range 100–1000 seconds.

A subsample of the ZZ Ceti stars shows amplitude variations on time scales of the order of one month. These variations could be driven by nonlinear phenomena.

One of these potentially non-linear pulsators, GD154, is on the red edge of the ZZ Ceti instability strip. It was first observed on May 1977 (Robinson et al. 1978). They obtained a power spectrum dominated by one mode

Astrophysics and Space Science **210**: 201–204, 1993.
© 1993 *Kluwer Academic Publishers.*

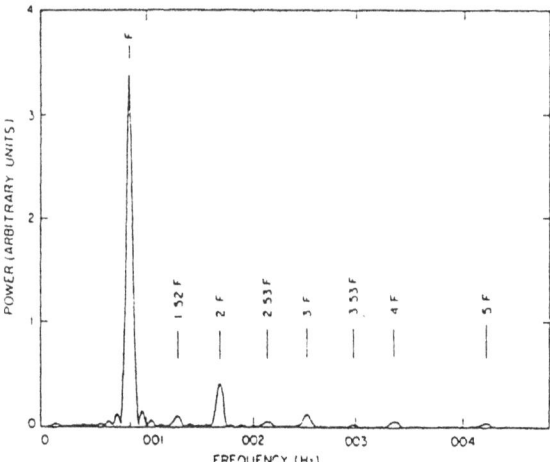

Fig. 1.   The power spectrum of the light curve of GD154 on the night of 1977 May 17 (UT). The peak labeled F is the fundamental frequency of the light curve.

(frequency F), its harmonics (nF) and "half integer modes" (($2n + 1)F/2$) (Fig. 1). Such a spectrum could be characteristic of resonant mode coupling or of a system on the way to chaos. Accordingly, a Whole Earth Telescope campaign (Nather et al. 1990) was organized on May 1991 to study GD154. The goal was to look for the (($2n + 1)F/2$) modes and amplitude variations.

Eight telescopes at different longitudes worked together to collect continuous data (Fig. 2). The coverage was about 50% from May 13 to 25 (Fig. 3). The resolution was about 10^{-6} Hz (1/12 days).

The Fourier Transform of GD154 light curve was computed (Fig. 4). A comparison of this new power spectrum with the previous one and shows that:

— there were no "half integer modes" during the W. E. T. campaign
— the light curve is dominated by 3 frequencies: $f1$ (0.842 mHz), $f2$ (0.918mHz) and $f3$ (2.484mHz).
— most of the other peaks are linear combinations or harmonics of those frequencies: $2f1$ (1.683mHz). $f1 + f2$ (1.761mHz), $2f2$ (1.837mHz), $f3 + f1$ (3.326mHz), $f3 + 2^{*}f1$ (4.169mHz).

Follow up observations were obtained from June 6 to June 12 from two telescopes (in Arizona and Canaries). Fig. 5 shows the Fourier Transform of these data. The same frequencies $f1$, $f2$ and $f3$ and their linear combinations and harmonics are still the only features. The main difference with the previous spectrum is the amplitude of the $f2$ mode: it is now three times the previous value.

During the W. E. T. campaign and the subsequent follow up observations, the "half integer modes" were not detected. Nonlinear phenomena could

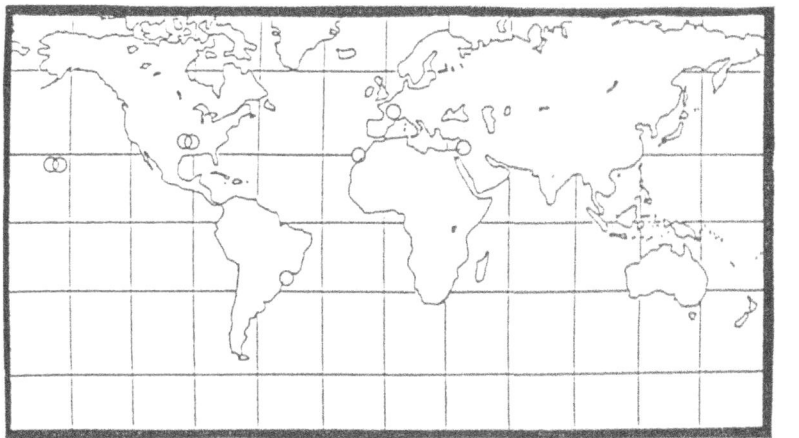

Fig. 2. During the W.E.T. campaign, eight telescopes were observing all around the world: Mauna Kea (C. F. H. and 24 inch), Texas (Mc Donald), Brazil (Itajuba), Canaries (La Palma). France (O. H. P.) and Israel (Wise).

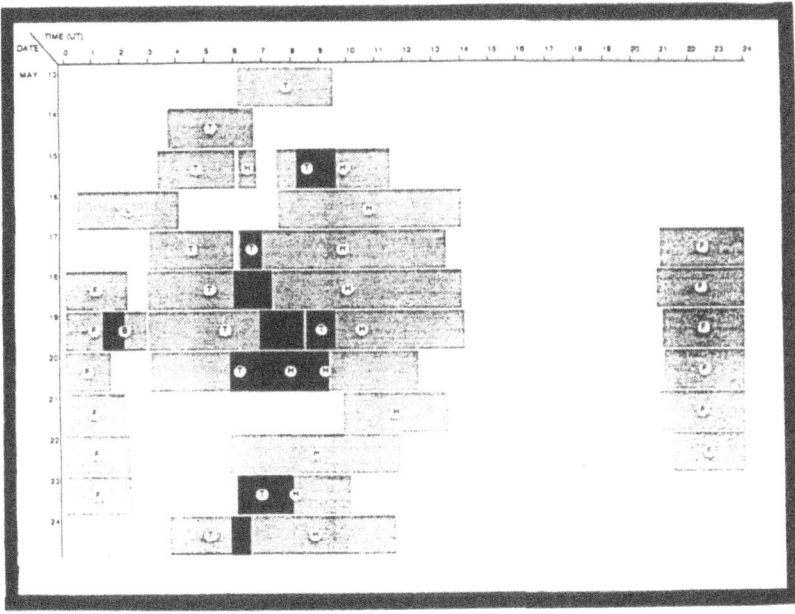

Fig. 3. W.E.T. coverage of GD154. On the vertical axis we plotted the days (13–24) and on the horizontal axis the U.T. Light grey corresponds to one telescope observing, dark grey to two telescopes and black to three. The letters indicate the sites: H for Hawai, T for Texas, B for Brazil. L for La Palma and F for France.

however take place in GD154: mode coupling could explain the observed amplitude variations and linear combinations of frequencies. The data are studied in more details to look for such a coupling.

References

Nather, R. E., Winget, D. E. Clemens, J. C.. Hansen, C. J., and Hine, B. P.: 1990, *Astrophysical Journal* **361**, 309

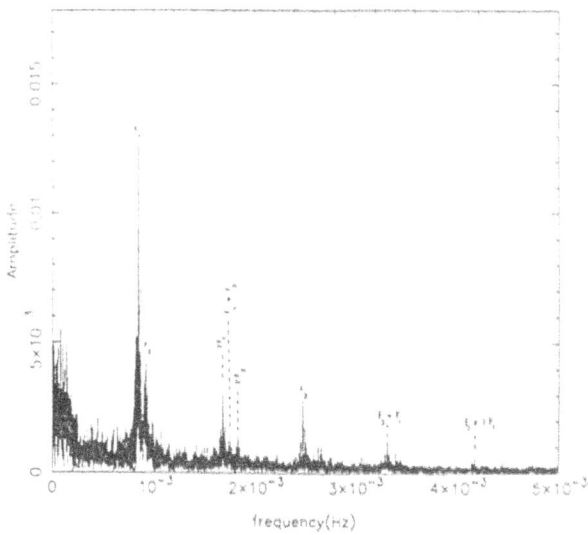

Fig. 4.   Power spectrum of GD154 light curve from May 13 to May 24.

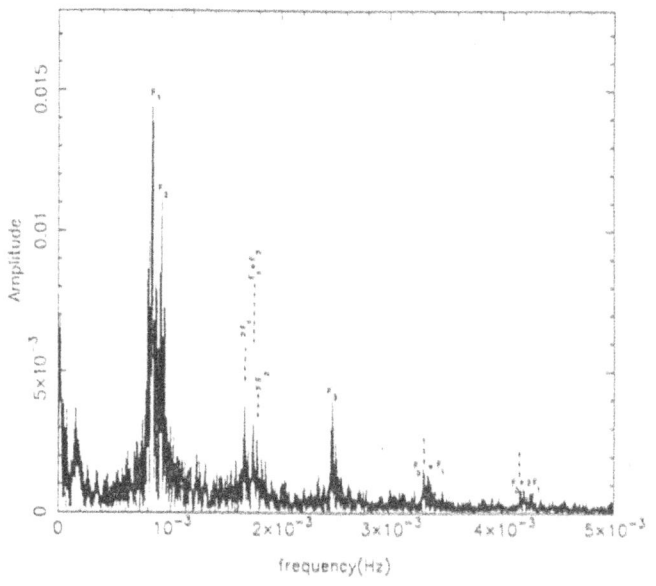

Fig. 5.   Power spectrum of GD154 light curve from June 16 to June 12.

Robinson, E. L., Stover, R. J., Nather, R. E., and McGraw, J. T.: 1978, *Astrophysical Journal* **220**, 614.

OBSERVATION AND STATISTICAL ANALYSIS
OF ZZ PISCIUM

H. SATOH

Institute of Astronomy, The University of Tokyo, Mitaka, Tokyo 181, Japan

K. SAIJO

Department of Science and Engineering, National Science Museum,
Taito-ku, Ueno Park, Tokyo 110, Japan

and

T. YANAGITA

The Institute of Statistical Mathematics, Minato-ku, Tokyo 106, Japan

Abstract. To examine the non-linear oscillation and period change of white dwarf variables, ZZ Psc is discussed observationally. As a result of data reduction we found two major periods, 680 seconds and 860 seconds, but could not find chaotic behavior of ZZ Psc.

1. Observations

Photoelectric observation of variable DA white dwarf ZZ Psc, one of the ZZ Cet type stars, was carried out for eleven nights from 1989 to 1991 using 91-cm telescope at Dodaria Station of the National Astronomical Observatory, Japan, as in Table I.

TABLE I
Journal of ZZ Psc observations

Date (U.T.)			Starting time (U.T.)	Integration time (s)	Number of points	Quality of data
1989	Dec.	15	10:50	55	80	poor
1990	Jan.	9	10:20	55	126	poor
1990	Nov.	5	10:25	55	1460	good
1990	Dec.	2	8:18	15	2090	poor
1990	Dec.	4	8:35	15	2265	poor
1990	Dec.	21	10:30	15	1250	poor
1990	Dec.	26	9:20	15	560	good
1991	Nov.	22	9:30	15	300	poor
1991	Nov.	24	13:12	15	50	poor
1991	Nov.	25	8:30	15	1560	good
1991	Nov.	26	9:40	15	650	good

All data were obtained by computer-controlled eight-channel ($\lambda\lambda$: 0.36–0.90 μm) photon-counting polarimeter cooled by dry ice and with cathode at

Astrophysics and Space Science **210**: 205–206, 1993.

−1.5 kV. The star SAO 128211. very near to ZZ Psc, was chosen as a comparison. In the observation, a pair of diaphragms is used for subtracting sky background effectively ("exchanging diaphragms method"). The aperture of the primary diaphragm (ϕ_1) used was $18''$ and the secondary diaphragm (ϕ_2) had a diameter about two times larger than primary one.

2. Data Analysis and Conclusion

Our present knowledge suggests that ZZ Psc is pulsating non-radially in g-mode. It has a multi-periodicity and among others has a photometric period of ∼13.0 minutes and an amplitude of 0.3 at most in visual magnitude. Furthermore, it is confirmed that ZZ Psc has large variation from one cycle to another in the amplitude and shape of the light curve even during a night. Series of data are analyzed after transformation to conventional UBV system. First data, taken on 15 December, 1989 was analyzed with Phase Dispersion Minimization (PDM) Method, which is known as a suitable reduction method in case of small number of data. As a result, we have obtained periods: 678 seconds in V, 665 seconds in B, and 668 seconds in U; so we concluded, that the period was 668 seconds on average. This is almost in good agreement with observational results of McGraw and Robinson (1975).

The purpose of our observation is to detect the variations of major periods, so we have tried to apply period analysis. The other data were preliminarily analyzed by FFT, AR (Auto-Regressive), AR-MA (AR-Moving Average) models and so on. In the observation, we were interfered by Moon light, so the quality of data is not good. Among them, we could get relatively good data on 25 November, 1991. From power spectrum by AR model, we obtained a period of 860 seconds (25 November, 1991). It corresponds to almost middle of f_2, f_3 periods of McGraw and Robinson (1975). We have examined whether there is chaotic behavior in the time variation of ZZ Psc by return map, but we have not found clear evidence for existence of chaotic behavior, because of limited amount of the data unequally due to missing observations and noise of the observational data.

We are going to continue observations on ZZ Cet type stars to examine the cause of light variation.

References

Landolt, A. U.: 1968, *Astrophysical Journal* **153**, 151.
McGraw, J. T. and Robinson, E. L.: 1975, *Astrophysical Journal* **200**, L89.
Saijo, K. and Satoh, H.: 1990, *Bull. of National Science Museum, ser. E.*, Vol. 13.
Stellingwerf, R. F.: 1978, *Astrophysical Journal* **224**, 953.

NONLINEAR, NONRADIAL PULSATION IN RAPIDLY

OSCILLATING AP STARS

D. W. KURTZ

Department of Astronomy, University of Cape Town

Abstract. The rapidly oscillating Ap stars pulsate in high-overtone, low degree p-modes with their pulsation axes aligned with their oblique magnetic axes. They show non-linearity in their pulsation in three ways:

1) The harmonics of the basic pulsation frequency are detectable.
2) The pulsation phase seems to vary stochastically on a time scale of days to years depending on the star.
3) The form of the nonradial surface distortion is not constant with time.

These three effects are illustrated with HR 3831, the best studied of the roAp stars. HR 3831 pulsates in distorted dipole mode which can be modelled as a linear sum of axisymmetric $l = 0$, 1, 2, and 3 spherical harmonics aligned with the magnetic axis. This gives rise to a 7-frequency multiplet split by exactly the rotation frequency. The form of the distortion shows small changes on a time-scale of years. HR 3831 shows a 5-frequency rotationally split first harmonic multiplet, a 3-frequency rotationally split second harmonic multiplet, and a single third harmonic frequency has probably been detected at an amplitude of 0.065 mmag. The first harmonic has changed its form significantly over the last 10 years. A technique for decomposing the fundamental frequency septuplet into its component spherical harmonics is used to fit the pulsation phase as a function of rotation phase. This allows a unique O-C to be defined for any length of light curve. The long term behaviour of the O-C diagram cannot be modelled adequately with a combination of periodic (Doppler shift) and quadratic (evolution) terms; there seems to be a significant stochastic component. The direction of the pulsation phase reversal at rotational phase 0.747 is indeterminate; sometimes it is a positive-going reversal, sometimes negative-going. At present it is not known whether this is a numerical artifact, or a physical effect in the star. If it is a physical effect, it means that small non-periodic differences in pulsation amplitude between the bipolar hemispheres have been detected.

1. Introduction to HR 3831

HR 3831 is an A7p SrCrEu magnetic star with an effective temperature of $T_{eff} = 8000 \pm 200$ K, a radius of $R = 1.9 \pm 0.1 R_\odot$ and an absolute magnitude of $M_{bol} = 2.0 \pm 0.3$, where the quoted errors are estimates. It has a polarity reversing magnetic field which ranges from about $+780$ G to -720 G with the rotation period of $P_{rot} = 2.851982 \pm 0.000005$ day. From $v \sin i = 33 \pm 3$ km s^{-1} and the radius estimate, $i > 38°$. The magnetic field variations require that either i or $\beta > 62°$. Mean light variations occur with the rotation period, but have extrema which lag behind the magnetic extrema. HR 3831 is a visual binary with a separation of 3.29 arcsec; the secondary is a main sequence G2 star. From the absolute magnitude of the secondary and from the parallax, the distance is estimated to be about 60 pc. (Kurtz *et al.* 1992; Kurtz, Kanaan and Martinez 1992).

Astrophysics and Space Science **210**: 207–214, 1993.
© 1993 *Kluwer Academic Publishers.*

HR 3831 is a singly-periodic rapidly oscillating Ap (roAp) star which pulsates with a frequency of $\nu = 1.4280128$ mHz ($P = 700.27$ s $= 11.67$ min). (See Kurtz 1990 for a review of the roAp stars.) Observed through a Johnson B filter, the semi-amplitude of the light variation associated with the pulsation ranges from a little over 4 mmag at the times of magnetic extrema to zero at one of the magnetic quadratures (Kurtz, Kanaan and Martienz 1992; Kurtz 1990; Kurtz, Shibahashi and Goode 1990; Kurtz and Shibahashi 1986; Kurtz 1982). The amplitude modulation period is the same as the rotation period, $P_{rot} = 2.851982$ day, and the times of pulsation amplitude maxima coincide with the magnetic extrema (Kurtz et $al.$ 1992).

Kurtz, Kanaan and Martinez (1992) found that HR 3831 pulsates in a single mode which is a distorted dipole with its pulsation axis aligned with the magnetic axis. Kurtz (1992) showed that the distortion can be modelled by a linear sum of axisymmetric spherical harmonic of degree $l = 0, 1, 2,$ and 3. The first second and third harmonics are also present. Table I gives the frequencies derived by Kurtz, Kanaan and Martinez (1992).

TABLE I

A least-squares fit of $\nu = 1.4280128$ mHz and its rotational sidelobes and harmonics to the 1991 data set.

Frequency name	Frequency mHz	Amp mmag	Phase rad
$\nu - 3\nu_{rot}$	1.4158380	0.243 ± 0.022	-3.0430 ± 0.0904
$\nu - 2\nu_{rot}$	1.4198962	0.267 ± 0.022	1.1426 ± 0.0811
$\nu - \nu_{rot}$	1.4239545	1.974 ± 0.022	2.6014 ± 0.0110
ν	1.4280128	0.481 ± 0.023	0.5805 ± 0.0483
$\nu + \nu_{rot}$	1.4320710	1.640 ± 0.022	2.6013 ± 0.0133
$\nu + 2\nu_{rot}$	1.4361293	0.085 ± 0.022	0.0084 ± 0.2549
$\nu + 3\nu_{rot}$	1.4401875	0.126 ± 0.022	-2.9291 ± 0.1742
$2\nu - 2\nu_{rot}$	2.8479090	0.113 ± 0.022	-2.4037 ± 0.1918
$2\nu - \nu_{rot}$	2.8519673	0.128 ± 0.022	2.7427 ± 0.1698
2ν	2.8560255	0.404 ± 0.021	-2.3549 ± 0.0511
$2\nu + \nu_{rot}$	2.8600838	0.054 ± 0.022	1.3846 ± 0.4032
$2\nu + 2\nu_{rot}$	2.8641420	0.121 ± 0.022	-2.3580 ± 0.1788
$3\nu - \nu_{rot}$	4.2799800	0.121 ± 0.020	-0.0447 ± 0.1698
3ν	4.2840383	0.064 ± 0.021	-2.2201 ± 0.3200
$3\nu + \nu_{rot}$	4.2880965	0.144 ± 0.020	-0.1701 ± 0.1426
4ν	5.7120510	0.065 ± 0.023	1.7344 ± 0.3163

$t_0 = $ HJD2448312.24019
$\sigma = 1.7386$ mmag per observation

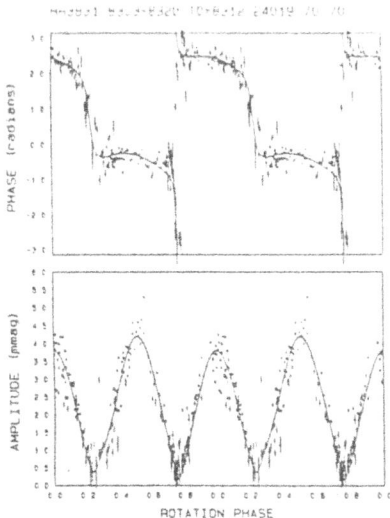

Fig. 1. This diagram plots the pulsation phase and amplitude as a function of the rotation phase for the 1991 data. The rotation phase is calculated from the time of magnetic maximum using the ephemeris given by Kurtz *et al.* (1992) with the rotation period $P_{rot} = 2.851982$ day. Two rotation cycles are plotted. Each point in the diagram has been calculated by fitting the frequency $\nu = 1.4280128$ mHz to 4 cycles (46.685 min) of the high-speed photometric data by linear least-squares. The theoretical lines show the best fit to the low frequency septuplet given in Table I assuming a pulsation mode which can be described by the sum of $l = 0, 1, 2$ and 3 axisymmetric spherical harmonics.

2. Theoretical Fits to the Phase and Amplitude Modulation Curves

Figures 1 to 5 show the theoretical fits of a sum of axisymmetric spherical harmonics of degree $l = 0, 1, 2$ and 3 to the 1981, 1985, 1986, 1990 and 1991 HR 3831 data sets. There appear to be real changes in the surface distortion from year-to-year. This is better illustrated in figures 6 and 7 which are schematic amplitude spectra. See Kurtz, Kanaan and Martinez (1992) and Kurtz (1992) for a complete discussion of the theory and observations which were used to construct these diagrams.

3. The O-C Diagrams

Figures 8 and 9 show the O-C diagrams for the entire data set and for the yearly data sets respectively. The long-term behaviour of these diagrams cannot be adequately modelled with periodic and quadratic terms, indicating that either the pulsation frequency or phase is not constant. Again, see Kurtz, Kanaan and Martinez (1992) for a more detailed discussion.

Other roAp stars also show changes in their pulsation frequencies, ampli-

D. W. KURTZ

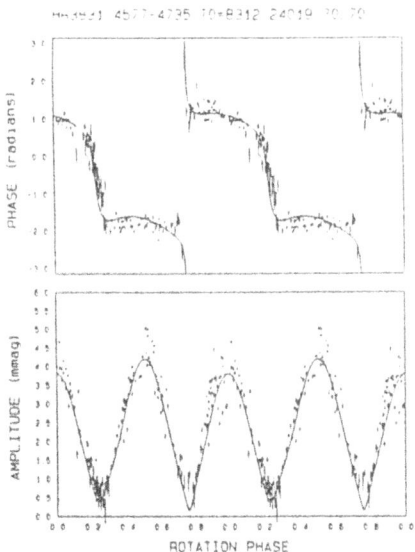

Fig. 2. This diagram plots the pulsation phase and amplitude as a function of the rotation phase for the 1981 data. The fitted curves are from the 1991 data and hence are the same as those in Fig. 1. The phase curve has been adjusted vertically to minimize the residuals to the fit.

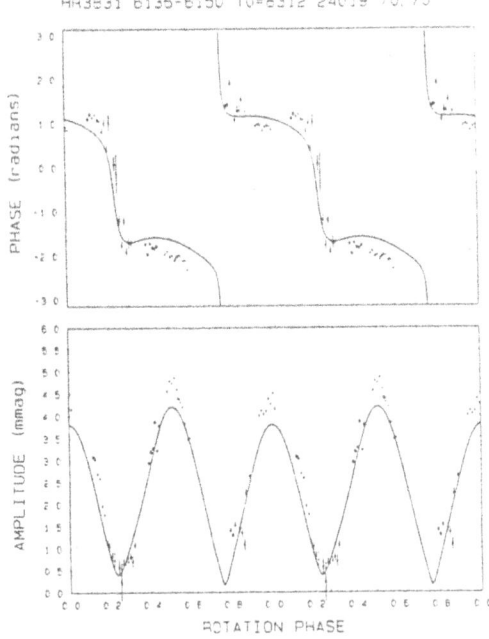

Fig. 3. This diagram plots the pulsation phase and amplitude as a function of the rotation phase for the 1985 data. The fitted curves are from the 1991 data and hence are the same as those in Fig. 1. The phase curve has been adjusted vertically to minimize the residuals to the fit.

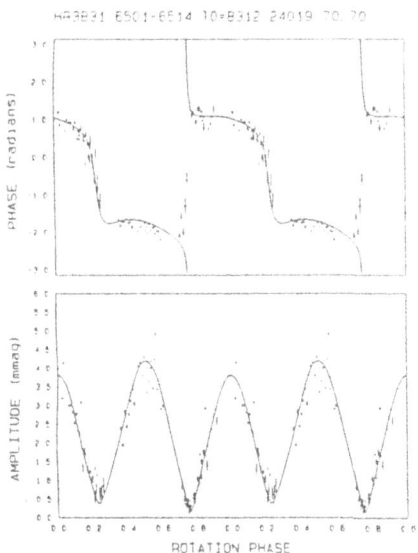

Fig. 4. This diagram plots the pulsation phase and amplitude as a function of the rotation phase for the 1986 data. The fitted curves are from the 1991 data and hence are the same as those in Fig. 1. The phase curve has been adjust vertically to minimize the residuals to the fit.

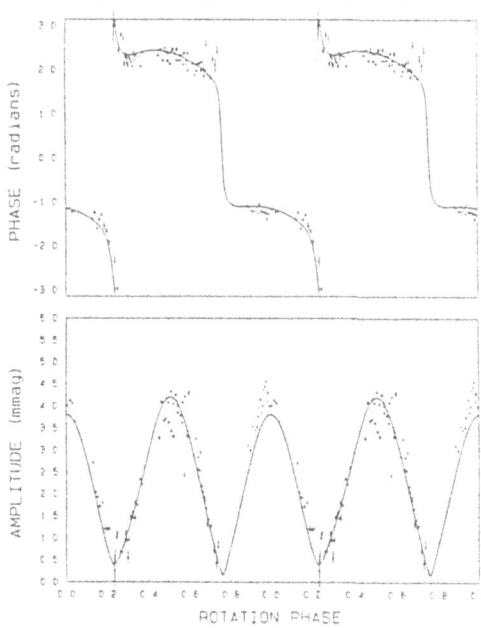

Fig. 5. This diagram plots the pulsation phase and amplitude as a function of the rotation phase for the 1990 data. The fitted curves are from the 1991 data and hence are the same as those in Fig. 1. The phase curve has been adjusted vertically to minimize the residuals to the fit.

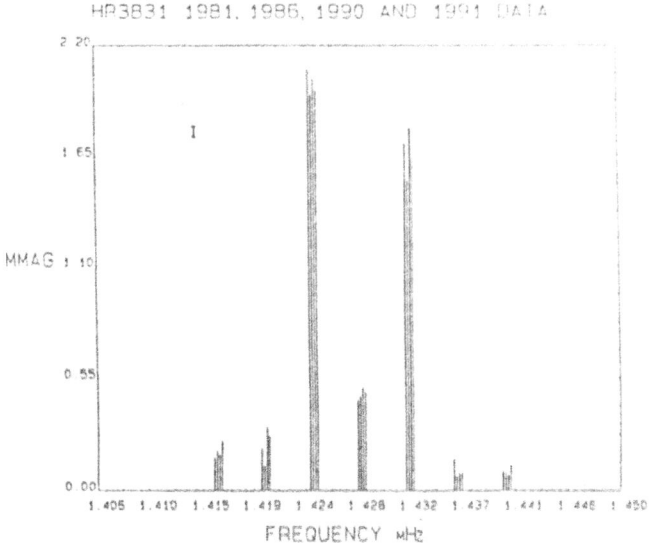

Fig. 6.   A schematic amplitude spectrum for the fundamental frequencies showing a linear least-squares fit of the frequency septuplet ν, $\nu \pm \nu_{rot}$, $\nu \pm 2\nu_{rot}$ and $\nu \pm 3\nu_{rot}$ to four data sets: JD2444577-4735 (1981), JD2446501-6514 (1986), JD2447931-7963 (1990) and JD2448303-8320 (1991). The amplitude spectrum for each data set has been shifted by a small frequency to display all four sets on the same diagram. At each frequency the left peak is 1981, the second peak 1986, the third 1990 and the right peak is 1991. *The separation of the four components in frequency is for display only; all four peaks coincide; the left peak is actually plotted at the correct frequency.* It is suggested that some of the apparent variation in the amplitudes of the components is real, but that the basic form of the rotational amplitude modulation has remained the same for 10 years. The error bar is ±0.05 mmag which is twice the internal error.

tudes and/or phases: HD 60435 shows amplitude modulation on a time-scale of days (Matthews, Kurtz and Wehlau 1986, 1987); HD 24712 (HR 1217) on a time-scale of weeks (Kurtz *et al.* 1989); HD 101065 on a time-scale of days or weeks (Martinez and Kurtz 1990); and HD 217522 on time-scales of days and years (Kreidl *et al.* 1991). It is not yet clear what governs the time-scale of the non-periodic component of the pulsation in roAp stars. There is the impression that the fewer the number of pulsation modes, the longer the life-times of the modes, but the statistics are inadequate as yet to be certain of this. More of these stars need to be discovered, and better studies are needed of those already known.

References

Kreidl, T. J., Kurtz, D. W., Kuschnig, R., Bus, S. J., Birch, P. B., Candy, M. P., and Weiss, W. W.: 1991, *Monthly Notices of the RAS* **250**, 477.
Kurtz, D. W.: 1982, *Monthly Notices of the RAS* **200**, 807.
Kurtz, D. W.: 1990, *Annual Review of Astronomy and Astrophysics* **28**, 607.

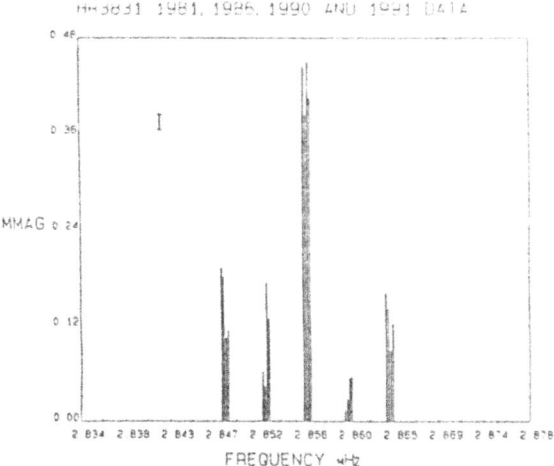

Fig. 7. A schematic amplitude spectrum for the first harmonic frequencies showing a linear least-squares fit of the frequency quintuplet 2ν, $2\nu \pm \nu_{rot}$ and $2\nu \pm 2\nu_{rot}$ to four data set: JD2444577-4735 (1981), JD2446501-6514 (1986), JD2447931-7963 (1990) and JD2448303-8320 (1991). At each frequency the left peak is 1981, the second peak 1986, the third 1990 and the right peak is 1991. The amplitudes are significantly different for the different data sets. The error bar plotted is ±0.02 mmag.

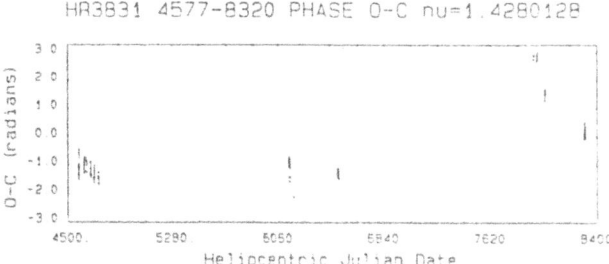

Fig. 8. This is the pulsation phase O-C diagram. The ordinate value of each point is the phase shift necessary to bring the 1991 theoretical phase curve shown in Fig. 1 into agreement with the 4-cycle linear least-squares phases for one night of data. This phase shift can be seen in the fit of the 1991 curve to the yearly data sets in Figs. 2 to 5, but here each point represents just one night.

Kurtz, D. W.: 1992, *Monthly Notices of the RAS*, submitted.
Kurtz, D. W., Kanaan, A., and Martinez, P.: 1992, *Monthly Notices of the RAS*, submitted.
Kurtz, D. W., Kanaan, A., Martinez, P., and Tripe, P.: 1992, *Monthly Notices of the RAS*, in press.
Kurtz, D. W., Matthews, J. M., Martinez, P., Seeman, J., Cropper, M., Clemens, J. C., Kreidl, T. J., Sterken, C., Schneider, H., Weiss, W. W., Kawaler, S. D., Kepler, S. O., van der Peet, A., Sullivan, D. J., and Wood, H. J.: 1989, *Monthly Notices of the RAS* **240**, 881.
Kurtz, D. W., and Shibahashi, H.: 1986, *Monthly Notices of the RAS* **223**, 557.
Kurtz, D. W., Shibahashi, H., and Goode, P. R.: 1990, *Monthly Notices of the RAS* **247**, 558.

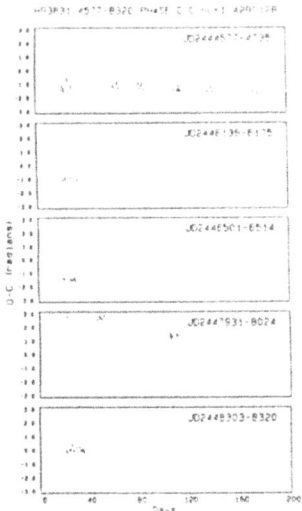

Fig. 9.   The phase O-C diagrams for the yearly data sets. Each panel is a blow up of a section of Fig. 8. The top panel is for the 1981 data; the second panel is for the 1985 data; the third panel is for the 1986 data; the fourth panel is for the 1990 data; and the bottom panel is for the 1991 data. All panels are plotted to the same scale in both coordinates. The 1991 data points are flat about O-C = 0 because the fitted frequency $\nu = 1.4280128$ mHz was derived from those data, as was the theoretical curve in Fig. 1 which defines the zero point of the O-C scale.

Martinez, P., and Kurtz, D. W.: 1990, *Monthly Notices of the RAS* **242**, 636.
Matthews, J. M., Kurtz, D. W. and Wehlau: 1986, *Astrophysical Journal* **300**, 348.
Matthews, J. M., Kurtz, D. W. and Wehlau: 1987, *Astrophysical Journal* **313**, 782.

PERIOD VARIATION OF BW VUL

JIANG SHI-YANG

Beijing Astronomical Observatory, Chinese Academy of Sciences

Due to the large fluctuation of the time of light maxima, in each observational season only very few observations is very dangerous for period variation research. Especially when the observation is not accurate enough such as for the early radial velocity data. Of course parts of the fluctuation are real which are caused by the chaotic behavior of the complicated pulsation. Generally the pulsation is quite periodic but is not like a timer such as the oscillation of the quartz or the rotating of the pulsar, all of these as much simple.

The period increasing rate of BW Vul from our results is 0.34 micron days per year which makes a evolution time scale about 3 million years. For a star with mass of 10 solar mass, the life age on the main sequence is about 10 million years. So BW Vul should not be very far away from the main sequence.

Our formula for calculate the time of light maxima is

$$T_{\max} = \text{HJD}2428802.7250 \underset{\pm 16}{} +0.201027291 \underset{56}{} +9.35 \underset{5}{} \times 10^{-11}E^2$$

This formula to fit the parabolic curve in figure 1 is almost exactly the same for to fit that in figure 2. Because after 1967 all the observations are very good, the situation in figure 2 should be very reliable. So the figure 1 also should be very reliable.

In figure 3 there are 6 points rather far away from the parabolic curve. The first one was given by Huffer in 1937 with very low time resolution so the light curve shape is quite different with our recent observation. So the reliability is very bad. The 3 points given by Cester in 1957 and 2 points given by McNamara in 1960 are not only have rather larger errors but really is too few points in each season. So we can not use them as a evidence to show some kind of special period variation.

Astrophysics and Space Science **210**: 215–217, 1993.
© 1993 *Kluwer Academic Publishers.*

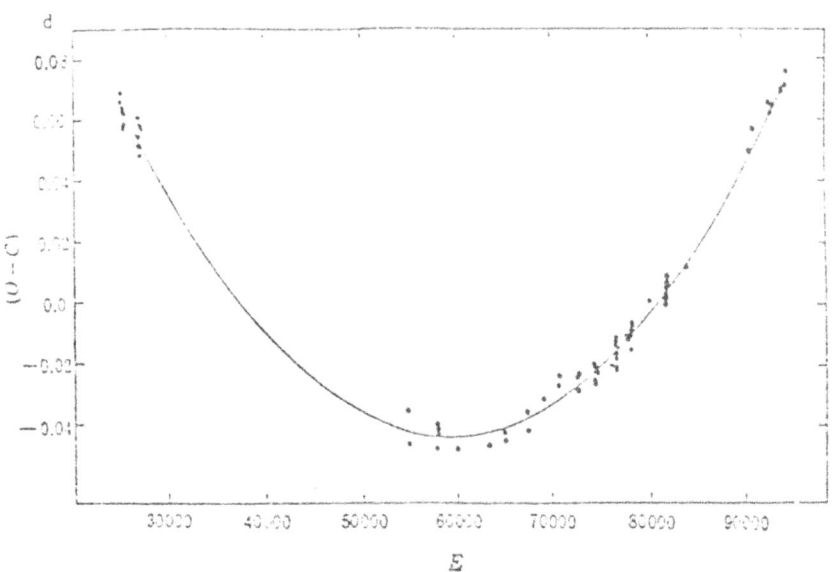

Fig. 1. The $(O-C)$ Curve for all the reliable photoelectric measures time of light maxima.

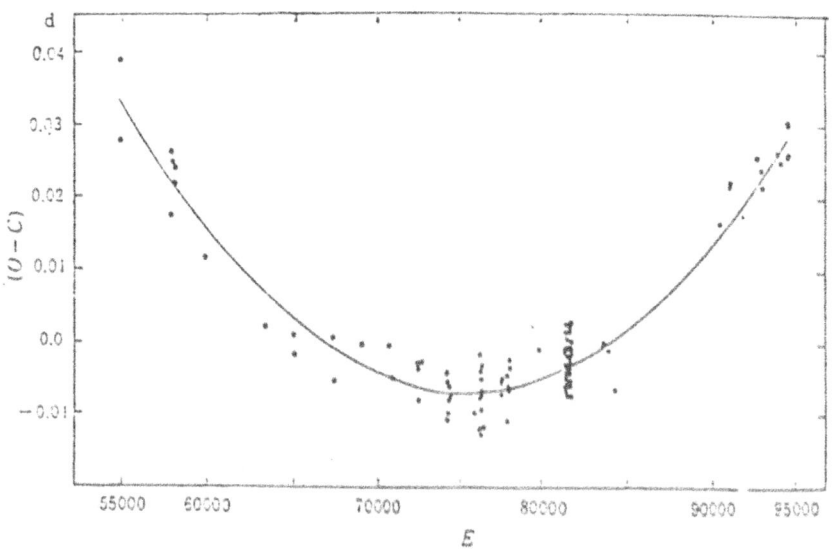

Fig. 2. The $(O - C)$ Curve for all the time of light maxima after 1967.

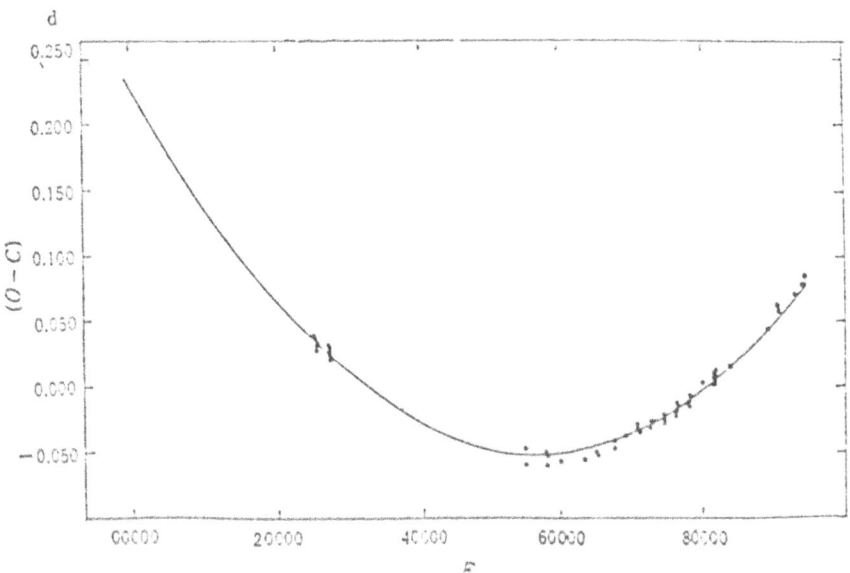

Fig. 3. The $(O - C)$ Curve for all the time of light maxima except that calculated from the radial velocity data.

NONRADIAL OSCILLATIONS IN ZETA OPHIUCHI IN 1991

E. KAMBE

Department of Geoscience, National Defense Academy, Yokosuka, Kanagawa 239, Japan

H. ANDO

National Astronomical Observatory, Mitaka, Tokyo 181, Japan

and

R. HIRATA

Department of Astrophysics, Kyoto University, Kitashirakawa, Kyoto 606, Japan

The line-profile variations have been found in many early-type stars surrounding the classical β Cephei variables in the H-R diagram. The feature of the variations has not been known enough mainly because of their periods of about one day. To specify the cause of the variations, however, it is necessary to make clear their features such as the multi-periodicity, the relation between the photometric variations and the spectroscopic ones, and the cause of existence or non-existence of mode (or period) switching (Balona 1991, Smith 1991). In Be stars, the correlation of the amplitude of the variations with their emission cycle is also important (Ando 1986).

TABLE I
Estimated (Period, m, k) in temporal domain

Candidates of periods (hr)	m	k	Amplitude ($\times 10^{-3}$)
2.14(2.43)±0.04	6~7	0.1±0.1	~4.8
2.49(3.33)±0.08	5~6	0.1±0.1	~4.5

We have been extensively monitoring the line-profile variations in a Be star, ζ Oph, a prototype of the line-profile variables to investigate their features (Vogt and Penrod 1983, Harmanec 1989). We have already shown that the variations could be reproduced well by two NRPs throughout our observations between 1987 and 1990. The correlation of the amplitudes of NRPs with its Be emission cycle is also found in the star. In this paper, we discuss the NRPs in May 1991, about one year after its latest emission episode.

The short-term variations in the He I $\lambda6678$ absorption line were monitored using the coude spectrograph of the 1.88-m telescope at the Okayama Astrophysical Observatory (Kambe 1991). The spectra thus obtained are shown in figure 1 as the 3-D contour map.

We use the mode identification method which was first developed by Gies and Kullavanijaya (1988) and later extended by Kambe et al. (1990) to analyze the observed line-profile variations. Our results are listed in table

Astrophysics and Space Science **210**: 219–221, 1993.

I. In this season, a mode with a period of 2.15 hr and a mode with a period
of 2.49 hr are detected; taking into account aliasing due to the sparse of
the data, these mode could correspond to modes with periods of 2.43 hr
and 3.33 hr which have been detected in all of our previous observations.
The m-value for a mode with a period of 2.49 (3.33) hr is slightly larger
than that of previous observations, but it might be superficial. Multi-site
simultaneous observations could be preferable way to increase the quality of
the data.

The amplitudes of the two modes seem to be smaller compared to those
in the midst of its last emission episode (1990 May), but slightly larger than
previous ones (figure 2).

Though the results have much ambiguity due to the shortage of the data,
we could confirm the previous results of the correlation between the amplitudes of the variations with its Be cycle; that is, the amplitudes of NRPs
increase toward the emission episode, reach their maximum at the episode
and decrease after that. However, the role of NRPs in the mass loss activity
from Be stars are still uncertain.

Numerical calculations were performed with Image Processing Facility
(VAX 4000 computer) at the National Defense Academy.

References

Ando, H.: 1986, *Astronomy and Astrophysics*, **163**, 97.
Balona, L. A.: 1991, *Be Newsletter No. 24*, ed. G. Peters, p. 5
Gies, D. R. and Kullavanijaya, A.: 1988, *Astrophysical Journal*, **326**, 813.
Harmanec, P.: 1989, *Bull. Astron. Inst. Czechosl.*, **40**, 201.
Kambe, E., Ando, H., and Hirata, R.: 1990, *Publications of the ASJ*, **42**, 687.
Kambe, E.: 1991, *Doctor Thesis, University of Tokyo*.
Smith, M. A.: 1991, *Be Newsletter No. 24*, ed. G. Peters, p. 8
Vogt, S. S. and Penrod, G. D.: 1983, *Astrophysical Journal*, **275**, 661.

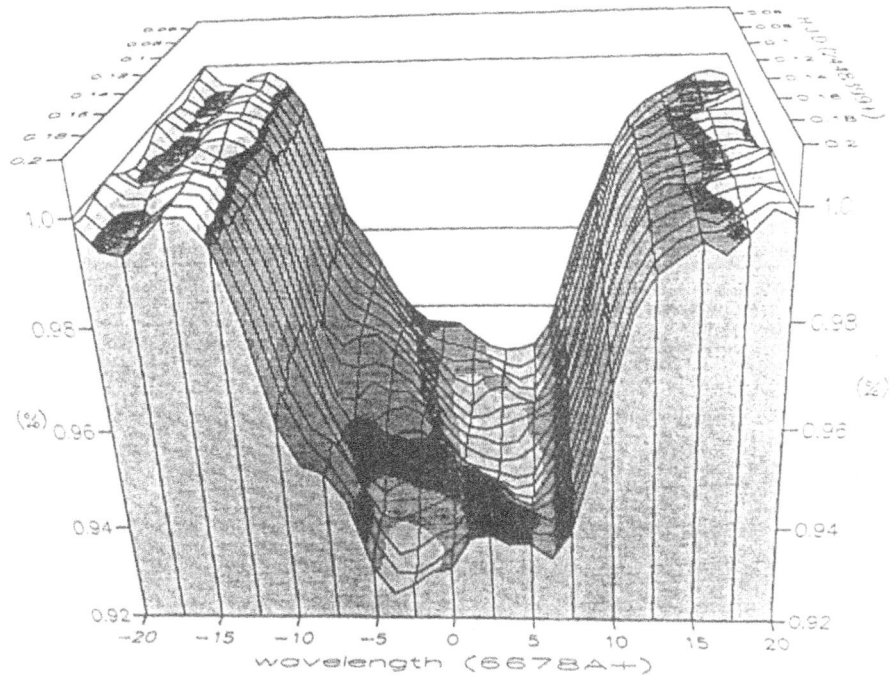

Fig. 1.   3-Dimensional view of observed profiles (He I λ6678) of May 22, 1991.

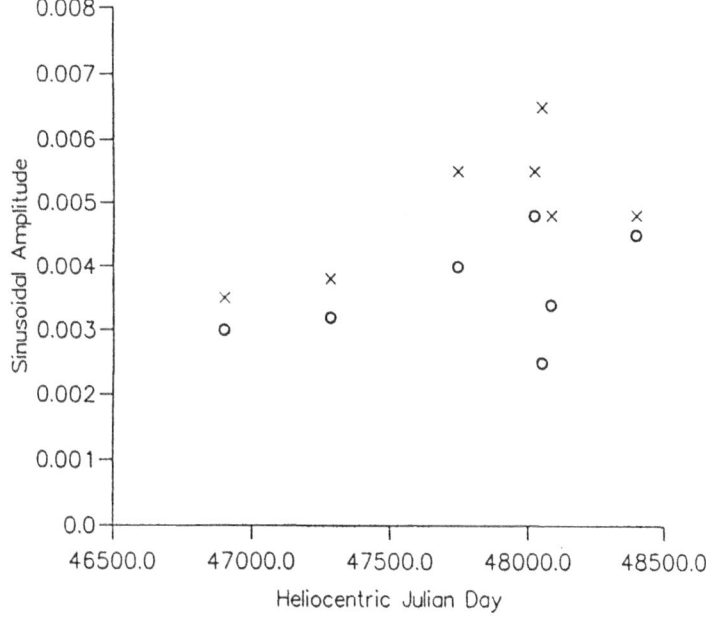

Fig. 2.   The amplitudes of two main sinusoidal waves detected during our observations of 1987-1991 (o; a $l == m = 4$ mode with a period of 3.33 hr, ×; a $l = -m = 7$ mode with a period of 2.43 hr). The abscissa is the Heliocentric Julian Day with an offset of 2400000.

THE ONSET OF QUASI-PERIODIC VARIATIONS IN Be STARS

T. KOGURE and M. MON

Department of Astronomy, Faculty of Science, Kyoto University, Sakyo-ku, Kyoto 601-01, Japan

and

M. SUZUKI

Kanazawa Institute of Technology, Kanazawa 921, Japan

We present some evidence of the quasi-periodic long-term variations (QP LV) in the violet-to-red ratio of double-peaked emission lines (V/R variation) and/or in the radial velocities of shell absorption lines for some Be stars. Although the V/R variations are rather prevailing phenomena among Be stars, the QPLV is remarkable by the following characteristics:

(1) The QPLV appears as a sudden onset of repeated V/R variations after a long (10 years), almost stable period, and it persists for a few or several periods in ten or more years.

(2) The period and amplitude of V/R variations change from cycle to cycle and from star to star. The variations of radial velocities (RV) of shell absorption lines are usually nearly parallel with the V/R variations.

(3) The QPLV appears usually in early type Be stars with large rotational velocities, regardless whether the stars are normal Be or shell stars, and whether they are close-binaries or single stars.

Table I gives a list of Be stars which showed the onset of the QPLV in recent decades. In column 5, distinction is given for the binarity (S= single, VB & SB= visual and spectroscopic binaries). The X-ray Be star γ Cas is supposed to have a compact companion as the X ray source. Column 6 and 7 give $\bigcirc$ marks when the QPLV is observed in the RV or V/R, and column 8 shows the periods of recurrent V/R variations observed in each term of the QPLV. Figure 1 exhibits two examples of the QPLV for ζ Tau (Ref. 3) and γ Cas (Ref. 1)

Two models have so far been proposed on the origin of the V/R variations. One is the rotation-pulsation of the envelope, and the other is the elliptical disk (or ring) model. There is no conclusive arguments as yet on these models, though the elliptical disk model has been found as more preferable in some close binary Be stars by the synchronization of the V/R variation with the orbital motion. The observed behaviors of the QPLV infer that the onset of these phenomena should be originated from some non-linear processes in stars, and taken place as a result of its interaction with the surrounding envelopes. Further studies from observational as well as theretical sides are desirable.

Astrophysics and Space Science 210: 223–225, 1993.
© 1993 *Kluwer Academic Publishers.*

TABLE I
List of Be Stars that showed Quasi-Periodic Long-Term Variability

Name	Sp.	Vsin i km/s	Be/ Shell	Binarity	Quasi-Periodic Var. RV	V/R	Quasi-P(yrs)	Ref.
γ Cas	B0IVpe	300	Be	X-ray Be	×	○	P=5,6,7,-	1
1H Cam	B2Ve	351	S/Be	SB	○	○	P=6?,6,6,6,	2
ζ Tau	B2IVpe	310	S	SB P=132.9d	○	○	P=7,4,4,7?	3
β¹Mon	B3Ve	346	S	VB,SB	○	○	P~12.5	4
48 Lib	B3IIIp	405	S	S	○	○	P=9,11,13,6,-	5,6
ε Cap	B3Vpe	293	S	VB, SB	×	(○)	P~8?	7
π Aqr	B1Ve	278	S	S	○	○	P=6,2,3,(5),-	8
EW Lac	B2IIIpe	350	S	S	○	○	P~7	9,10

References:
1. Horaguchi, Kogure, Hirata, Kawai, Matsuoka, Murakami, Doazan, Slettebak, C. C. Huang, Cao, Guo, L. Huang, Tsujita, Ohshima, & Ito (in preparation).
2. McLaughlin: 1963, *ApJ*, **137**, 1085.
3. Mon, Kogure, Suzuki, & Singh,: 1992 *PASJ*, **44** (in press).
4. Cowley & Gugula,: 1973, *A&A,*, **22**, 203.
5. Merrill & Sanford,: 1944, *ApJ*, **100**, 14.
6. Aydin, & Faraggiana,: 1978, *A&A, Sup.*, **34**, 51.
7. Mennickent & Vogt,: 1991, *A&A*, **241**, 159.
8. McLaughlin,: 1962, *ApJ, Sup.*, **7**, 65.
9. Hubert, Floquet et al.: 1987, *A&A. Sup.*, **70**, 443.
10. Kogure & Suzuki,: 1987, in *Physics of Be Stars*, eds. Slettebak & Snow, p. 192.

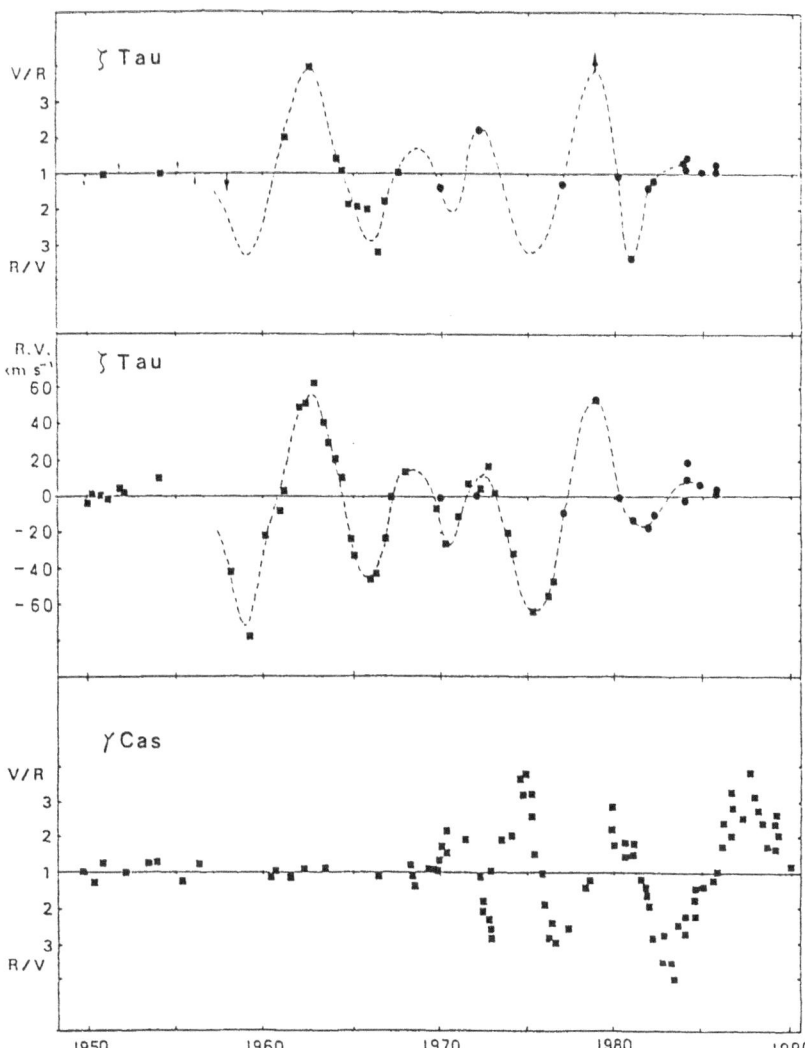

Fig. 1.   The QPLV in ζ Tau (V/R and RV) and γ Cas (V/R). Filled circles in ζ Tau denote our original data (Ref. 3). The RV cureve is given after subtracting the binary motion of ζ Tau. The observational data for γ Cas are taken from Horaguchi et al. (Ref. 1).

OPTICAL OBSERVATIONS OF FY PERSEI

AKIRA OKAZAKI

Department of Science Education, Gunma University, Maebashi, Japan

Abstract. The results are presented of multi-channel photometry and spectroscopy of the suspected nova-like variable FY Per. These results suggest that FY Per may be a Herbig Ae/Be star rather than a nova-like variable.

1. Introduction

The light variability of FY Persei was first detected by Morgenroth (1936). Richter (1964) examined photographic data of FY Per from 1935 to 1963, and found that the light of FY Per varies irregularly between 11.2 and 14.7 mag. Subsequently, Shugarov (1980) found tentative evidence for periodicity on a few hour time scale from his UBV photoelectric observations. He suggested that FY Per may be a nova-like variable. Ritter (1990) also listed this star in his catalogue of cataclysmic binaries.

The nature of FY Per, however, still remains an open question, because neither spectroscopic studies nor further photometric ones of FY Per have been published.

2. The Observations

2.1. MULTI-CHANNEL PHOTOMETRY

We observed FY Per at the Dodaira Station of the National Astronomical Observatory of Japan (NAOJ), with the 0.91-m telescope equipped with the eight-channel polarimeter (ch.1: $\lambda_{\rm eff} = 360$nm, ch.2: 410nm, ch.3: 460nm, ch.4: 530nm, ch.5: 650nm, ch.6: 700nm, ch.7: 760 nm, ch.8: 880nm. see Kikuchi, 1988), on 13 nights from October 1989 to February 1991. An integration time in each observation was 36s. We chose two nearby stars as the comparision stars.

We obtained $\sim$2400 individual observations in each channel (The data in ch. 8 have not been used in this study because of their poor S/N). The observed light curve of FY Per (in ch.4) shows irregular variation with an amplitude of $\sim$0.5 mag, as presented in Fig. 1, where each cross is a normal point consisting of 30 individuals (the estimated error of each normal is 0.01–0.02 mag). It is noticed that FY Per occasionally exhibit rapid light variation, as mentioned by Richter (1964). For example, on January 24, 1990, FY Per became fainter by 0.15 mag for a few hours.

We also found that the color of FY Per becomes redder when it dims as far as the observed magnitude range. For example, FY Per moves almost

Astrophysics and Space Science **210**: 227–229, 1993.

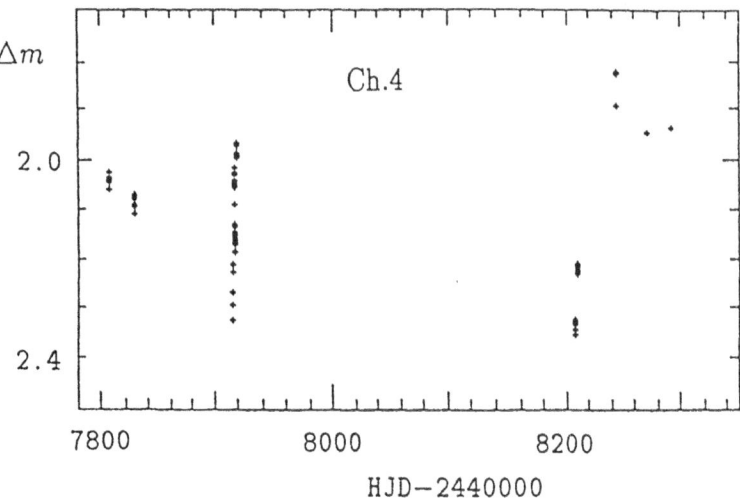

Fig. 1.   The observed light curve of FY Per in ch. 4.

along the line represented by $\Delta m_4 / \Delta(m_3 - m_4) \approx 10$ in the color-magnitude diagram.

2.2. SPECTROSCOPY

We obtained three spectrograms with a dispersion of $\sim$220 Åmm^{-1} on baked IIa-O emulsion with exposure times of 50–65 min on February 14, 1989, using the Cassegrain image-tube spectrograph of the 1.88-m telescope at the Okayama Astrophysical Observatory of NAOJ.

We found that FY Per shows a rather narrow Hα emission line (corresponding to a velocity dispersion of 150–200 km s^{-1}), though other members of Balmer series are absorption lines which are considered to originate from an early-type star.

3. Discussion

Shugarov (1980) found tentative evidence for periodicity on a few hour time scale in the light of FY Per on the basis of his 23 observations. He deduced several periodicity candidates such as 0.0702 d and 0.4407 d. Then, we made a period analysis for all the $\sim$2400 individual observations (the combined light of Ch.2, 3 and 4) using the PDM method (Stellingwerf, 1978). We found, however, no definite evidence for such short periodicity.

As mentioned above, (1) FY Per shows irregular light variation, (2) the color of FY Per becomes redder when it dims, and (3) FY Per has a rather narrow Hα emission line which is superimposed on an early-type star spectrum. It is also noticed that FY Per is located near by the dark nebula complex B20.

These observational facts suggest that FY Per may be a Herbig Ae/Be star rather than a nova-like variable.

Acknowledgements

We would like to thank Dr. S. Kikuchi for helpful discussions. We are also grateful to Mr. T. Kitazume and the staffs of the Dodaira Station and the Okayama Astrophysical Observatory of NAOJ for their assistance with the observations.

References

Kikuchi, S.: 1988, *Tokyo Astron. Bull.*, second ser., 3267.
Morgenroth, O.: 1936, *Astron. Nachr.*, **261**, 261.
Richter, G.: 1964, *Mitt. Veränderl. Sterne*, **2**, No.4, 79.
Ritter, H.: 1990, *Astronomy and Astrophysics, Supplement Series*, **85**, 1179.
Shugarov, S. Yu.: 1980, *Astron. Tsirk.*, No. 1119, 3.
Stellingwerf, R. F.: 1978, *Astrophysical Journal*, **224**, 953.

SOME INSIGHTS INTO STELLAR STRUCTURE FROM NONLINEAR PULSATIONS

M. J. GOUPIL

Observatoire de Paris, DASGAL, URA CNRS 335, France

1. Introduction

Efficient tools of investigation of stellar pulsation are the integral relations which link oscillation frequencies to the static structure of stellar models, as provided by the linear theory of pulsation (for a review, see Saio, this conference).

Similarly, oscillation amplitudes and phases, which arise from nonlinear processes, can be related to the stellar structure by means of amplitude equation formalisms (for a review, see Buchler, this conference).

For the simple case of a monoperiodic oscillation, involving only one unstable marginal mode, such a formalism shows that the (limit cycle) radius variations, at time t and mass level m, can be approximated, up to second order of approximation, (Buchler and Goupil, 1984; Buchler and Kovàcs, 1986) by:

$$\frac{\delta r}{r}(m,t) = 2A|\xi_r(m)|\cos \Omega_{nl}t + 2A^2|C_1(m)|\cos(2\Omega_{nl}t+\phi) + A^2C_0(m) \quad (1a)$$

$$A^2 = -\kappa/Q_r; \quad \Omega_{nl} = \Omega(1+\Delta\Omega/\Omega); \quad \Delta\Omega/\Omega = A^2Q_t \quad (1b)$$

where $A, R, \Omega, \kappa, \xi_r(m)$ respectively are the amplitude, stellar radius, linear nonadiabatic frequency, growth rate, radius eigenfunction. Second order nonlinearities generated first harmonic oscillations and change in equilibrium radius about which the star oscillates, as represented by the last two terms in (1a) respectively. Analogous expressions are obtained for velocity and light variations, that can be compared with observations.

The nonlinear, nonadiabatic coefficients, C_1, C_0, ϕ, Q_r, Q_t, are integrals over mass of kernels which depend on eigenfrequencies, eigenfunctions, on second and third order Taylor quantities from the equations modelling the star. They can either be computed from static models (Klapp et al., 1985) or obtained by numerical fits of hydrodynamical results (Kovàcs and Buchler, 1989).

2. Nonlinearities as Probes of Stellar Structure

When local quasiadiabaticity is assumed, approximated expressions for the eigenfunctions and their adjoints in terms of adiabatic ones can be used to

Astrophysics and Space Science **210**: 231–233, 1993.
© 1993 *Kluwer Academic Publishers.*

simplify the nonlinear coefficients entering (1).

The amplitude and $\Delta\Omega/\Omega$ are then found to be composed of two types of competiting contributions. The first one comes from nonlinearities of the restoring forces and energy transfer. The radius variations, initially sinusoidal (small amplitude), thereby become nonlinear i.e. are contaminated by harmonics (as in (1a)). This nonlinear contamination changes the magnitude of the restoring forces and energy transfer. This, in turn, modifies the radius variations. This feedback gives rise to the second contribution.

Further simplification (taking the adiabatic exponent as $\Gamma_1 \sim 5/3$ and neglecting spatial derivatives of $\xi(m)$ in linear variations of densities) leads to

$$\frac{\Delta\Omega}{\Omega} = A^2 \int dm \; 18\xi_r^3(3\Gamma_1 - 4)(\frac{5}{4}F_1 - \xi_r(m)) \tag{2}$$

For a fundamental mode, $\xi_r \geq 0$. Then, each stellar region contributes negatively or positively to (2), according to its position with respect to the critical value $5/4F_1$, which depends on the whole stellar structure (consequence of the aforementioned feedback). To F_1's value, however, mainly contributes the exterior where $\xi_r^3(m)dm$ peaks. Roughly, $F_1 \sim 1$ for a fundamental mode, therefore $\Delta\Omega/\Omega > 0$. The nonlinear period is longer than the linear one, in agreement with results from numerical models. The discussion must be slightly changed when stable modes are taken into account.

Contributions to the amplitude, $A^2 = (\mathcal{G}_2 + \mathcal{G}_3 + \mathcal{G}_4 + \mathcal{G}_{st})^{-1}$, come from eigenvectors nonadiabaticity, nonlinear dependence of pressure on entropy, nonlinear energy gains and losses and indirect influence of stable modes. Saturation effects that contribute to the existence of a limit cycle ($A^2 > 0$) can be locally discussed as above. In particular, energy nonlinearities play an important role in modifying the impact on the pulsation of the nonlinear variation of pressure with entropy.

To lowest quasiadiabatic order, the nonlinear change in equilibrium radius is $A^2C_0 = 2A^2F_1\xi(m)$. It increases towards the exterior with ξ_r as it is observed in numerical models (Fadeyev, this conference). In quasiadiabatic regions, a phase lag of π exists between the main and first harmonic oscillations and their amplitude ratio, $(1/3)F_1A$, is independent of mass level. Existence of stable modes introduces a slight dependence with mass level.

In conclusion, the above work can be extended to more realistic cases involving mode interactions. Though the necessary assumptions (local quasiadiabaticity, simplified boundary conditions) limit the discussion to qualitative behaviors, they provide simplified expressions of nonlinear coefficients which enable to investigate which processes, in what stellar regions, affect finite amplitude pulsations.

Acknowledgements

I am grateful to J. R. Buchler for providing useful suggestions about this work, during this conference.

References

Buchler, J. R. and Goupil, M. J.: 1984, *Astrophysical Journal*, **279**, 394.
Buchler, J. R. and Kovàcs, G.: 1986, *Astrophysical Journal*, **303**, 749.
Klapp, J., Goupil, M. J., Buchler, J. R.: 1985, *Astrophysical Journal*. **296**, 514.
Kovàcs, G. and Buchler, J. R.: 1989, *Astrophysical Journal*, **346**, 898.

CAPABILITIES OF THE OPTICAL MONITOR FOR THE RESEARCH IN X-RAY SOURCE AND STELLAR VARIABILITY

E. ANTONELLO
Osservatorio Astronomico di Brera, Via E. Bianchi 46, 22055 Merate, Italy

L. MARASCHI
Dipartimento di Fisica, Via Celoria, 20131 Milano, Italy

and

O. CITTERIO
Osservatorio Astronomico di Brera, Via E. Bianchi 46, 22055 Merate, Italy

1. Introduction

The project of an Optical Monitor (OM) for X-ray satellites, in particular the JET-X (Joint European Telescope for X-ray astronomy) experiment (Wells et al., 1991), derives from the scientific need of having complete data coverage at various wavelengths, UV and optical, of the observed X-ray sources, because these data are essential for a deeper understanding of the various classes of objects. When studying variable sources and/or transient astronomical phenomena, one needs that the multifrequency observations be performed essentially at the same time, because it is the knowledge of the simultaneous optical and X-ray behaviour of a source which contributes substantially to the clarification of its nature. In principle optical observations simultaneous with X-ray ones can be performed from ground based telescopes. However the complexity of satisfying the constraints typical of the optical telescopes (weather conditions, source observability) and of the X-ray instrumentation (e.g. orbital constraints) lead inevitably to a substantial loss of observing time. Therefore the only practical way of having an optimal utilization of the time available for X-ray observations, together with the wealth of scientific potential of simultaneous UV-optical observations, is to have a small telescope to be part of the same space mission.

2. JET-X Optical Monitor and Some Scientific Applications

JET-X is an experiment developed in the framework of a collaboration among United Kingdom, Italy, Germany and USSR, and is a part of the SPECTRUM X-GAMMA satellite, which will be launched by the beginning of 1995. The JET-X OM (Antonello et al., 1990) is a small Ritchey-Chretien optical reflector of 26 cm, equipped with two frame transfer CCD detectors,

Astrophysics and Space Science **210**: 235–237, 1993.

one designed to observe a small field (SF) of 8 arcmin with a resolution of 1.67 arcsec, the other to observe a wide field (WF) of 30 arcmin with lower angular resolution (6.2 arcsec). The WF of the OM coincides approximately with the total field of view covered by the X-ray detectors. In the main operating mode of the OM, the SF camera will observe the optical counterparts of X-ray sources, while the WF camera will observe the bright stars falling in the field of view. In the following, the expected performances of OM are briefly discussed in the context of some possible scientific applications.

X-ray Binaries. The basic criterion for reconstructing the geometry and the emission region(s) in these sources is the accurate study of the light curves at different frequencies. Irregular variability is observed in X-rays on timescales from seconds to weeks. With the JET-X telescope and OM the light curves of AM-Her or DQ-Her like systems (binaries containing a magnetic white dwarf) with $m_v \sim 15$ could be sampled with a resolution of 10 s in X-rays and few seconds in the UV and visual bands. Taking these values as typical for a distance of 100 pc, at a distance larger by a factor of three, X-ray, visual and UV light curves could still be measured with a resolution of minutes. In the low mass X-ray binaries, which consist of a collapsed object (neutron star or black hole) accreting matter through a disk from a low mass companion, besides rapid irregular variability, intensity dips with duration of minutes are observed in X-rays, while the orbital periods, difficult to detect due to the large intrinsic variability, are of several hours. A source such as Cyg-X2 can be monitored in the optical band with a time resolution of few seconds. Finally, the identification of the optical counterpart of the sources responsible for the so-called transients will be easily achieved with the OM if the source happens to be in the field of view.

AGNs. The study of variability of AGNs has long been recognized as one of the most powerful probes for understanding what occurs in the regions close to the massive black hole, thought to be at the nucleus, and in the gas surrounding it. The OM will be able to detect AGNs as faint as $m_v \sim 23$ in a reasonable exposure time (about 1000 s). Moreover, it will be able to yield the optical light curve of an AGN with $m_v \sim 15$ with a time resolution of about 10 s and an accuracy better than 10%, while for an AGN with $m_v \sim 18$ the same accuracy will be obtained with a resolution of about 100 s.

Stellar Variability. The study of all the variable celestial objects, from stars to AGNs, will take advantage of the observation from space over that from the ground, because the problems related to atmospheric phenomena such as transparency variations and scintillation are excluded. and it is possible to perform a continuous monitoring for some days. These advantages are vital for the development of stellar seismology. With the OM it will be possible to perform accurate differential photometry of bright stars and to detect short period variations with very small amplitude, less than 10^{-5} mag for $m_v \leq 7$. This kind of observations will be done in serendipity mode for the bright stars falling in the field of view of OM when JET-X will perform long surveys of X-ray source variability.

References

Antonello, E., Citterio, O., Mazzoleni, F., Mariani, A., Pili, P., and Lombardi, P.: 1990, in *Instrumentation in Astronomy VII*, SPIE **1235**, 867.

Wells, A., Stewart, G. C., Turner, M. J., Watson, D. J., Whitford, C. H., Antonello, E., Citterio, O., Cropper, M. S., Curtis, W. J., Peskett, S., Eyles, C. J., Goodall, C. V., Mineo, T., Sacco, B., and Terekhov, O.: 1991, in *Multi-layer and Grazing Incidence, X-ray/EUV Optics*, SPIE **1546**, in press.

WIND VARIABILITY OF LBV STARS

G. L. ISRAELIAN

Byurakan Astrophysical Observatory, Rep. of Armenia

Abstract. General properties of Luminous Blue Variables (LBV) have been reviewed by Lamers(1987). The LBV's are all close to the Humphreys-Davidson luminosity upper limit. The semi periods of the photometric microvariations with $\Delta V \sim 0.1^m$ are about twice as large as for normal supergiants of the same L and T_{eff}, and 4–20 times larger than the fundamental mode for radial pulsations. So, it is likely due to nonradial pulsations. During shell ejections, which accompanied by moderate photometric variations of $\Delta V \sim 1^m$, L_{bol} remains constant. This fact has been explained by the quasiperiodic variations of T_{eff} and the radius of pseudo-photosphere (Lamer 1986). LBV's are less stable than normal supergiants so that any internal instability has more effect on their envelopes. The nature of these instabilities remains unknown. We have found some interesting peculiarities which can throw light to this problem.

By the comparison of P Cyg CCD spectra obtained by Stahl *et al.* (1990) with that of Johnson *et al.* (1978), we have found that all 11 comparatively strong emission lines of FeII observed in 1975, disappeared, but lines of FeIII became more intensive. Emission wings of Balmer and some HeI lines also became stronger (Israelian and de Groot 1991a). Obviously, the excitation of FeII atoms for the most part is due to collisions with the free electrons. Lines of [FeII] were stronger in 1975, however, they still exist in 1990 (Stahl *et al.* 1990). Same kind of things have been observed in stars S Dor (Stahl and Wolf 1982) and R 71 (Wolf *et al.* 1981), and looks like an expansion of shock waves in the envelopes of long-period variables (Gorbatski 1958). Variations in the ionization degree of metals have been predicted by Pouldrach and Puls (1990), but for a variety of reasons we turn to the shock waves.

A comprehensive analysis of P Cyg CCD spectra permitted to identify 15 from 45 weak hitherto unidentified emission lines (Johnson et al. 1978, Israelian and Nicoghossian 1991). The majority of them belongs to multi-ionized atoms of metals. (Table I). Therefore an anomalous heating of some layers of the extended envelope can be expected to occur. Some of these unidentified spectral lines exists also in spectra of Sco X-1, γ Cas, RR Tel. V 1016 Cyg.

Let us draw a parallel between the evolution of absorption dips (or shells) of H_α and propagation of shock waves in the stellar wind of P Cyg. Absorption dips started at 0.4–0.5 V_s and accelerated with a mean value of $a = 0.6$ cm/s^2 until they reach $V = -220$ km/s. This evolution is well approximated with formulae $M = \text{const } \rho_0^{-1/\alpha}$ (Castor 1987), where $\alpha = 4.9$, M is Mach value, and $\rho(r)$ accepted from Pouldrach and Puls (1990). It has been shown that if the ratio of the intrinsic line width to the sound speed is large than

TABLE I
Hitherto Unidentified Emission Lines in P Cyg CCD
Spectra (Israelian and Nicoghossian 1991).

Line		Line		Line	
He II	6890	[Ni IV]	5959	[Fe VII]	5159
N III	6465	[Ni IV]	5363	[Fe VII]	4988
N II	6173	[Ni IV]	5289	[Fe IV]	5228
He II	6170	[Ni IV]	5040	[Ca VII]	5619
Ni I	5847	[Ca V]	5309	[K V]	4168

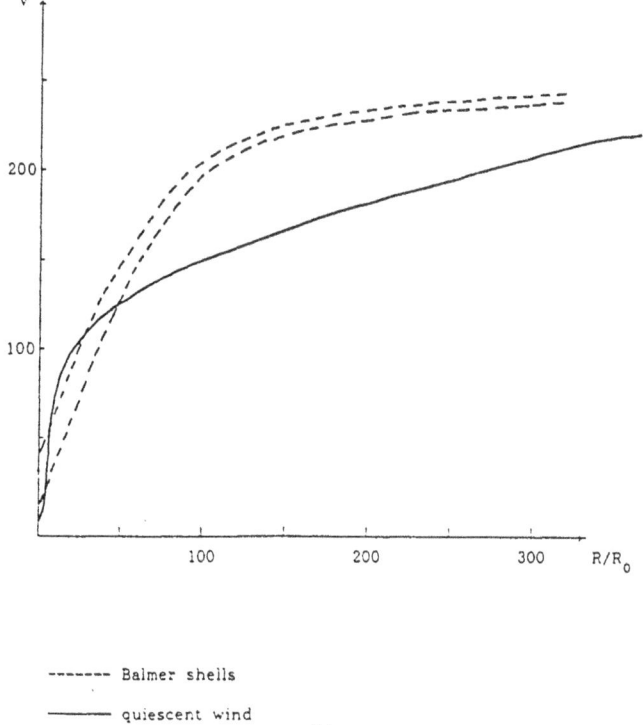

Fig. 1.

0.5, only a negligible wind results; when the ratio is smaller than 0.35 (which
is the case of P Cyg), the wind executes permanent self-excited oscillations;
in an intermediate range the wind is globally stable, but acts as a powerful
wave amplifier (Castor 1990). It is interesting to note that if we compare
the velocity law of the Balmer shells (Lamers 1986) with that of the quiescent wind (Pouldrach and Puls 1990, Israelian and de Groot 1991b), we can
see that the accelerations for the shells are at least equal (for smaller V) or
slightly higher (Fig. 1). It means that these shells are not driven by optically
thick lines in the Balmer continuum and there must be another mechanism
of the shells acceleration.

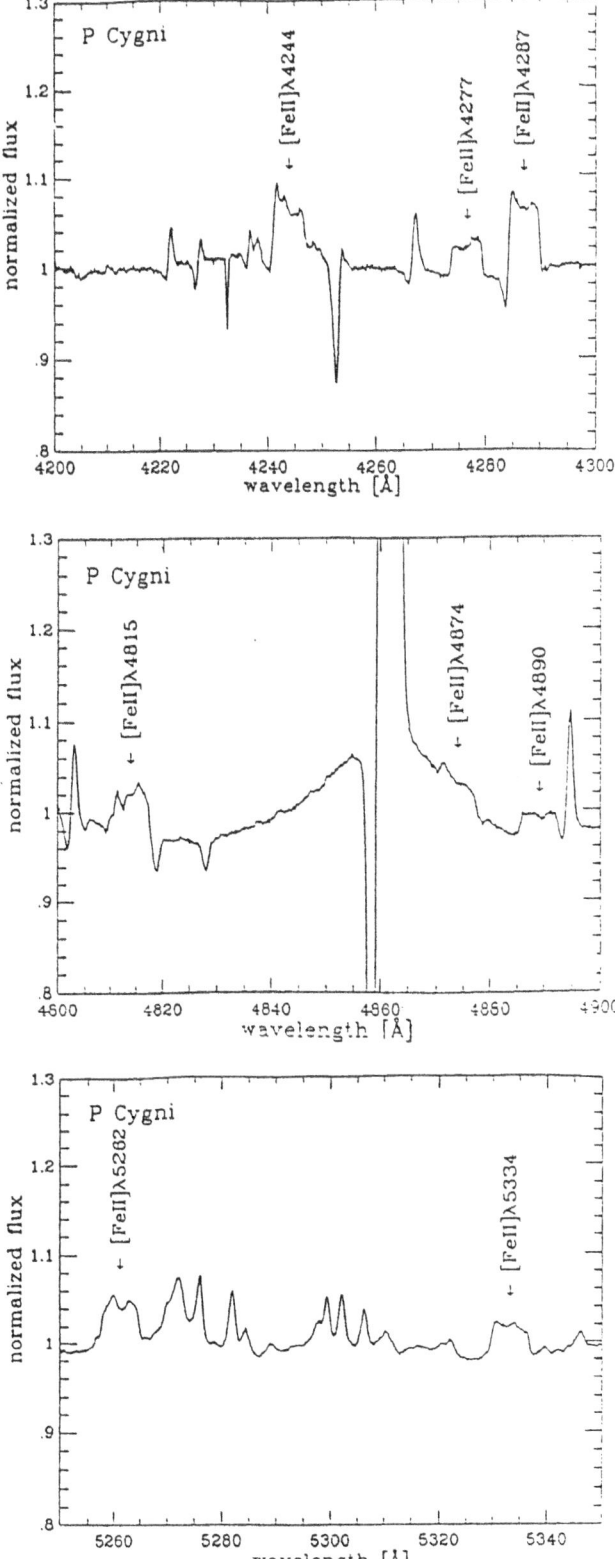

Fig. 2.

Post-shock radiation can be absorbed in pre-shock region, so the number of free electrons in that region is much more higher and proportional to V_s. For the optical depth τ for electron scattering of pre-shock region we have $\tau = 14c\ln(T_2/T_1)/3V_s$ (Klimishin 1984) where T_2 and T_1 are the post and pre-shock temperatures respectively. In case of adiabatic shock $T_2 \sim V_s^2$ and $T_2 \gg T_1$, but if the post-shock radiation pressure exceeds an adiabatic one, such that $T_2 \sim V_s^{-0.5}$ (Klimishin 1984), then will be $\tau < 1$. Expansion of the shock must be accompanied by considerably increasing strengths and extensions of emission line wings. Photometric variations due to electron scattering effects can be expected to occur.

Not long ago Wolf and Stahl (1990) found a change for most of the lines of singly ionized metals from normal to inverse P Cygni type profiles in spectrum of S Dor. We suppose, that this inverse absorption forms in dense post-shock region, as the velocity of this region is less than V_s (in an adiabatic case it's equal to $(3/4)\,V_s$), and the parcel of material is in free fall between shock passages, as the envelope has a small value of V_{esc}.

At the end, we want to draw attention on emission peaks of [FeII] flat-topped profiles (Fig. 2) (Johnson et al. 1978). Obviously these features exist in principle as the CCD spectra has a high ratio $S/N \sim 500$. Much more likely they represent some puffs or blobs of outflowing matter, but there is also the possibility due to thin rotating disk (Underhill and Nemec 1989). Hayes (1985) observed numerous polarization episodes in P Cygni in which Q and U fluctuated irregularly by 0.1-0.5%. These variations were found not to be colinear in Q and U plane, suggesting the existence of anisotropic mass flows in the wind of P Cygni (Underhill and Nemec 1989). This is also supported by last observations held in Byurakan observatory. We have found an increase of H alpha strength during a 5 days. Another LBV star η Car shows extended jet with the multigraded structure, which also reminds the shock waves (Hayes 1985).

We are grateful to Prof. A. Underhill for helpful discussion and Dr. O. Stahl for an opportunity given by him to analyze his spectra.

References

Castor, J. I.: 1987, in *Instabilities in Early Type Luminous Stars*, eds. H. J. G. L. M. Lamers and C. W. H. de Loore (Reidel, Dordrecht), p. 159.

Castor, J. I.: 1990, in NATO Workshop on *Stellar Atmospheres*, eds. Hubeny and Crivellari (Trieste).

Gorbatski, V. G.: 1958, *Soviet Astronomy* 35, 5.

Hayes, D. P.: 1985, *Astrophysical Journal* 258, 639.

Israelian, G. L. and de Groot, M.: 1991a, *Astrofis.* 34, 3.

Israelian, G. L. and de Groot, M.: 1991b, *Astron. Tsirk.* 1548, p. 15

Israelian, G. L. and Nicoghossian, A. G.: 1991, *Astron. Tsirk.*, in press .

Johnson, H. L., Wiisniewski, W. Z., and Fay, T. D.: 1978, *Rev. Mex. de Astron. Astrof.* vol 2, no 4, p. 273.

Klimishin, I. A.: 1984, *Shock Waves in Stellar Envelopes* (Moscow, 'Nayka'), p.75.

Lamers, H. J. G. L. M.: 1986, *Astronomy and Astrophysics* **159**, 90.

Lamers, H. J. G. L. M.: 1987, in *Instabilities in Luminous Early Type Stars*, eds. H. J. G. L. M. Lamers and C. W. H. de Loore (Reidel, Dordrecht), p.99.

Pouldrach, A. W. A. and Puls, J.: 1990, *Astronomy and Astrophysics* **237**, 409.

Stahl, O., Mandel, H., Szeifert, Th., Wolf, B. and Zhao, F.: 1990, *Astronomy and Astrophysics* submitted for publication.

Stahl, O. and Wolf, B.: 1982, *Astronomy and Astrophysics* **110**, 272.

Underhill, A. B. and Nemec, A. F. L.: 1989, *Astrophysical Journal* **345**, 1008.

Wolf, B., Appenzeller, I., and Stahl, O.: 1981, *Astronomy and Astrophysics* **103**, 94.

Wolf, B. and Stahl, O.: 1990, *Astronomy and Astrophysics* **235**, 340.

VARIABILITY IN ACTIVE GALACTIC NUCLEI

M. MATSUOKA

Institute of Physical and Chemical Research (RIKEN),
Hirosawa, Wako-shi, Saitama 351-01, Japan

Abstract. AGN (Active Galactic Nuclei) have their profound time variability over a wide range of time scales. Although many results of AGN variability have been provided from wide band wavelength observations, I would like to concentrate the recent problems concerning a nearby region of their central engine based on the X-ray observations which are most efficient to investigate this region. In this paper we will investigate mainly the result of Seyfert galaxies which would be generalized to other AGN.

1. Introduction

Various variabilities of AGN have given much information to investigate the structure of AGN. Prior to description of the variability in AGN I would like to introduce a recent progress of the structure of AGN which has been obtained from spectrum observations as well as time variabilities.

A schematic overall spectrum of AGN is given by Sanders *et al.* (1989) who have derived from their investigations of wide band continuum observations of radio to X-ray for more than 100 AGN. The radio is emitted by non-thermal high energy electrons produced often in relation to a jet-like non-thermal phenomenon in AGN. It has been well known that AGN have generally BLR (broad line region) and NLR (narrow line region) in optical emissions (e.g. Netzer *et al.* 1990). In 1980 era remarkable UV bump was discovered by IUE observations (Malkan and Sargent 1982). This UV bump is represented by a black body spectrum with a temperature of several tens times 10^3 K. On the other hand, it has been known that in X-ray region overall continuum spectra of AGN are represented by a power law function.

Recently an evidence of cold thick gas, so called "accretion or molecular torus" with H_2-molecule is pointed out by IR H_2-line ($\sim 2\mu$) observations (Kawara *et al.* 1989; Kawara *et al.* 1990). This accretion torus would reflect somewhat covering thick matter related to a discovery of polarized broad line emissions in the Seyfert 2 galaxy NGC1068 (Antonucci and Miller 1985) and very thick absorption gas in Seyfert 2 galaxies made by X-ray observations (Koyama *et al.* 1989; Awaki *et al.* 1990). Consequently, the unified model of Seyfert 1 & 2 has been emphasized (Miller and Goodrich 1990), but the unification of AGN can not be so simply accepted (Mulchaey *et al.* 1992).

On the other hand a theoretical investigation has been in favour of an accretion disk surrounded a black hole. So called "UV bump" is considered to be related to such an accretion disk (Malkan and Sargent 1982; Sun and Malkan 1989), while recently Ginga discovered an X-ray hard bump above 10 keV (Matsuoka *et al.* 1990; Pounds *et al.* 1990; Piro *et al.* 1990), which

Astrophysics and Space Science **210**: 245–257, 1993.
© 1993 *Kluwer Academic Publishers.*

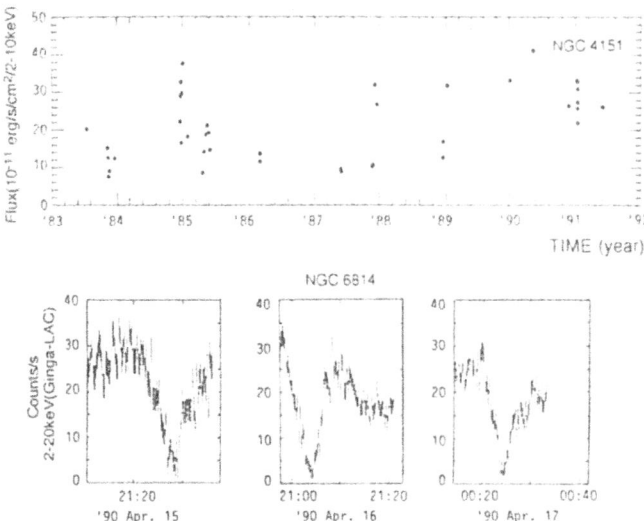

Fig. 1. X-ray light curves of NGC4151 (Yaqoob *et al.* 1992) and NGC6814 (Kunieda *et al.* 1991).

probably is related to this accretion disk or UV bump. Consequently it has been considered that general X-ray spectra of AGN exhibit a composite spectrum consisting of the direct and reflection components (Lightman and White 1988; Guilbert and Rees 1988).

In the following I would like to mention about model structures and X-ray variabilities near the AGN central engine. It is noted that most figures in this paper are modified from the original ones without essential change.

2. Variabilities of AGN

2.1. CONTINUUM VARIABILITY

Firstly I would like to show several snaps of X-ray variabilities of AGN. Fig. 1 are examples of the light curves in two Seyfert galaxies which show a long term X-ray variability of NGC4151 (Yaqoob *et al.* 1992) and a short term X-ray variability of NGC6814 (Kunieda *et al.* 1991). As shown in this figure the time variabilities are seen over several years to several tens sec. Fig. 2 shows the X-ray variabilities from the AGN which are observed by *EXOSAT*. A Scale of 10 hour-time is shown in each figure (Pounds and McHardy 1990; McHardy 1989).

When we performed a standard power spectrum density (PSD) analysis, we obtained that these AGN gave a significant power suggesting a rapid variability as shown in Fig. 3. The PSD is extended to a frequency of 10^{-2} Hz or higher. Generally we can describe the X-ray PSD of AGN as a function

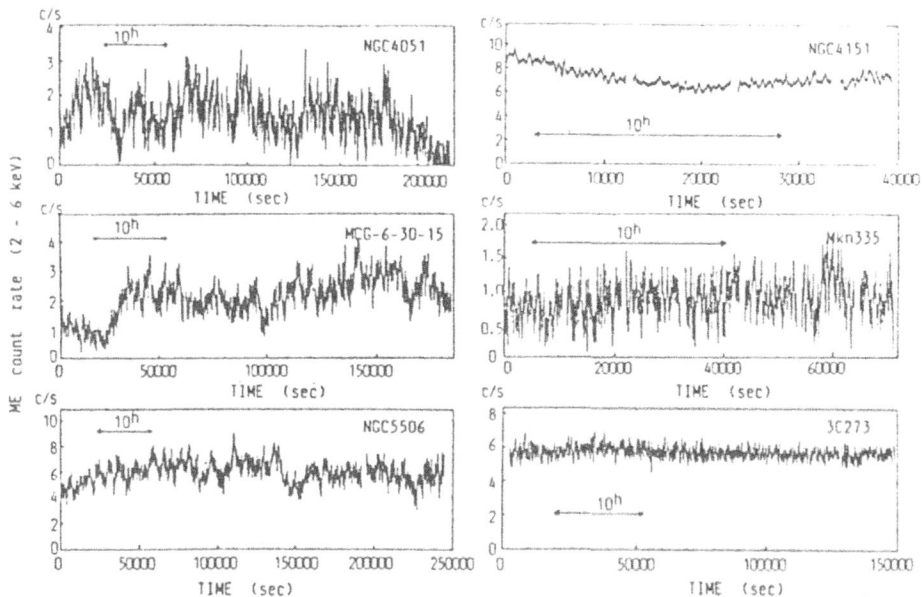

Fig. 2. Examples of X-ray light curves observed by EXOSAT (Pounds and McHardy 1988).

of $f^{-\alpha}$ for frequency f. The value of α is about 1~2 for high frequencies and then decreases to be flat towards low frequencies.

Mathematical ways of achieving this PSD are considered (Begelman and De Kool 1991); e.g. (1) one way in which a Poisson process can produce f^{-1} noise is to have a pulse shape with a very slow decay, (2) a second way to obtain a steeper power law is to superpose white noise spectra with a range of normalizations and high-frequency cut offs, and (3) a third way is to assume that there are correlations between the pulse height and either the decay time of the pulse or the mean rate of pulses. In any case the PSD of AGN X-ray intensity would reflect a chaotic central phenomenon of AGN. A simple interpretation of producing spectrally invariant intensity variations is released of stored energy in electron pairs or magnetic fields by shocks, but neither simple shot noise models nor models dominated by a simple cooling or heating time scale. However, the real physical activity in this most inside region is open to question.

2.2. IRON LINE VARIABILITY

Next we will consider some outside region away from the central engine. The K X-ray iron line from Seyfert galaxies, one of major groups of AGN, generally has been known to be produced by the fluorescent mechanism in the cold gas surrounding the central engine (e.g. Matsuoka *et al.* 1986).

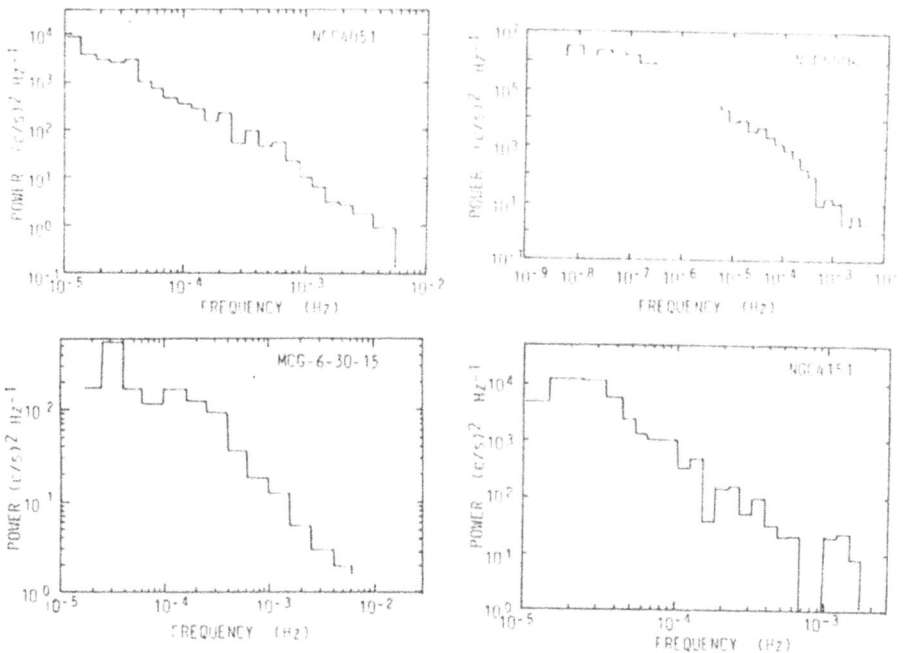

Fig. 3.   Power density spectra of X-ray intensities of AGN (Pounds and McHardy 1988),
where the power density spectrum of MCG-6-30-15 is obtained from Ginga data by Fiore
et al. (private communication, 1991).

Recently Ginga has discussed on exciting phenomenon that the ion K
X-ray flux of NGC6814 correlates with the continuum X-ray (Kunieda et al.
1990; Kunieda et al. 1991). The continuum and iron line intensities observed
by Ginga are shown in Fig. 4. This suggests that the fluorescence material
should be located at somewhat nearby region from the central engine. Thus
the cross correlation analysis has been done for the X-ray line and continuum
intensities (Done et al. 1992). Fig. 5 shows a time lag of iron line against the
continuum (3–5.5 keV) for NGC6814 (Done et al. 1992). The lag time was
about less than 200 sec. This suggests that the fluorescent matter is located
at the region less than 10^{13} cm from the central engine. Such matter would
be highly ionized by strong irradiation from the central source (Hayakawa
1991; Yamauchi et al. 1992).

2.3. REFLECTION SPECTRUM

I would like to mention the reprocessed region with very thick gas (dense
matter) derived from recent spectral analyses of X-rays. Prior to Ginga
observations, X-ray spectra of AGN had been described by the model of a
single power law plus an iron line emission (e.g. Mushotzky 1984; Turner
and Pounds 1989). However, detailed observations of the X-ray spectra of

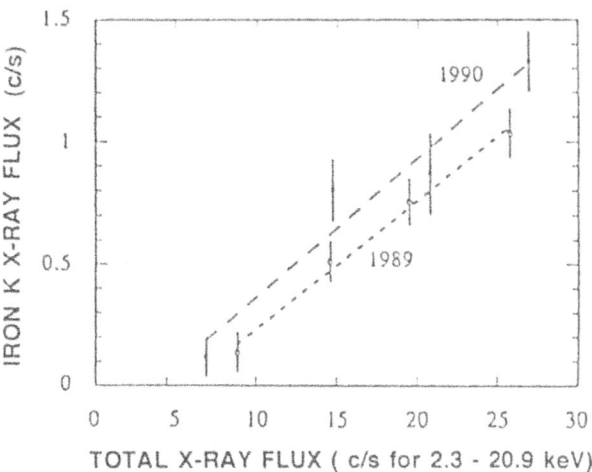

Fig. 4.   The X-ray continuum and iron line intensities of NGC6814 (Kunieda *et al.* 1991).

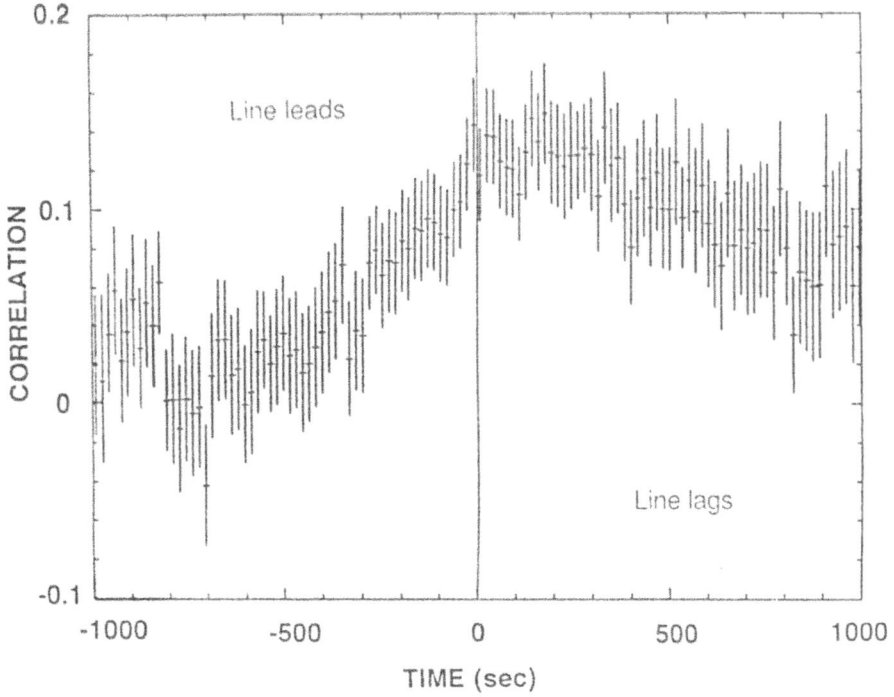

Fig. 5.   The cross correlation of the extracted iron K X-ray and 3–5.5 keV flux (Done *et al.* 1992).

Seyfert galaxies by Ginga have ruled out this model generally (Matsuoka *et al.* 1990; Piro *et al.* 1990). As aforementioned, some X-ray spectra require the reflection component in addition to a single power law plus an emission

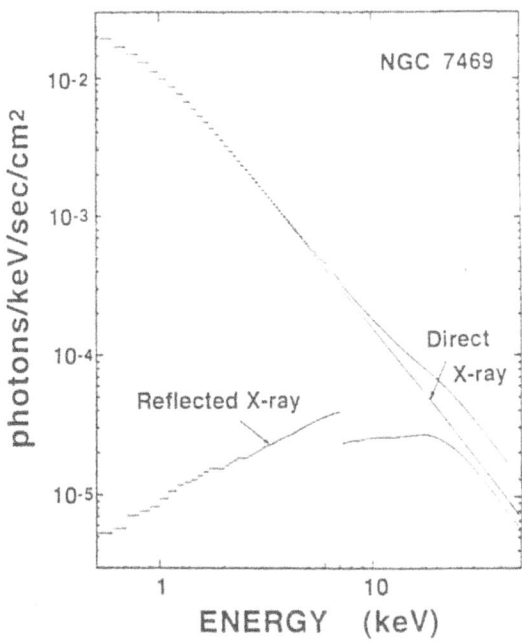

Fig. 6.   A reflection model of AGN spectrum; an example of NGC7469.

line.

Detailed observations of the X-ray spectra of AGN have proved that
the X-ray spectra of many AGN are well fitted to so called the "reflection
model", which consists of direct and reflection components as shown in Fig.
6. This is a leading model of X-ray spectra of AGN, and then this reflection
feature would be related to the UV bump; that is, the X-rays from the
central engine are reflected by some part of accretion disk. Thus so far the
X-ray spectra of a number of AGN observed by Ginga have been well fitted
to the reflection model (Seyfert galaxies: Matsuoka *et al.* 1990; Pounds *et
al.* 1990; Piro *et al.* 1990; Nandra *et al.* 1991, QSO's: Williams *et al.* 1992).

2.4. VARIABILITY OF SPECTRAL INDEX

The power law index in AGN X-ray spectra has been called as a "canonical"
index (Mushotzky 1984). If the spectra are fitted to a single power law model,
the canonical index is represented to be about 1.7 in photon index. But some
AGN X-ray spectra show that the power law index depends on the X-ray
intensity, if the spectra are fitted to a simple power law model (Halpern 1985;
Perola *et al.* 1986; Turner 1987; Matsuoka *et al.* 1990). However, actually
the X-ray spectra of many Seyfert galaxies are well fitted to the reflection
model. If the reflection model is applied to X-ray spectra of some AGN, it

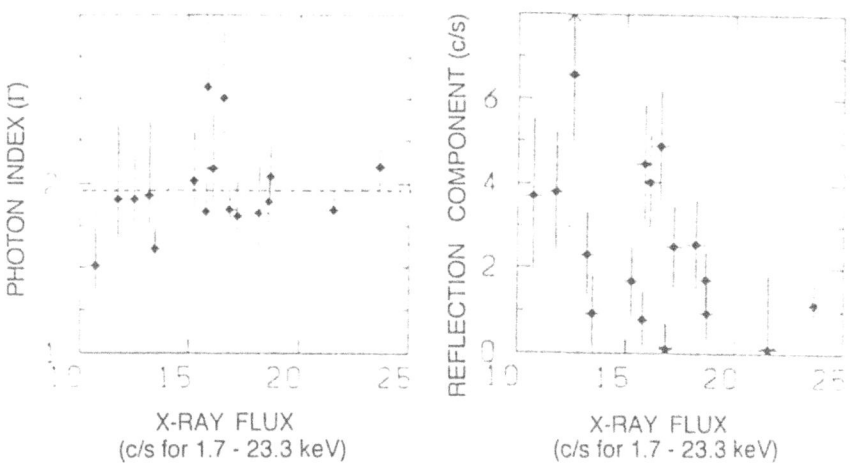

Fig. 7.   The photon-index and reflection component for MCG-6-30-15 (Fiore *et al.* 1992).

seems that the photon index is not variable for the X-ray intensity, while
the reflection component is variable for the X-ray intensity as shown in Fig.
7 (Fiore *et al.* 1992). If we use a single power law model for NGC5548, the
photon index is variable for the X-ray flux, while the photon index is constant
for the reflection model (Nandra *et al.* 1991). This is not common to AGN
X-ray spectra. In case of NGC4151 the results of Ginga observations indicate
the variation in the spectral index (Yaqoob and Warwick 1991; Yaqoob *et
al.* 1992), in which the spectral softening as the source brightens is due to
a correlation of the spectral index with 2–10 keV flux as shown in Fig. 8.
Furthermore, it is noted that the X-ray spectrum of NGC4151 is well fitted
by a partial covering model, but rules out the generally favoured reflection
model (Yamauchi *et al.* 1992).

   The flux vs. index correlation in NGC4151 is interpreted in terms of
the current "pair model" of X-ray emission in AGN which involves inverse
Compton scattering of UV photons on relativistic electrons (Yaqoob 1992;
also see Lightman and Zdziarski 1987).

2.5. SOFT EXCESS

Soft X-ray spectra of Seyfert galaxies are complex in form and that contributions from several different emission and absorption components make
up the soft spectra (Turner *et al.* 1991; Urry *et al.* 1989). Recent AGN survey by *ROSAT* shows that the soft excess below about 1 keV is common
to AGN spectra (Brinkmann 1991). Past observations suggest that at least
four types of soft excess exist in Seyfert galaxies as follows (Turner *et al.*
1991).

   (1) A steep, ultrasoft, and rapidly variable excess, which may be associ-

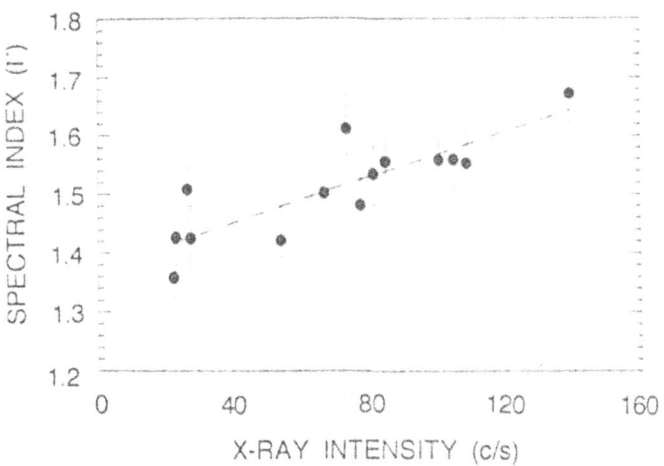

Fig. 8.   The photo-index - intensity relation of NGC4151 (Yaqoob *et al.* 1992).

ated with the inner region of an accretion disk (see Wilkes and Elvis 1987, Turner and Pounds 1989). This component has been characterized as a low-temperature ($T \sim 6$–10 eV) black body or thermal bremsstrahlung emission or a a steep second-power law. Most AGN spectra observed by *ROSAT* seem to be addressed to this type.

(2) Excess soft emission due to leakage of the hard X-ray continuum through a patchy absorber, i.e., the "partially covered" sources. NGC4151 is a typical example of this type (Holt *et al.* 1980; Yamauchi *et al.* 1992).

(3) An emission-line feature typically centered around 0.8 keV, whose possible explanation is a blend of soft X-ray lines.

(4) An extended soft excess component seen in *Einstein HRI* observations of NGC4151 (Elvis *et al.* 1983), and NGC2992 and NGC1566 (Fassnacht *et al.* 1990). This type of soft excess would not be variable in a short time scale.

Detailed observations of Seyfert galaxy NGC5506 were made by *Ginga* (Bond *et al.* 1992). Fig. 9 shows a spectral variability of NGC5506 over a full observation period of July 6–10, 1991 and on July 9, 1991. The soft (1.2–3 keV), intermediate (3–10 keV), and hard (10–28 keV) X-ray intensities are plotted for each orbit along with the softness and hardness ratio. This X-ray spectrum is not well fitted to a "reflection" model, remaining a soft excess. After investigation of several spectral models, we have achieved the composite model consisting of reflection and partial covering components in addition to the direct one as shown in Fig. 10. Thus this is a generalized model, in which the partial covering model and the reflection model are combined.

The spectra of NGC4151 as well as NGC5506 require the partial covering components. The results of a covering fraction derived from the spectra of these AGN give that the covering fractions normally have a value of 80–

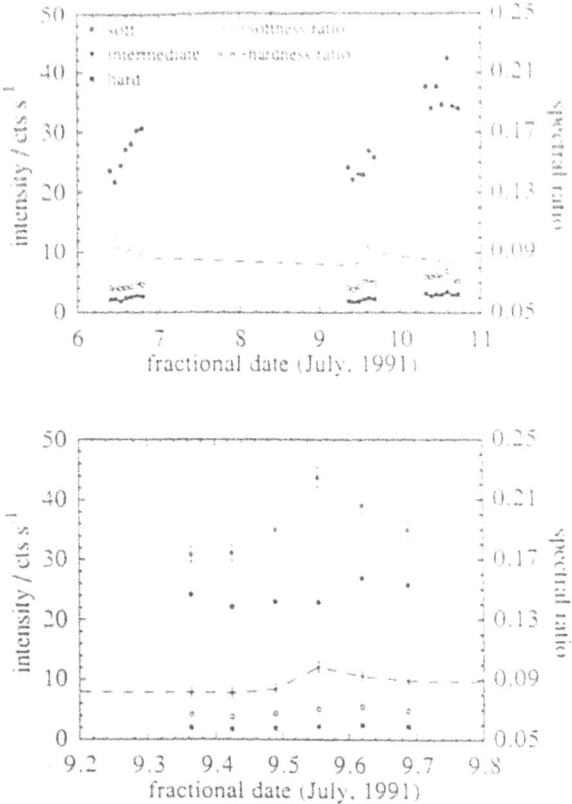

Fig. 9. A spectral variability of NGC5506 (Bond *et al.* 1992). The soft (1.2–3 keV), intermediate (3–10 keV), and hard (10–28 keV) X-ray intensities are plotted for each orbit along with the softness and hardness ratios.

90%, but one time declined to around 60% and 20% for NGC5506 and NGC4151, respectively, as shown in Fig. 11. The existence of a "canonical" value of the covering fraction would rule out the cloud interpretation of the partial covering model since it is hard to realized a situation where a cloud cover can re-configure itself to a normal state after some distribution such as a temporary break in the clouds (Bond *et al.* 1992; Yaqoob *et al.* 1992). Anyway very dense matter in the central engine of AGN have an important effect on the emission and absorption of the radiation (Celotti *et al.* 1992).

2.6. RAPID VARIABILITIES OF TWO AGN

Ginga detected a surprising flare from the QSO PKS0558-504, whose X-ray flux increased by 67% in the space of only 3 minutes (Remillard *et al.* 1991). This flare is one of the most dramatic examples of rapid variabilities from AGN. The X-ray rise of PKS0558-504 implies $dL/dt \sim 3.2 \times 10^{42}$ erg/sec^2.

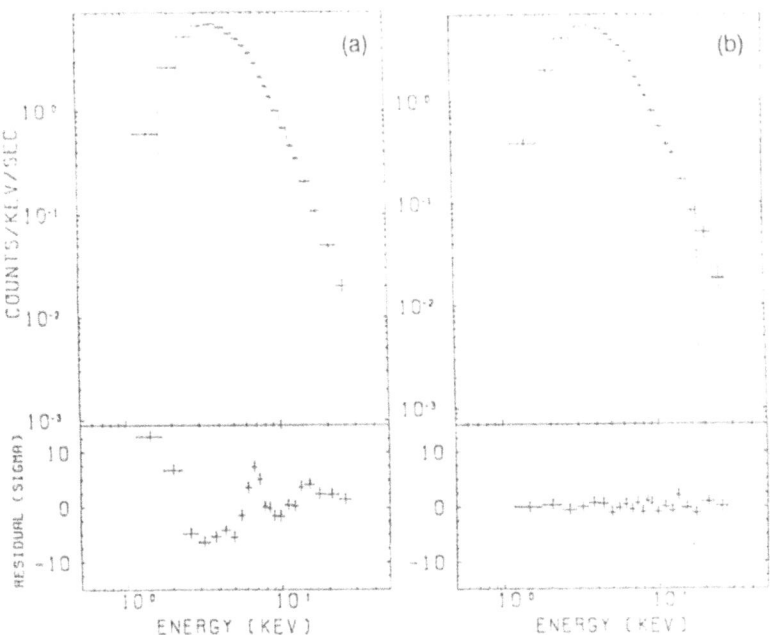

Fig. 10. The NGC5506 spectral fits are shown for two cases of a single power law model and a composite model consisting of "reflection" and "partial covering" models (Bond *et al.* 1992).

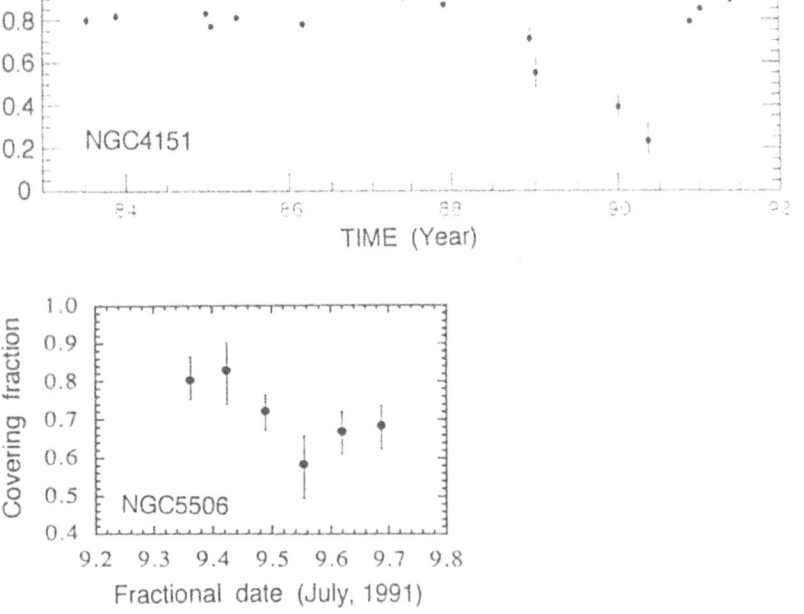

Fig. 11. Time variabilities of covering fractions for NGC5506 (Bond *et al.* 1992) and NGC4151 (Yaqoob *et al.* 1992).

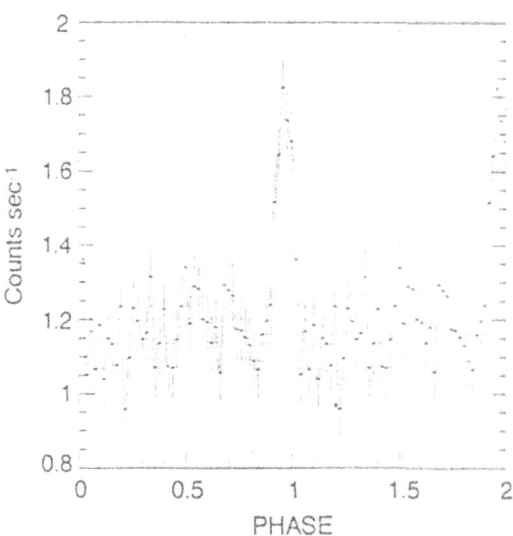

Fig. 12.   A light curve folded on the period of 12,200 sec for NGC6814 X-ray light curve (Fiore *et al.* 1991).

Ginga also detected that another QSO 3C279 exhibited a 20% increase in 45 minutes giving a change in luminosity of 2–10 keV, $dL/dt > 2.9 \times 10^{42}$ erg/sec^2 (Makino *et al.* 1990).

In the interpretation of AGN variability time scales an efficiency limit (η) is derived for the rate of conversion of gravitational potential energy into X-ray emission, assuming that the photons must escape through the accreting matter, and that the photon opacity is dominated by Thomson scattering. The derived limit is $dL/dt < 2 \times 10^{42} \eta$ erg/sec^2 in a spherical symmetric model (Cavallo and Rees 1978; Fabian 1979). It is unlikely that η exceeds 0.1. The results of both PKS0558-504 and 3C279 suggest that the apparent luminosity must be enhanced by relativistic beaming (Remillard *et al.* 1991). Such as rapid variabilities in X-ray luminosity have been attained by BL Lac objects (Barr and Mushotzky 1986).

Further surprising result of a periodicity of AGN is discovered in Seyfert galaxy, NGC6814, on a time-scale of approximate 12,000 sec during long observations by *EXOSAT* (Mittaz and Branduardi-Raymont 1989; Fiore *et al.* 1992) as a light curve folded on the period of 12,200 sec shown in Fig. 12. It has been interpreted that a bright spot is moving surrounding the central object (Abramowicz *et al.* 1992), or some peculiar dynamic oscillation of the disk (Honma *et al.* 1992) and so on, but the further explanation is open to question.

3. Conclusion

Recent results of X-ray time variabilities as well as X-ray spectra have made a great progress for a structure of a nearby region of their central engine. The reprocessed region as well as direct X-ray region has an important role in AGN. Thus we could inspect most nearest region in AGN by X-ray observations. For further future observations with *ASTRO-D* and *SPECTRUM-X* it is important to investigate the variability of X-ray intensity in connection with the X-ray spectral structure. Especially, the observation of the energy shift of X-ray lines is one of the most important subjects. Systematic investigation of intrinsic absorption column densities and soft excesses in X-rays will also provide a great progress for AGN.

References

Abramowicz, M. A., Bao, G., Fiore, F. *et al.*: 1992, *Nature*, submitted.
Antonucci, R. R. J. and Miller, J. S.: 1985, *Astrophysical Journal* **297**, 621.
Awaki, H. *et al.*: 1990, *Nature* **34**, 544.
Barr, P. and Mushotzky, R. F.: 1986, *Nature* **320**, 421.
Begelman, M. C. and De Kool, M. 1991, *Variability of active galactic nuclei*, eds. H. R. Miller and P. J. Wiita (Cambridge Univ. Press.), p. 198.
Bond, I. A., Matsuoka, M. and Yamauchi, M.: 1992, *Astrophysical Journal*, submitted.
Brinkmann, W.: 1991, MPE conference
Cavalo, G. and Rees, M. J.: 1978, *Monthly Notices of the RAS* **183**, 359.
Celloti, A., Fabian, A. C. and Rees, M. J.: 1992, *Monthly Notices of the RAS* **255**, 419.
Done, C., Madejski, G. M., Mushotzky, R. F., Turner, T. J. Koyama, K., and Kunieda, H. 1992, *Frontiers of X-ray Astronomy*, eds. Y. Tanaka and K. Koyama (Univ. Academy Press), p. 525.
Elvis, M., Briel, U. G. and Henry, J. P.: 1983, *Astrophysical Journal* **268**, 105.
Fabian, A. C.: 1979, *Proc. Roy. Soc.* **336**, 449.
Fabian, A. C., George, I., Miyoshi, S. and Rees, M. J.: 1990, *Monthly Notices of the RAS* **242**, 14.
Fassnacht, C., Elvis, M., Wilson, A. S. and Briel, U.: 1990, *Astrophysical Journal* **361**, 459.
Fiore, F., Massaro, E. and Barone, P.: 1991, *Astronomy and Astrophysics*, submitted.
Fiore, F., Perola, G. C., Matsuoka, M., Yamauchi, M. and Piro, L. :1991, *Variability of Active Galactic Nuclei*, eds. H. R. Miller and P. J. Wiita (Cambridge Univ. Press), p. 273.
Fiore, F., Perola, G. C., Matsuoka, M., Yamauchi, M. and Piro, L. : 1992, *Astronomy and Astrophysics*, to be appeared.
Guilbert, P. W. and Rees, M. J.: 1988, *Monthly Notices of the RAS* **233**, 475.
Halpern, J. P.: 1985, *Astrophysical Journal* **290**, 130.
Hayakawa, S.: 1991, *Nature* **351**, 214.
Holt, S. S., Mushotzky, R. F., Becker, R. H., Boldt, E. A., Serlemitsos, P. J., Szymkowiak, A. E. and White, N. E.: 1980, *Astrophysical Journal, Letters to the Editor* **241**, L13.
Honma, F., Matsumoto, R. and Kato, S.: 1992, preprint.
Lighman, A. J. and White, T. R.: 1988, *Astrophysical Journal* **335**, 57.
Lightman, A. P. and Zdziarski, A. A.: 1987, *Astrophysical Journal* **319**, 643.
Kawara, K., Nishida, M. and Gregory, B.: 1989, *Nature* **341**, 27.
Kawara, K., Nishida, M. and Gregory, B.: 1990, *Astrophysical Journal* **352**, 433.
Koyama, K., Inoue, H., Tanaka, Y., Ohashi, T. and Matsuoka, M.: 1989, *Publications of the ASJ* **41**, 731.

Kuieda, H. *et al.*: 1990, *Nature* **235**, 786.

Kunieda, H. *et al.*: 1991, *Iron Line Diagnostics in X-ray Sources*, eds. A. Treves, G. C. Perola and L. Stella (Spring-Verlarg), p. 241.

Malkan, M. A. and Sargent, W. L. W.: 1982, *Astrophysical Journal* **254**, 22.

Makino, F. *et al.*: 1990, *Astrophysical Journal, Letters to the Editor* **347**, L9.

Matsuoka, M., Ikegami, T., Inoue, H. and Koyama, K.: 1986, *Publications of the ASJ* **38**, 285.

Matauoka, M., Piro, L., Yamauchi, M. and Murakami, T.: 1990, *Astrophysical Journal* **361**, 440.

McHardy, J. M. 1989, *23rd ESLAB Symposium X-ray Astronomy. 2. AGN and the X-Ray Background*, eds. by J. Hunt and B. Battrick (ESA-France), p.111.

Mittaz, J. P. D. and Branduardi-Raymont, G.: 1989, *Monthly Notices of the RAS* **238**, 1029.

Mulchaey, J. S., Mushotzky, R. F. and Weaver, K. A.: 1992, *Astrophysical JournalLetters*, in press.

Mushotzky, R. F.: 1984, *Adv. Space Res.* **3**, No. 10-12, p.153.

Nandra, K., Pounds. K. A. *et al.*: 1991, *Monthly Notices of the RAS* **248**, 760.

Netzer, H.: 1990, "Active Galactic Nuclei" eds. by R. D. Blaudford, H. Netzer and L. Woltjer (Springer-Verlag), p.57.

Piro, L. Yamauchi, M. and Matsuoka, M.: 1990, *Astrophysical Journal, Letters to the Editor* **360**, L35.

Pounds. K. A. and McHardy, I. M.: 1988, *Physics of Neutron Stars and Black Holes*, Ed. by Y. Tanaka (Universal Academy Press. Inc.), p.285.

Pounds, K. A., Nandra, K., Stewart, G. C., George, I. M. and Fabian, A. C.: 1990, *Nature* **344**, 132.

Remillard, R. A. *et al.*: 1991, *Nature* **350**, 589.

Sanders, D. B., Phinnery, E. S., Neugebauer, G., Soifer, B. T. and Mathews, K.: 1989, *Astrophysical Journal* **347**, 29.

Sun, H-W. and Malkan, M. A.: 1989, *Astrophysical Journal* **346**, 68.

Turner, T. J., Weaver, K. A., Mushotzky, R. F., Holt, S. S. and Madejski, G. M.: 1991, *Astrophysical Journal* **381**, 85.

Turner, T. J. and Pounds, K. A.: 1989, *Monthly Notices of the RAS* **240**, 833.

Urry, C. M., Arnaud, K. A., Edelson, R. A., Kruper. J. S. and Mushotzky. R. F.: 1989. *Proceedings of the 23rd ESLAB Symp.*, eds. J. Hunt and B. Battrick (ESA publications), p. 789.

Wilkes, B. and Elvis, M.: 1987, *Astrophysical Journal* **323**, 243.

Williams, O. R., Turner, M. J. L., Stewart, G. C. *et al.*: 1992, *Astrophysical Journal*, in press.

Yamauchi, M., Matsuoka, M., Kawai, N. and Yoshida, A.: 1991, *Astrophysical Journal*, in press.

Yaqoob, T.: 1992, *Monthly Notices of the RAS* in press (March 1992).

Yaqoob, T. *et al.*: 1992, in preparation.

Yaqoob, T. and Warwick, R. S.: 1991, *Monthly Notices of the RAS* **248**, 773.

III. MODELS

FOURIER ANALYSIS OF THE HYDRODYNAMIC
LIMIT-CYCLE MODELS OF PULSATING STARS

YU. A. FADEYEV
Institute for Astronomy of the Russian Academy of Sciences
and
Institute for Mathematics of the Vienna University

Abstract. The pulsation motions of the limit-cycle model can be described as a superposition of the Fourier harmonics, in the adiabatic layers each harmonics being identified with the corresponding standing wave. Near the resonance $\Pi_0/\Pi_l = k$ the harmonics of order k is also identified with the overtone of order l. The spectra of the oscillatory moment of inertia obey to the power dependence on the Fourier harmonics order k. In cepheids with periods shorter than 9 days the bump is due to the wave packet generated by the second overtone, whereas at periods longer than 10 days the bump feature is due to the traveling pulse reflected off the stellar core.

1. Introduction

The Hertzsprung progression of Classical cepheids is one of the most conspicuous features which is unmistakably reproduced in hydrodynamic calculations on the theoretical light and radial velocity curves. At the same time the nature of the secondary bump is still unknown because the long competition between two alternative hypotheses on the nature of the secondary bump has not been ended yet. The first of these hypotheses considers the bump feature as the traveling pulse reflected off the stellar core (Christy, 1968; 1975), whereas the second one proposed by Simon and Schmidt (1976) assumes that the bump is due to the resonance between the second overtone and fundamental mode. The attempts to reconcile both these hypotheses also did not reach their logical completion (Whitney, 1983; Aikawa and Whitney, 1985). Below we try to shed the light onto the problem of the Hertzsprung progression using the fact of the strict repetition of the pulsation motions in Classical cepheids. This allows us to calculate the Fourier coefficients for the main hydrodynamic variables at each mass zone of the limit cycle model. Together with the problem of the Hertzsprung progression we briefly discuss also some other features of pulsating stars. In more detail some preliminary results of this study are given by Fadeyev and Muthsam (1992).

2. Radial Properties of Fourier Harmonics

The most conspicuous feature of the Fourier harmonics of hydrodynamic variables is that their radial dependence are very similar to the eigenfunctions of the linear equation of stellar pulsation. For example, shown on the upper panel of Fig. 1 are the normalized amplitudes of the three lowest

Astrophysics and Space Science **210**: 261–267, 1993.
© 1993 *Kluwer Academic Publishers.*

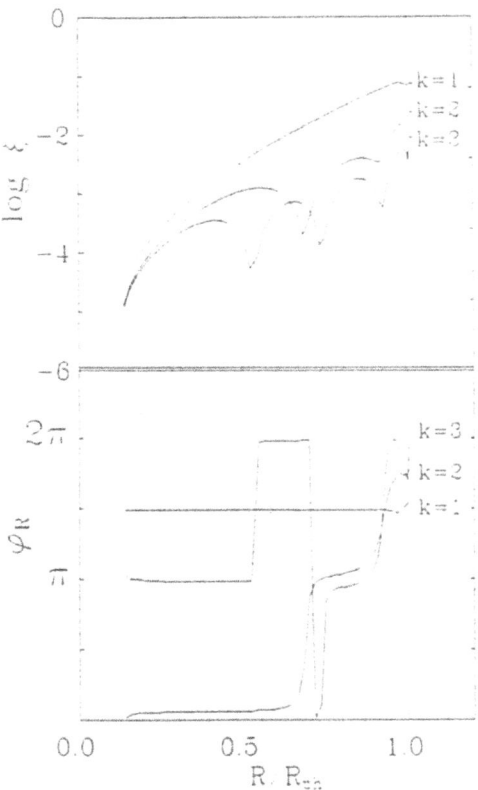

Fig. 1.   The normalized amplitudes (upper panel) and the phases (lower panel) of the
Fourier harmonics of order $1 \leq k \leq 3$ as a function of the radial distance.

Fourier harmonics of the radial displacement. This similarity is most prominent in the adiabatically pulsating layers where the difference between the
adjacent maxima and minima is largest. Moreover, the phase of the Fourier
harmonics is constant between two adjacent amplitude minima, whereas at
the minimum the phase abruptly changes by π radian. Such a behaviour of
the Fourier harmonics is recognized up to the order of $k = 8$. For higher
Fourier harmonics this conclusion becomes uncertain due to the limited
space resolution of the hydrodynamic models.

So, in the adiabatic layers the pulsation motions can be represented as
a superposition of the standing waves. However in the radiative damping
region as well in the outer layers where effects of nonadiabaticity become
perceptible the amplitude minima and the corresponding phase changes become shallower and smoother, respectively. This implies that in the nonadiabatic layers the pulsation motions cannot be described in the terms of the
pure standing waves due to the presence of the progressive wave component.

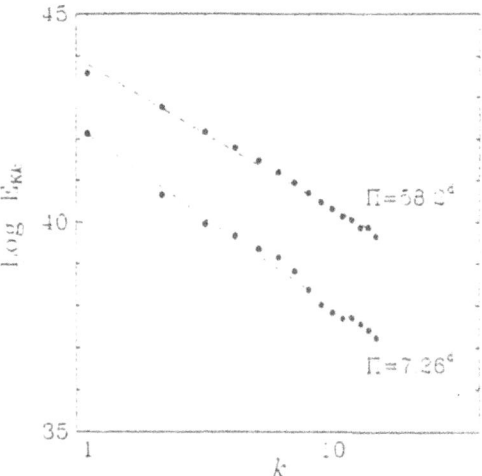

Fig. 2.   The spectra of the kinetic energy E_{Kk} of the models of Classical cepheids with pulsation periods 7.26 and 58.2 days.

Comparison with the eigenfunctions of the linear equation of stellar pulsation shows that the Fourier harmonics of order k can be identified with the overtone of order l if the period ratio is $\Pi_0/\Pi_l = k$. When the period ratio is not integer and is in the range from k to $k + 1$, the harmonics of order k reveals the features typical for both eigenfunctions of order l and $l - 1$, respectively. When the period ratio becomes closer to one of these integer values, the properties of one of the overtones escape, whereas the properties of another overtone enhance.

Using the Fourier coefficients of radius and velocity we can calculate the oscillatory moment of inertia J_k and kinetic energy E_{Kk} for each Fourier harmonics of order k. Another conspicuous feature of the Fourier harmonics is that in all hydrodynamic limit-cycle models the pulsation spectra of J_k and E_{Kk} obey to the power law: $J_k = J_1 k^{-\nu_J}$ and $E_{Kk} = E_{K1} k^{-\nu_E}$ (see Fig. 2). Increase of nonadiabaticity in the envelope is accompanied by the decreasing slope of the spectrum, so that there is a correlation between the spectrum index ν_J (or ν_E) and the parameter related to nonadiabaticity (e.g. the growth rate of pulsation instability or mass to radius ratio). This implies that increase of nonadiabaticity is accompanied by the redistribution of the pulsation energy among higher harmonics.

3. Wave Packets

The compression wave propagating inwardly from the He^+ zone is easily traced using the temporal velocity dependence (see, e.g. Christy, 1975), whereas the outwardly propagating pulse reflected off the core is often lost

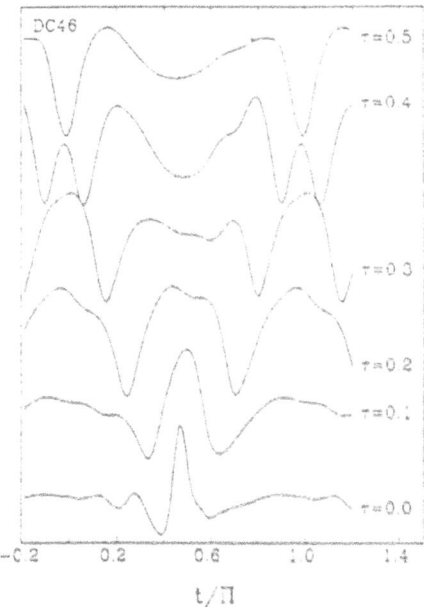

Fig. 3.   The temporal dependence of the wave packets consisted of Fourier harmonics of order $2 \leq k \leq m = 15$. The dependencies are arbitrarily normalized, τ is acoustic coordinate of the layer.

in the helium ionizing region due to its relatively small amplitude. So, it is instructive to remove the influence of the high amplitude oscillations of the low-order harmonics and to consider the propagation of the wave packets consisted of Fourier harmonics of order $k_L \leq k \leq m$.

Shown in Fig. 3 are the temporal dependence of the velocity wave packets consisted of Fourier harmonics of order $k \geq k_L = 2$. For the sake of the graphic representation all these dependence are arbitrarily normalized. As is seen, the wave packets reveal the presence of the both inwardly and outwardly propagating pulses. However, though this method of the pulse tracing has a certain advantage, the procedure nevertheless remains rather cumbersome. In order to avoid such shortcoming, we considered the acoustic coordinates of most prominent maxima and minima of the wave packets. Fig. 4 shows the typical acoustic coordinate - time diagram for the cepheid model with the pulsation period of 8.5 day. On this diagram the minima and maxima of the velocity wave packets are shown as filled and open circles, respectively, the larger circles corresponding to the most prominent maxima and minima. So, the traveling pulse can be traced as a sequence of minima or maxima located along the characteristics ($|d\tau/dt| = 1$).

As is seen from Fig. 4, the inwardly propagating pulse is created in the instability excitation region at maximum compression (the phase $t/\Pi \simeq -0.1$). Below the instability excitation region the pulse propagates along the charac-

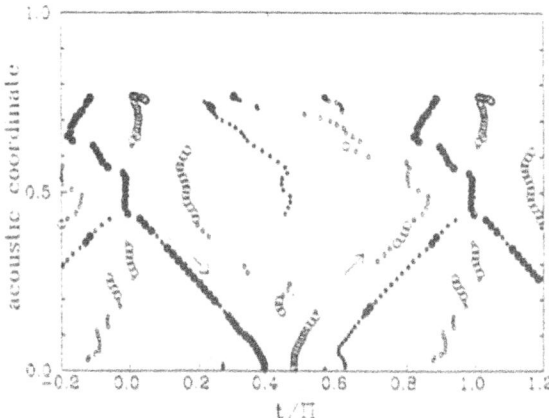

Fig. 4.   The acoustic coordinate - phase diagram for the Classical cepheid model with period of 8.5 day. The filled and open circles show the minima and maxima of velocity, respectively. The largest circle corresponds to the most prominent maximum or minimum of the wave packet.

teristics and his trajectory is broken only at the acoustic midpoint ($\tau = 0.5$) due to interaction with the outwardly propagating pulse reflected off the stellar core. The reflected pulse also propagates along the characteristics but as is seen from Fig. 4, this pulse cannot be responsible for the secondary bump since his arrival at the photosphere nearly coincides with the main maximum.

The secondary bump appears due to the wave packet generated in the ionizing region at phases from 0.5 to 0.7 (see Fig. 4). The maximum of this wave packet coincides with the phase of the maximum of the second Fourier harmonics. Calculations show that the bump location changes in phase with the maximum of the second Fourier harmonics of the velocity. This implies that the secondary bump at periods shorter than 9 days is generated by the second Fourier harmonics and the nature of the Hertzsprung progression is tightly related to the second Fourier harmonics identified with the second overtone. Shown in Fig. 5 are the radial dependence of the phase difference between the fundamental mode and the second Fourier harmonics. This phase difference becomes close to $\pi/2$ radian near the resonance center. It is interesting to compare the change of the phase difference $\varphi_{U1} - \varphi_{U2}$ with the corresponding dependence of the secondary bump on the pulsation period. According to Fadeyev (1982), in the period range from 6 to 9 days the phase change of the secondary bump is $d\varphi/dlg\Pi = -1.66$. According to our models, the corresponding phase change is $d\varphi/dlg\Pi = -1.54$. There are hopes that the agreement will be improved when the more extended grid of the models is considered.

At periods longer than 10 days the secondary bump is due to the out-

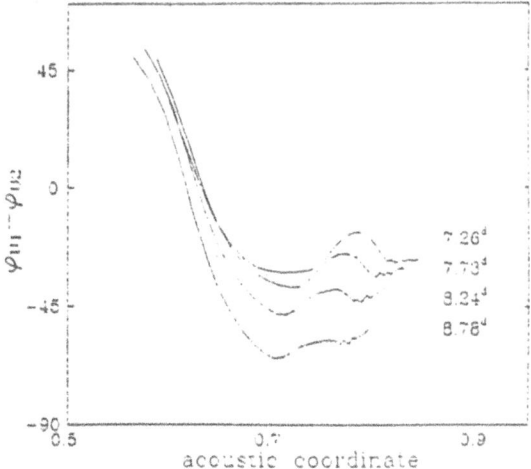

Fig. 5.   The phase difference between the fundamental mode and second Fourier harmonics as a function of acoustic coordinate. The numbers near dependencies show the corresponding pulsation period.

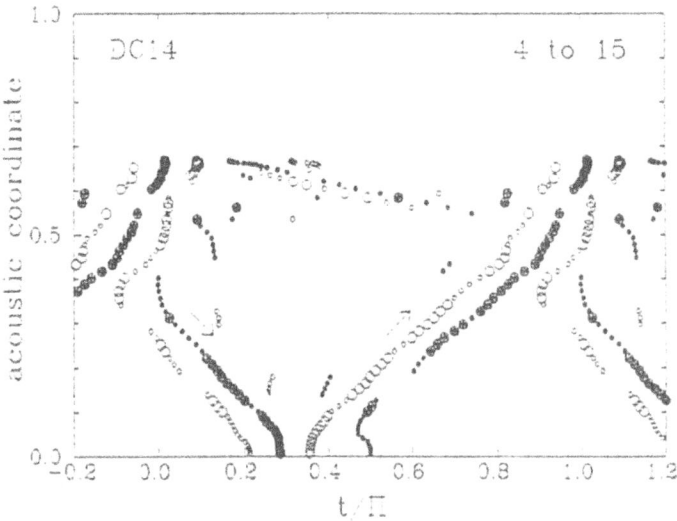

Fig. 6.   The acoustic coordinate - time diagram for the Classical cepheid model with period of 17.4 day.

wardly propagating pulse reflected off the core (see Fig. 6). The energy of this pulse is distributed among the harmonics of order $k > 2$, i.e. the bump observed before the main maximum is not related to the second Fourier harmonics. As in observed light and radial velocity curves, the phase of the traveling pulse appearance does not change with the period.

4. Concluding Remarks

Now we can certainly assert that transfer of the pulsation energy from the instability excitation region into the inner layers of the envelope is due to the traveling pulse created as a superposition of the standing waves. In population II cepheids the pulsation spectra are not so steep as in Classical cepheids so that the kinetic energy of the second Fourier harmonics identified with the first overtone is nearly a quarter of the kinetic energy of the fundamental mode. This implies that the second Fourier harmonics might be responsible for the alternating oscillations in RV Tau stars.

Acknowledgements

This work was supported in part by the Hochschüljübiläumsstiftung der Gemeinde Wien and by the Jubiläumsfonds der österreichisches Nationalbank, proj. 3376.

References

Aikawa, T., and Whitney, C. A.: 1985, *Astrophysical Journal* **296**, 187.
Christy, R. F.: 1968, *Quarterly Journal of the RAS* **9**. 13.
Christy, R. F.: 1975, *Mem. Soc. Roy. Sci. Liege*, 6 ser., tome **VIII**, 173.
Fadeyev, Yu. A.: 1982, Ph. D. (Moscow).
Fadeyev, Yu. A., and Muthsam, H.: 1992, *Astronomy and Astrophysics*, in press.
Simon, N. R., and Schmidt, E. G.: 1976, *Astrophysical Journal* **205**, 162.
Whitney, C. A.: 1983, *Astrophysical Journal* **274**, 830.

BIFURCATION IN HYDRODYNAMIC MODELS OF
STELLAR PULSATION

T. AIKAWA

Tohoku Gakuin University, Faculty of Liberal Arts,
Izumi-ku, Sendai 981-31, Japan

Abstract. Phenomena of bifurcation in hydrodynamic stellar models of radial pulsation
are reviewed. By changing control parameters of models, we can see qualitatively different
pulsation behaviors in hydrodynamic models with transitions due to various types of
bifurcation.

In weakly dissipative models (classical Cepheids). the bifurcation is induced by modal
resonances. Two types of the modal resonances found in models are discussed: The higher-
harmonic resonances of the second overtone mode in the fundamental mode pulsator and
of the fourth overtone mode in the first overtone pulsator are relevant to observations. The
subharmonic resonance between the fundamental and first overtone modes is confirmed in
classical Cepheid models.

In strongly dissipative models (less-massive supergiant stars), the bifurcation of nonlinear pulsation is induced by the hydrodynamics of ionization zones as well as modal
resonances. The sequence of the bifurcation sometimes leads to chaotic behaviors in nonlinear pulsation. The transition routes from regular to the chaotic pulsations found in
models are discussed with respect to the theory of chaos in simple dynamical systems:
The cascade of period-doubling bifurcation is confirmed to cause chaotic pulsation in W
Virginis models. For models of higher luminosity, the tangent bifurcation is found to lead
intermittent chaos.

Finally, hydrodynamic models for chaotic pulsation with small amplitudes observed in
the post-AGB stars are briefly discussed.

1. Introduction

We can guess stellar parameters and physical properties of pulsating stars
with comparison of observational data with corresponding outputs of hydrodynamic models. This is an important purpose for hydrodynamic modeling
of pulsating stars. In linear models, only pulsation periods are used for these
comparisons, while pulsation periods give us rich information in multimode
pulsators, particularly in nonradial pulsators (Saio, 1992). In nonlinear models, however, we can use the time variations of magnitude and velocity, although observations for these quantities are time consuming and the outputs
for these quantities from hydrodynamic models depend on parameters for
numerical treatments and assumptions used for simplifying hydrodynamic
models.

The pulsation behaviors in hydrodynamic models are changed qualitatively with transitions due to bifurcation of nonlinear pulsation. In my review, we discuss these transitions in hydrodynamic models of radial pulsation for classical Cepheids and less-massive supergiant stars. Finally we
discuss chaotic pulsations with small amplitudes in hydrodynamic models
for the post-Asymptotic Giant Branch Stars.

Astrophysics and Space Science **210**: 269–280, 1993.

2. Bifurcation in Weakly Dissipative Models (Cepheid Models)

Bifurcation in nonlinear pulsations of weakly dissipative models (Cepheid models) is induced by modal resonances. There are two types of resonance: higher harmonic resonance and subharmonic resonance.

2.1. HIGHER HARMONIC RESONANCES

Modal resonance with the damped second overtone in the fundamental pulsator is well-known for a mechanism of the features in 10 days bump Cepheids (Simon and Schmidt, 1976). The higher harmonic component of the fundamental pulsation is induced in non-linear pulsation by this resonance. Many hydrodynamic models have been built to reproduce the feature of bump (Simon and Davis, 1983; Takeuti et al., 1983; Aikawa, 1987; Carson and Stothers, 1988, Fadeyev, 1992). Recently, Buchler et al. (1990) have made an intensive survey of hydrodynamic models for bump features. They have reproduced the Hertzsprung progression of velocity curves observed in bump Cepheids nicely.

Moskalik and Buchler (1990) have reported that other higher harmonic resonances of damped modes in the fundamental pulsator with $P_f = mP_n$ (where P_f and P_n are periods of the fundamental and n-th overtone modes, and m is an integer) are realized in their Cepheid models.

According to analytical theories for these resonances (e.g., Takeuti and Aikawa, 1981), bifurcation of unstable pulsation from stable pulsation with the bump feature is predicted near the center of the resonance. Thus, we include this type resonance as a phenomenon of bifurcation.

2.2. SUBHARMONIC RESONANCES

The Floquet stability analysis (e.g., Iooss and Joseph, 1980) of limit cycles in Cepheid models has been performed by Moskalik and Buchler (1990). They confirm effects of subharmonic resonances between the fundamental mode and overtone modes with the condition $P_f = m/2 \, P_n$ where m is an odd integer, for instance 3 or 5. In particular, they have found that subharmonic bifurcation with period doubling is induced by the subharmonic resonance between the fundamental and first overtone modes with the condition $P_f = 3/2P_1$, when both the modes have unstable limit cycles. Instead of limit cycles, the model has stable periodic oscillation with a period of twice the fundamental period. This type of the subharmonic resonance is found in other Cepheid models (Moskalik and Buchler, 1991).

2.3. FIRST OVERTONE PULSATOR

Antonello, Poretti and Redzzi (1990) have suggested that s-Cepheids are first overtone pulsators, and moreover the progression of light curves which is demonstrated as a trend in Fourier components obtained for light curves.

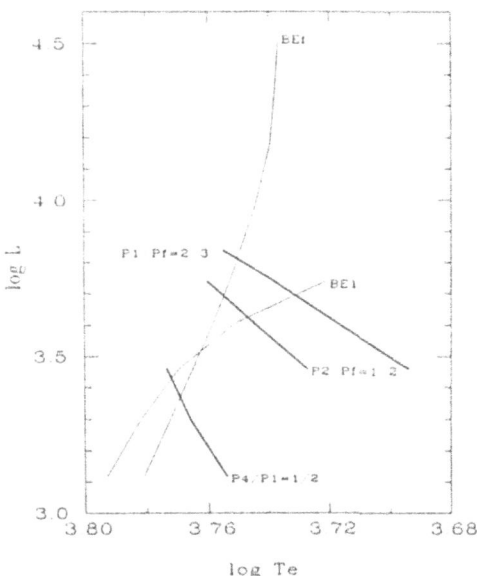

Fig. 1.   Resonance lines in the HR diagram. Resonance lines for the fundamental and first overtone pulsators are drawn with labels of resonance conditions. The blue edges of the instability strip for the fundamental (BEf) and the first overtone mode (BE1) are also plotted.

Recently Antonello (1991) has suggested that modal resonance of the first overtone pulsator with the fourth overtone mode for the mechanism of the progression in s-Cepheids. We examine this possibility. Fig. 1 displays the location of the resonance in the HR diagram using the Mass-Luminosity relation adapted from Chiosi (1990) and the augmented metal opacities (Simon, 1982). Other resonance lines for the fundamental pulsator are also displayed in the figure. The factor for the augmented metal opacities is set so that the resonance for dump cepheids with $P_f = 2P_2$ is realized in the fundamental pulsator with periods around 10 days.

The non-linear simulations for the first overtone pulsator with the suggested resonance are performed by TGRID hydro code (Simon and Aikawa, 1986) for models with $M = 4.0\ M_\odot$, $L = 1316\ L_\odot$, and the effective temperatures are chosen for models to cover the resonance.

Fig. 2 demonstrates light curves of the models as a function of effective temperature and they show a systematic trend, as shown in Fig. 3 for ϕ_{21} and R_{21} of the Fourier decomposition (see Simon and Lee, 1981 for definition of ϕ_{21} and R_{21}). We conclude that the higher harmonic resonance in the first overtone pulsator can make the features of the resonance in light curves.

This resonance with the damped fourth overtone mode in the first overtone pulsator is interesting also to the problem of double-mode Cepheids

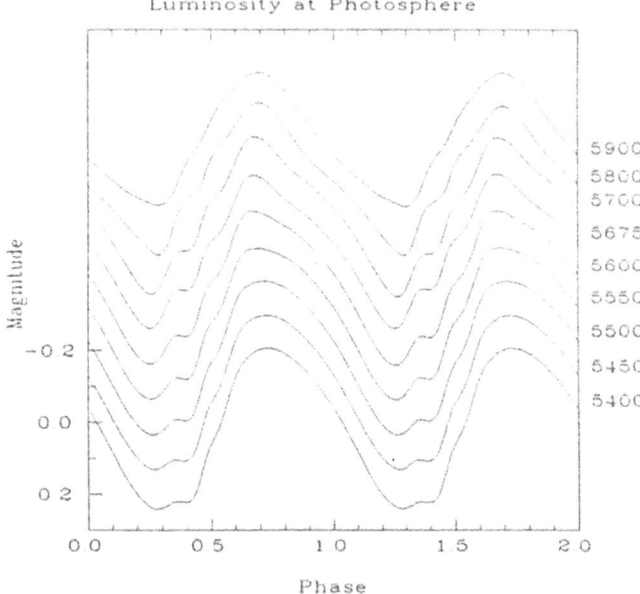

Fig. 2.   Light curves for the limit cycle models of the first overtone pulsators in the resonance. The effective temperature of models is labeled for each curve which is shifted vertically.

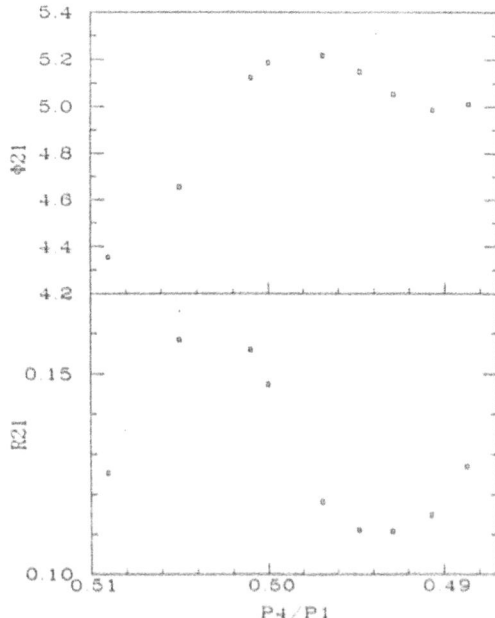

Fig. 3.   The Fourier phases ϕ_{21} and Fourier amplitude ratios R_{21} versus linear period ratio, P_4/P_1.

(Balona, 1985) which have similar periods with s-Cepheids. Buchler et al. (1990) found a Cepheid model which has unstable limit cycles for the fundamental and first overtone modes, and confirmed a persistent beat behavior. In their models, the fundamental mode is in a very close to the higher harmonic resonance of damped second overtone mode, and they have suggested that the resonance lowers the fundamental mode amplitude and leads to destabilization of the limit cycle. If this reasoning is correct, the present resonance also will work for destabilization of the first overtone limit cycle.

3. Bifurcation in Strongly Dissipative Models
(less massive supergiant stars)

Pulsations in strongly dissipative models (less-massive supergiant stars) are characterized by appearance of irregular behaviors (e.g. Tuchman et al., 1979; Fadeyev, 1982).

Nakata (1987) has shown that a sequence of less-massive supergiant models has a transition from regular to irregular pulsations, when the mass of the models is reduced step by step. Recently, the irregular pulsation has been investigated with respect to deterministic chaos. We concentrate in this section the bifurcation which leads irregular pulsations. It is well-known that there are universal routes from regular to irregular behaviors represented by simple mathematical models and confirmed by computer simulations and experiment measurements (Schuster, 1988). So far, there have been found two typical routes of transitions from regular to irregular pulsations in less-massive supergiant star models: the period doubling cascade (Buchler and Kovács, 1987) and the intermittency (Aikawa, 1987; Buchler et al., 1987).

3.1. PERIOD DOUBLING CASCADE

Buchler and Kovács (1987) have shown that a model sequence for Population II Cepheids has transitions from regular to irregular pulsations through the period-doubling cascade, one of universal routes of transitions from regular to irregular behaviors in deterministic chaos. Models of the sequence have the following stellar parameters: $M = 0.6M_\odot$, $L = 500L_\odot$, and chemical composition: $x = 0.745$, $z = 0.005$. The effective temperature is a control parameter of the model sequence. The models with higher effective temperature have stable limit cycles. They, however, show that the sequence suffers a subharmonic bifurcation, decreasing the effective temperature of models, step by step. At this point, the limit cycle becomes unstable, but the model has a stable periodic oscillation with a period of 2^1 times the period of unstable limit cycle. The sequence suffers another subharmonic bifurcation again, in the course of decrease of the effective temperature, and then has models with 2^2 periods. The subharmonic bifurcation repeats indefinitely (the period doubling cascade) and finally leads to chaotic pulsation.

Aikawa (1990) has demonstrated using TGRID hydro code that the period doubling cascade in one of model sequences studied by Buchler and Kovács (1987) is interrupted by the intermittency. Goupil et al. (1991) have suggested that analysis with higher dimension rather than the conventional first one-dimensional maps may reveal the nature of chaotic pulsation in this model.

3.2. INTERMITTENCY

Aikawa (1987) performed more luminous model sequence of less-massive supergiant stars. The model parameters are: $L = 3200_{\odot}$, $Te = 5300$ K, and chemical composition: $x = 0.70$, $z = 0.02$. The mass is a control parameter of the sequence. The results with LNA analysis are tabulated in Table I. The sequence shows different type of the transition from limit cycles to chaos, as shown in Fig. 4.

TABLE I

Model	$M/M_{\odot}$	P_f^a	η_f^b	P_1	η_1
a	1.50	27.755	0.417	15.038	−0.752
b	1.46	28.329	0.421	15.271	−0.786
c	1.45	28.476	0.422	15.331	−0.795
d	1.43	28.775	0.424	15.453	−0.812
e	1.42	28.927	0.424	15.515	−0.821
f	1.40	29.236	0.425	15.640	−0.840

a) The periods in units of days.
b) The growth rates are defined as $-4\pi\omega_i/\omega_r$.

We shall make the one-dimensional return map for the hydrodynamic pulsation models. We pick up values of total pulsational kinetic energies at their maxima. During one oscillation, the quantity has two maxima at the expansion and contraction phases, and we pick up one at the expansion phase. Fig. 5 shows these return maps for chaotic pulsation in a model as well as stable limit cycles.

It is shown from Fig. 5 that in chaotic pulsation with parameters close those to stable limit cycles, nonlinear pulsation stays at a 'ghost' limit cycle for a while, but gradually obtains pulsationally kinetic energies and finally moves away from the 'ghost' limit cycle, causing an outburst. By dissipation of the kinetic energies by shock waves, the pulsation is suddenly quenched. Then, the model repeats the previous process in a similar fashion. We identify this transition from limit cycles to chaotic oscillations in the hydrodynamic models as the type I intermittency, another universal route in simple dynamical models with dissipations (Pomeau and Manneville, 1980).

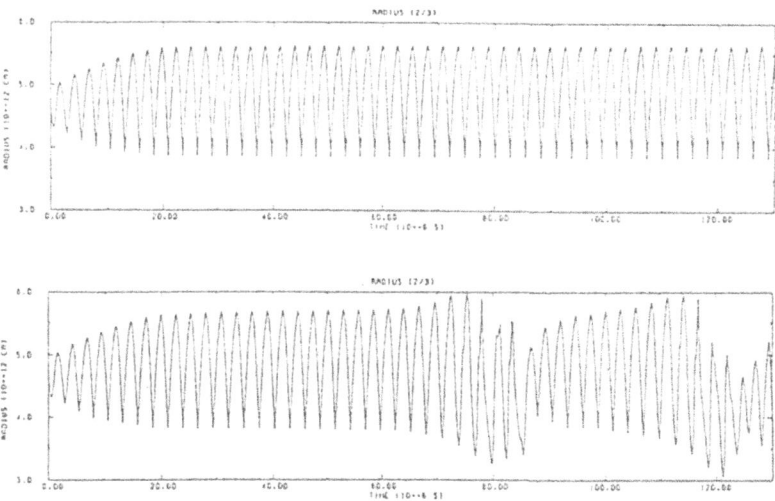

Fig. 4. Variations of the radius at the photosphere for model c (above) and d (below). It is noted that model d shows irregular oscillations with outbursts at times, while model c has a stable limit cycle.

Buchler et al. (1987) have demonstrated the intermittent chaos in Pop II Cepheid models.

3.3. DISSIPATION STRUCTURE

We shall discuss physical mechanisms for the transition. Aikawa (1987) has shown that the model which has stable limit cycles in the vicinity of the transition to the intermittent chaos has another unstable fixed point beyond a stable fixed point, and the transition from regular to irregular pulsation may be connected with disappearance of these fixed points (Tangent bifurcation). Aikawa (1988) shows these characteristics directly from the work integral of the nonlinear pulsations in one of models in question. It is shown that pulsation driving at the hydrogen ionization zones is strongly enhanced at amplitudes beyond the stable limit cycle. This additional driving makes the models to have another unstable fixed point.

It is suspected that the dissipation structures at the ionization zones with shock wave dissipation, as discussed in Aikawa (1988) may be also responsible for chaotic pulsations with other types of transition in less-massive supergiant models (see, Takeuti, 1987).

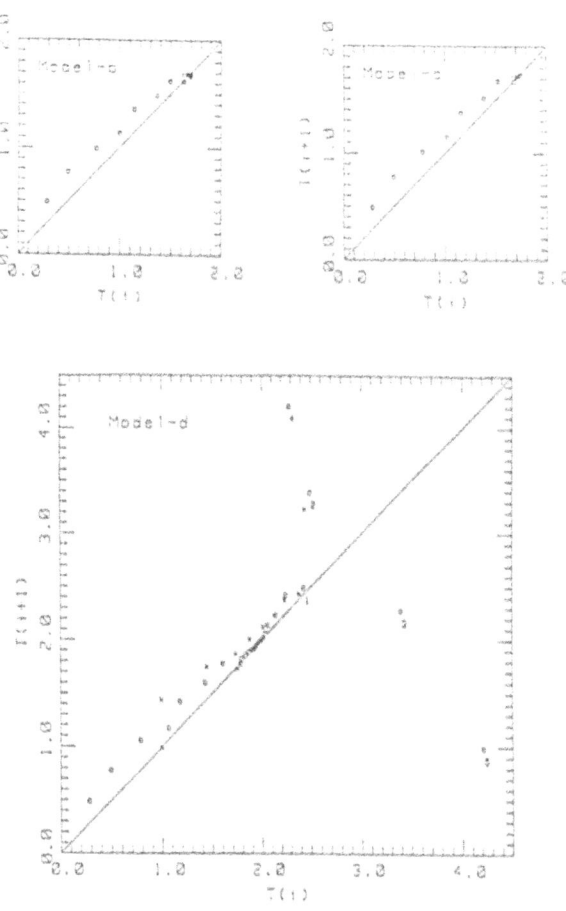

Fig. 5. The one-dimensional return maps for limit cycles (above) and irregular pulsation of model d (below). The latter model stays the bisectrix for a long time, following an outburst of large amplitudes. The oscillation is then suddenly quenched and restarts with a small amplitudes. The symbol (◯) and (×) are for first and second cycles of this process. The data points at the outburst are numbered in chronological order.

3.4. DYNAMIC REGIMES IN THE HR DIAGRAM

Kovács and Buchler (1988) have demonstrated with investigation of transitions from regular to irregular pulsations of model sequences of different luminosities. The transition in lower luminosity sequences is induced by the period doubling cascade, and on the other hand, the intermittency for higher luminosity models.

4. Bifurcation in Pulsation Models for The Post-AGB

4.1. PULSATION IN YELLOW SUPERGIANT STARS

Pulsation in yellow supergiant stars (e.g. 89 Her and HD 161796) has the following characteristics:

(1) Irregular pulsation with time scale of about 40 days,

(2) small amplitudes about 0.2 mag., which is contrasted with irregular variables in red giant stars,

(3) Some of them are suggested the post-asymptotic Giant Branch (post-AGB) stars.

The time variations of magnitude in 89 Her and HD161796, the prototypes of this class have been obtained by Fernie through the Automatic Photometric Telescope (APT) Service since Fernie (1983). Recently, Zsoldos and Sasselov (1991) have claimed that complicated light variation in UU Her which used to be classified as this class can be explained by superpostion of two linear oscillators with slowly modulated amplitudes. Thus, we need careful examinations for analysis of the variations.

4.2. LINEAR MODELS

The observed effective temperature of these yellow supergiant stars indicates that these stars are located at bluer region outside the conventional instability strip for pulsating stars in the HR diagram. Aikawa (1991) has shown that less-massive supergiant stars (e.g. $M < 1M_\odot$) with appropriate luminosities are unstable by overstable oscillations with higher overtone modes, while the modes have different properties from ordinary modes and must be related with strange modes (Gautschy and Glatzel, 1990), but the physical interpretation on the driving of the mode is unclear.

4.3. PULSATIONS IN NONLINEAR REGIME

Nonlinear simulations are performed for models of $M = 0.8\ M_\odot$ with sequences of models with different values of the effective temperature ($Te = $ 6000 K, 6300 K, 6600 K, 6900 K, 7200 K and 7500 K) to cover the region of the HR diagram for F type supergiant stars. The luminosity is a control parameter in this study and is changed with the range of 3500 $L_\odot < L <$ 7000 $L_\odot$.

For all model sequences except for the sequence of $Te = 6000$ K, models with higher luminosities show chaotic pulsation with small amplitudes, while pulsation in lower luminosity models is rather regular. It is noted that subharmonic components of the driving mode are strongly enhanced in nonlinear pulsation in chaotic regime, and time variations with much longer time scale are induced in chaotic oscillations, compared with the linear period of about 10 days.

We conclude that chaotic pulsation with small amplitudes can be gener-

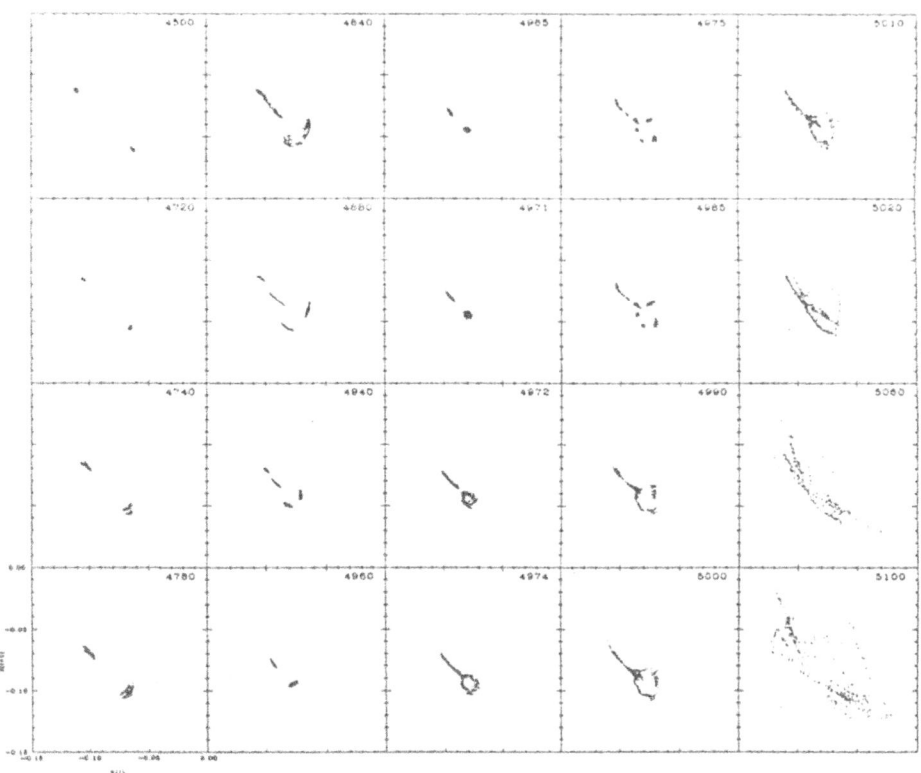

Fig. 6. The series of the one-dimensional return map for model sequence $Te = 6300$ K. Only higher luminosities part is drawn and luminosities of models are labeled for each map. Data of the Poincare section are made from data on magnitude at the photosphere at the time of maximum expansion of the same place.

ated robustly in pulsation of less-massive supergiant stars located apparently at bluer region outside the conventional blue edge.

The model sequence of $Te = 6300$ K is studies in detail to find out nature of chaotic pulsation. The model sequence shows much complicated transitions from regular to chaotic pulsations, as demonstrated in Fig. 6 as a series of the first return maps for the higher luminosity part in the model sequence.

Plotting data of the Poincare section with 3D, we can see more clearly the structure of transition, as shown in Fig. 7 and so the transition may be understood in higher dimensional mappings. The properties are often appeared in the transitions to chaos in which universal routes are interrupted by periodic oscillations and other type routes (Arneodo et al., 1983).

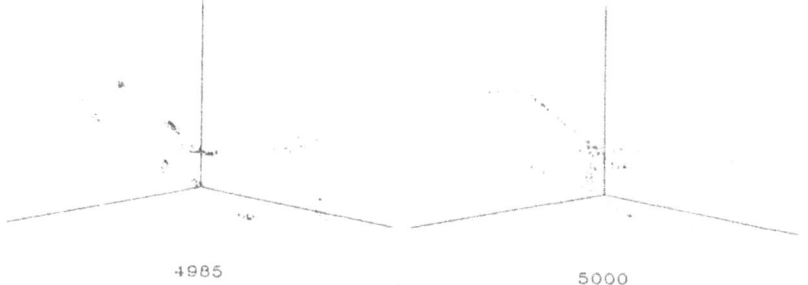

4985   5000

Fig. 7.   3D plots of the Poincare section data. Three successive data are plotted in three
dimensional space. For model $L = 4985 L_\odot$, we can see a periodic oscillation with 9 period,
and this periodic oscillation becomes divergent in model $L = 5000 L_\odot$.

5. Conclusions

Phenomena of bifurcation in non-linear pulsation in hydrodynamic models are reviewed. It is pointed out that qualitatively different behaviors in
nonlinear pulsation are realized with transitions due to bifurcation in hydrodynamic models. We may summarize as follows:

(1) In weakly dissipative models (Cepheid Models), bifurcation in nonlinear pulsation is induced by modal resonances in the fundamental and first
overtone pulsators. Some of them are responsible for observed properties.

(2) In strongly dissipative models (less-massive supergiant stars), there
have been found two systematic routes of transition from regular to chaotic
pulsations in hydrodynamic models: the period-doubling cascade and the
tangential bifurcation. These transitions are induced by the dissipation structure of the ionization zones including shock waves generated in the region.

(3) Models for chaotic pulsation with small amplitudes in the post-AGB
stars are proposed. There are transitions from regular to irregular pulsations with small amplitudes in a wide range of stellar parameters. Thus,
the transitions are robust for generation of chaotic pulsations with small
amplitudes.

References

Aikawa, T.: 1987, in *Stellar Pulsation*, eds. A. N. Cox, W. M. Sparks and S. G. Starrfield
      (New York: Springer), p. 175.
Aikawa, T.: 1987, *Astrophysics and Space Science*, **139**, 281.
Aikawa, T.: 1988, *Astrophysics and Space Science*, **149**, 149.
Aikawa, T.: 1990, *Astrophysics and Space Science*, **164**, 295.
Aikawa, T.: 1991, *Astrophysical Journal*, **374**, 700.
Antonello, E.: 1991, private communication.
Antonello, E., Poretti, E., and Reduzzi, L.: 1990, *Astronomy and Astrophysics*, **236**, 138.
Aeneodo, A., Coullet, P., Tresser, C., Libchaber, A., Maurer, J. and d'Humieres, D.: 1983,
      Physica, **6D**, 385.

Balona, L. A.: 1985. in *Cepheids: Theory and Observations*, ed. B. F. Madore (Cambridge: Cambridge Univ. Press), p. 17.

Buchler, J. R. Goupil M. -J., and Kovács. G.: 1987, *Phys. Lett.*, **A126**, 177.

Buchler, J. R. and Kovács, G.: 1987, *Astrophysical Journal, Letters to the Editor*, **320**, L57.

Buchler, J. R. Moskalik, P. and Kovács, G.: 1990, *Astrophysical Journal*, **351**, 617.

Carson, T. R. and Stothers, R. B.: 1988, *Astrophysical Journal*, **328**, 196.

Chiosi, C.: 1990, in *Confrontation between Stellar Pulsation and Evolution*, eds. C. Cacciari and G. Clementini (ASP Conf. Ser. vol.11), p. 158.

Fadeyev, Y. A.: 1982, *Astrophysics and Space Science*, **86**, 143.

Fadeyev, Y. A.: 1984, *Astrophysics and Space Science*, **100**, 329.

Fadeyev, Y. A.: 1992, this conference.

Fernie, J. D.: 1983, *Astrophysical Journal*, **265**, 999.

Gautschy, A. and Glatzel, W.: 1990, *Monthly Notices of the RAS*, **245**, 597.

Goupil, M. -J., Auvergne. M. and Serre, Th.: 1991, in *Applying Fractals in Astronomy*, eds. A. Heck and J. M. Perdang (New York: Springer), p. 43.

Iooss, G. and Joseph, D. D.: 1980, *Elementary Stability and Bifurcation Theory* (New York: Springer).

Kovács, G. and Buchler, J. R.: 1988, *Astrophysical Journal*, **334**, 971.

Moskalik, P. and Buchler, R. J.: 1990, *Astrophysical Journal*, **355**, 590.

Moskalik, P. and Buchler, R. J.: 1991, *Astrophysical Journal*, **366**, 300.

Nakata, M.: 1987, *Astrophysics and Space Science*, **132**, 337.

Saio, H.: 1992, this Conference.

Schuster, H. G.: 1988, *Deterministic Chaos: an Introduction* (Weinheim:VCH).

Simon, N. R.: 1982, *Astrophysical Journal, Letters to the Editor*, **260**, L87.

Simon, N. R. and Aikawa, T.: 1986, *Astrophysical Journal*, **304**, 249.

Simon, N. R. and Davis, C.: 1983, *Astrophysical Journal*, **266**, 787.

Simon, N. R. and Lee, A. S.: 1981, *Astrophysical Journal*, **248**, 291.

Simon, N. R. and Schmidt, E. G.: 1976, *Astrophysical Journal*, **205**, 162.

Takeuti, M.: 1987, *Astrophysics and Space Science*, **136**, 129.

Takeuti, M. and Aikawa, T.: 1981, *Science Rep. Tôhoku Univ.*, *8-th ser.*, **2**, 106.

Takeuti, M. Uji-iye, K, and Aikawa, T.: 1983, *Science Rep. Tôhoku Univ.*, *8-th ser.*, **4**, 129.

Tuchman, Y., Sach, N. and Barkat, Z.: 1979, *Astrophysical Journal*, **234**, 217.

Zsoldos, E. and Sasselov, D. S.: 1991, preprint (to appear to Astronomy and Astrophysics).

DOUBLE-MODE STELLAR PULSATION

G. KOVÁCS

Department of Physics, University of Florida, Gainesville
and
Konkoly Observatory, Budapest, Hungary

Abstract. The current observational and theoretical status of the double-mode variables is reviewed. Focusing mostly on the RR Lyrae stars, we address the question of the observational evidence of *modal stability*. The problem of stationarity is a crucial issue in the modelling of these stars.

We mention past efforts in hydrodynamical and analytical modelling together with a detailed discussion of some very recent results. It is suggested that stochastic forcing due to turbulent convection may play a crucial role in exciting some marginally stable modes in the limiting pulsation. The latest hydrodynamical results first demonstrate that purely *radiative models* are able to show *permanent* double-mode behavior in the relevant period regime of RRd stars. The reason for the previous lack of double-mode behavior is attributed to the large dissipation, *i.e.* artificial viscosity, generally used in the codes to ensure numerical stability and to obtain amplitudes comparable to the observed ones.

We think that better models should include some physical dissipation, most probably turbulent convection, and a more accurate numerical treatment of the radiative hydrodynamics.

1. Introduction

Double-mode variables represent perhaps the 'cleanest' example of multimode stellar pulsation. Their light- (or velocity-) variations are described usually within the observational errors as a result of a Fourier-sum of two non-commensurable frequency components and their linear combinations. The two principal components are identified with some *low-order*, usually the fundamental and first overtone *radial* modes of pulsation. Simple linear pulsational models allowed already some 20 years ago to estimate the masses of the double-mode variables directly from their observed frequencies of pulsation. The development and perfection of this method and its application to multimode variables is called *stellar seismology* nowadays. The power of this method led to the initiative of revising the older stellar opacities and providing not only a better modelling of the atomic physics but also resolving the long-standing mass-discrepancy problems between the evolution and pulsation theories.

Based on the reasonable assumption of radial stellar pulsation, we can try to model finite amplitude double-mode pulsation. It is the basic *nonlinear* nature of double-mode pulsation and its modelling which are our main concerns in this review.

Numerical modelling of double-mode pulsation is important for three main reasons: (1) double-mode variables are hoped to be modelled by $1 - D$ hydrodynamical codes, unlike the majority of the multimode pulsators which

Astrophysics and Space Science **210**: 281–300, 1993.
© 1993 *Kluwer Academic Publishers*.

are non-radially pulsating stars; (ii) modelling means fine tuning the physics and the numerical methods we use in solving the physical problem, and perhaps also narrowing down further the model parameters; (iii) successful numerical modelling is the *only* way to reach an understanding of the underlying physics of double-mode stellar pulsation and to build to a simple model of it.

As it is well known, so far, the hydrodynamical models have failed to give permanent double-mode behavior in the observed ranges of periods. One of the main purposes of this paper is to summarize the results of some very recent tests which show that purely radiative RR Lyrae models do exhibit sustained double-mode pulsation in the right period range if the artificial viscosity is reduced enough.

The outline of this paper is the following. After a brief review of the current observational status of the double-mode variables in Section 2, we mention the effect of the new opacities on the period ratio masses in Section 3. Some possible phenomenological models with the implication of the dynamical effects of noise are discussed in Section 4. The past hydrodynamical simulations together with the very recent ones are reviewed in Section 5. Finally, in Section 6 we summarize our conclusions and highlight the important questions.

2. Observations

Since the primary purpose of this review is to deal with the theoretical aspects of double-mode pulsation, here we merely mention the main observational features which are important for the nonlinear modelling.

2.1. CLASSICAL CEPHEIDS

The observational properties of the beat Cepheids have been reviewed recently by Balona (1985) and Szabados (1988). It is remarkable that about 30% of the observed galactic Cepheids are double-mode in the period range of 2–4 days, whereas no definite identification of this type of variables has yet been made in any extragalactic objects (see however Andreasen 1988). There is no preferred range in mode amplitude ratios, although most of the beat Cepheids have large fundamental than first overtone amplitudes. There is no convincing evidence for amplitude change on a long-time scale (Balona 1985). Even more, some stars show remarkably stable light variation on a 50–70 years time base (Jerzykiewicz 1988).

2.2. DWARF CEPHEIDS

The large-amplitude pulsators situated in the lower portion of the Cepheid instability strip close to the main sequence constitute an inhomogeneous group of stars, often called dwarf Cepheids. Some of them are very similar to the δ Scuti stars, but there are a few with distinctly Pop II characteristics. For a recent observational review of mostly the single-mode Pop II dwarf Cepheids we refer to Nemec and Mateo (1990), and for a more general review to Fitch (1980). The two principal components of the pulsation are generally assumed to be the radial fundamental and first overtone modes. The important question of amplitude stability has not yet been seriously addressed for these variables. There are, for example irregularities which are not explained yet (*e.g.* Fernley *et al.* 1987). It is possible, that some higher order overtones are somehow excited with low amplitudes which cause the seemingly irregular behavior. The exciting new analysis of AI Vel by Walraven, Walraven and Balona (1992) seems to support this view. They identify altogether four (or possible five) modes of pulsation. The newly discovered modes have very small ($\lesssim 10\%$) amplitudes compared to the principal ones. There are also some fluctuations in the amplitudes of these modes. The discovery of these modes is very significant, especially if some of them will be proved to be radial modes.

2.3. RR Lyrae stars

Until 1983, the field star AQ Leo and the M3 variables V67 and V87 were the only known double-mode RR Lyrae (RRd) stars (Jerzykiewicz and Wenzel 1977; Goranskij 1981). The discovery of a large number of RRd stars in the globular cluster M15 by Cox, Hodson and Clancy (1983) initiated a number of subsequent investigations in other clusters and galaxies. According to the periods, the RRd stars form two distinct groups, following largely the overall metal abundance of the cluster. The low-metal (Oosterhoff II) RRd stars have $P_0 = 0.55 \pm 0.02$ day, $P_1/P_0 = 0.746 \pm 0.001$, whereas those in the high-metal (Oosterhoff I) clusters have $P_0 = 0.48 \pm 0.005$ day, $P_1/P_0 = 0.7444 \pm 0.0004$. There are altogether 16 Oo I and 24 Oo II RRd stars securely identified and some more suspected. A distinct property of all the RRd stars is that their first overtone amplitudes are a few times larger than those of the fundamental. The only exception is variable V68 in M3, where the two amplitudes are about the same (Nemec and Clement 1989). There have been no significant long-term amplitude changes reported in any RRd star (Kovács, Shlosman and Buchler 1986; Nemec and Clement 1989; Jurcsik and Barlai 1990). AQ Leo, the only RRd star observed photoelectrically, showed remarkable stability ($\Delta A \leq 0.005$ mag.) during three years of observation (Jerzykiewicz and Wenzel 1977).

One great mystery about the RRd stars is their occurrence and frequency. Many careful analyses have been made during the past several years based mostly on old photographic materials. The searches have largely been unsuccessful (see Clement and Walker 1991 and references therein). The complete absence of RRd stars in the cluster ω Cen which has 155 known RR Lyrae stars, or the only 2 RRd stars found among the 180 RR Lyrae stars of M3, indicate that double-mode variability is not all common among the R R Lyrae stars and may actually be a very delicate phenomenon, which depends on some, so far unknown fine details of stellar structure and evolution.

For a more detailed review of the observational aspects of the R Rd stars we refer to Szeidl (1988).

3. Period Ratio Masses

The very powerful method of mass determination with the aid of the observed periods of double-mode variables (*e.g.* Petersen 1973) has triggered a lot of discussions and arguments on the discrepancies between the pulsation and evolution theories (*cf.* Cox 1987). Among the many, sometimes exotic proposals to resolve this discrepancy, Simon's (1982) hypothesis on the possible underestimation of heavy element opacities has proved to be the most fruitful one. Due mainly to this idea, a substantial effort for updating the equations of state and opacities for stellar conditions has led to the first results in the last year (Iglesias and Rogers 1991; Rogers and Iglesias 1992). Now it seems that all major discrepancies between the evolution and pulsation theories have been eliminated. However, a stronger sensitivity to the heavy element abundance leaves us with fairly large uncertainties in some cases.

Moskalik, Buchler and Marom (1991) studied the beat and bump Cepheids with the new opacities of Iglesias and Rogers (see also Zalewski 1992). Moskalik *et al.* conclude that for beat Cepheids the period ratio masses are between $4 - 7M_\odot$, in essential agreement with the standard evolution theories. For bump Cepheids, the derived masses are still somewhat low, but only if we take some moderate heavy element abundance ($Z = 0.02$) and a standard evolutionary mass - luminosity ($M - L$) relation. We think that the errors both in Z and in the evolutionary $M - L$ relations are high enough to render the remaining small discrepancy insignificant.

As for the RRd stars, the new opacities completely ruined our previous picture about these stars. Even for the low Z Oo II variables the new opacities predict a mass of $\approx 0.77M_\odot$, which is $\approx 0.1M_\odot$ higher than the old value. Because of their higher Z, Oo I variables tend to have the same mass as the Oo II stars (Kovács, Buchler and Marom 1991; see also Cox 1991). In addition to the very strong Z dependence of the derived mass, there is a substantial sensitivity even to the detailed mixture of the heavy elements. In a recent paper Kovács *et al.* (1992) discuss this mixture dependence within the framework of the latest observational and theoretical results regarding the chemical compositions of Pop II stars (for a review, see Wheeler, Sneden and Truran 1989). They conclude that present inaccuracies in the observed chemical compositions and the lack of direct measurements on RRd stars, prevent us from estimating their masses on the basis of their periods more accurately than $\pm 0.1 M_\odot$. For this reason, period ratio masses are not very useful at the present moment for making a more thorough comparison with the evolution studies. More accurate direct chemical composition measurements are indispensable to make further progress here.

4. Phenomenological Models

The basic nonlinear behavior of a pulsating star is described by a set of ordinary differential equations which refer to the dynamical evolution of the amplitudes of the normal modes (*cf.* Buchler 1985). However, except for second order adiabatic coupling, the nonlinear parameters entering in the amplitude equations are very difficult to compute *ab initio* from the stellar models. Some information can be obtained through a comparison with the nonlinear hydrodynamical models, but in general the nonlinear coupling coefficients are regarded as 'almost' free parameters. Therefore, most of the studies on the amplitude equations can be considered as phenomenological. It is important however, that in some cases one can derive strict and general results without invoking the specific value of the coupling coefficients.

In the following we summarize the basic conclusions obtained from the amplitude equations relevant for the double-mode problem. First we review the cases when the system does not contain any stochastic forcing. The effect of the additive noise on the non-resonant pulsation is to be discussed in the second part of this section.

4.1. DOUBLY-PERIODIC SOLUTIONS IN NOISELESS SYSTEMS

(1) NON-RESONANT SYSTEM

Two modes coupled in a non-resonant way may settle down on a single- or double-mode state. The two possibilities are mutually exclusive in the sense that for a given set of parameters either one or two single-mode states or only one double-mode state can physically exist. More specifically, if we assume (supported by hydrodynamical results) that the nonlinear coupling coefficients are constants and negative across the instability strip, we can classify the solutions in two distinct groups, depending solely on the nonlinear coupling. Crossing the instability strip from the blue to the red, we have the following set of states.

Case (1): *first overtone only - either first overtone or fundamental - fundamental only*

Case (2): *first overtone only - double-mode only - fundamental only*

One of the major practical conclusion we can draw from these results is that non-resonant double-mode behavior is unique, there is no hysteresis which includes this state. For further discussion of the non-resonant mode coupling we refer to Dziembowski and Kovács (1984) and Buchler and Kovács (1986).

(2) RESONANT SYSTEM

We consider only the lowest order of resonances. Depending on the number of modes involved in the resonance there are two types of interaction.

(2A) TWO-MODE: $2\omega_0 \approx \omega_j$

The importance of the 2:1 resonance in stellar pulsation was first noticed by Simon and Schmidt (1976). In the case of classical Cepheids this type of resonance between the fundamental and second overtone modes gives rise to the specific progression of Fourier parameters (*i.e.* Hertzsprung progression, see Buchler, Moskalik and Kovács 1990). The possibility that the same type of resonance may also affect the stability of the single-mode state, thereby leading to double-mode pulsation, was recognized by Dziembowski and Kovács (1984).

The basic mechanism is the amplitude decreasing effect of the resonance. For example, if there is a resonance between the fundamental and a high overtone, the amplitude of the fundamental limit cycle decreases and reaches a minimum at or near the center of resonance. As a result of it, the stability of this limit cycle decreases, and may allow other, linearly unstable modes (most importantly the first overtone) to grow in the limit cycle. The decrease of the stability of the resonant limit cycle is a general consequence of the resonance and can be seen both in the analytical considerations and in the numerical results (Kovács and Buchler 1988). If the fundamental limit cycle becomes unstable and the resonant first overtone limit cycle is also unstable, there is a three-mode (but because of phase lock, doubly-periodic) state which is the *only* stable state of the system. In the general case of non-adiabatic coupling the situation in principle could be more complicated, but numerical results support the simple picture we described.

(2B) THREE-MODE: $\omega_0 + \omega_1 \approx \omega_j$

Simon (1979) suggested that this type of resonance may play a role in the double-mode pulsations of Cepheids and dwarf Cepheids. Subsequent hydrodynamical and analytical studies, however, did not support this hypothesis (Simon, Cox and Hodson 1980; Dziembowski and Kovács 1984). The following rigorous analytical results obtain.

Assuming that the fundamental and first overtone modes are linearly excited, while the resonant high overtone is damped and that there is resonant adiabatic coupling only, then, the three-mode resonance is never able to establish stable pulsation with constant amplitudes. When nonresonant coupling is also included, one of the single-mode states remains always stable while a stable three-mode state may occasionally exist simultaneously with the single-mode state for some very restricted range of parameters.

The general case (including non-resonant and non-adiabatic couplings) was discussed by Kovács and Kolláth (1988). Here we can observe the whole spectrum of nonlinear behavior of a dynamical system. Since there are no direct computations regarding the non-adiabatic effects in the resonant coupling and there are no hydrodynamical simulations which indicate the importance of three-mode resonance, we think that the role of this resonance remains hypothetical.

For completeness, we remark that the three-mode resonance may play an important role in the case of low-amplitude pulsators (δ Scuti, Ap stars, white dwarfs). In δ Scuti stars, the low frequency modes correspond to higher overtone g-modes which interact with the low-order p- or g-modes. This parametric excitation was investigated by Dziembowski and Krolikowska (1985), Dziembowski, Krolikowska and Kosovitchev (1988) and by Moskalik (1985).

4.2. Noise Generated Multimode States

There is no physical system which is free of random perturbations. This is particularly true for stars, where turbulent convection may exert a substantial effect not only on the static (*i.e.* average) structure, but on the pulsation too. There are two types of convective dynamical effects: (1) The part of the stochastic interaction which has non-vanishing ensemble average changes the dynamics in a deterministic way. The classical treatment of pulsation-convection interaction with the mixing length theory deals with this problem (*e.g.* Stellingwerf 1984); (2) The stochastic forcing which has zero ensemble average establishes a sustained perturbation of the system. One of the results of this is an excitation (usually at very small amplitudes) of all the normal modes of the system. This is what is claimed to happen in the case of the solar 5-min oscillations (Goldreich and Keeley 1977).

In this subsection we would like to elaborate further on the effects of stochastic mode excitation in the context of multimode nonlinear pulsation. Naturally, we are interested in systems which are close to some bifurcation, therefore, they are easy to influence even by a small amount of noise. We show, that for a sufficiently large number of stochastically excited modes the non-resonant interaction among these and the principal modes may lead to a genuine multimode state even if the noiseless system is unstable in that state. The fluctuation of the amplitudes of the principal modes can be arbitrary small depending on the number of modes included in the interaction.

Let us first consider the case when we omit all the modes except for the tow principal ones. Assuming additive noise in the original system and short noise correlation time, one can derive the following amplitude equations (Buchler, Goupil and Kovács 1992)

$$\frac{dA_0}{dt} = \kappa_0 A_0 + Q_{00} A_0^3 + Q_{01} A_0 A_1^2 + \frac{1}{2} \frac{S_0}{A_0} + \eta_0(t), \tag{1}$$

$$\frac{dA_1}{dt} = \kappa_1 A_1 + Q_{11} A_1^3 + Q_{10} A_1 A_0^2 + \frac{1}{2} \frac{S_1}{A_1} + \eta_1(t). \qquad (2)$$

Here $\eta_0(t)$, $\eta_1(t)$ are the fluctuating non-parametric noise components which are easily related to the ones appearing in the original system. S_0, S_1 are the spectral densities of $\eta_0(t)$, $\eta_1(t)$ at frequency zero. Some useful information can be obtained regarding the average values and the stability of the amplitudes by studying the system (1)–(2) with $\eta_0(t) = \eta_1(t) = 0$, but keeping the intensity terms. It follows immediately that we only have double-mode solutions because of the S_i/A_i terms. These are, however, no genuine double-mode solutions, because as one can show by an analysis of the amplitude fluctuations, one of the two modes remains of *precursor*-type, *i.e.* its average amplitude and the fluctuation around it will be of the same size. For high enough noise the two double-mode solutions (associated with the noiseless single-mode states) can merge together and form a unique noisy double-mode state with one single maximum in the probability distribution function (see Buchler 1992).

To illustrate the behavior of the stochastic double-mode solution, we integrate Eqs. (1)–(2) for a parameter set which corresponds to a 'first overtone only' case in the noiseless system. The empirical probability distribution functions are shown in Fig. 1. Because of the large dispersion, it is clearly seen that the low-amplitude mode is a noise generated one. The smaller the absolute value of the switching rate of this mode is, the larger is the excited amplitude at a constant noise level. It is important to remark that though the amplitude has a large scatter, the variation of the signal is smooth on a time scale of some fraction of the dynamical evolution (*i.e.* $1/\bar{\kappa}$, where $\bar{\kappa}$ is the switching rate).

Regarding the observed double-mode variables, the model discussed above is not acceptable, because of the large amplitude fluctuations of the precursor mode. It is clear, however, that the mechanism could be quite efficient to excite some other modes which are just marginally damped in the limit cycle or in the double-mode state. We conjecture that the observed irregularities in δ Scuti stars, dwarf Cepheids and maybe in other stars too, could quite possibly be attributed to noise excited higher order modes. We emphasize the fact that the amplitudes of these modes could be much larger than the ones maybe obtained by a solar analogy (Christensen-Dalsgaard and Frandsen 1983). The reason for this is that we deal with linearly excited modes which could be only very mildly damped in the limit cycle. The discovery of a few very low-amplitude modes in AI Vel (Walraven *et al.* 1992) is very important and deserves further attention from the theory of the stochastic mode excitation.

Let us now examine the case of multimode non-resonant interaction in the presence of noise. Our aim is to find situations in which the system bifurcates to a truly double-mode state in which the principal modes have very little

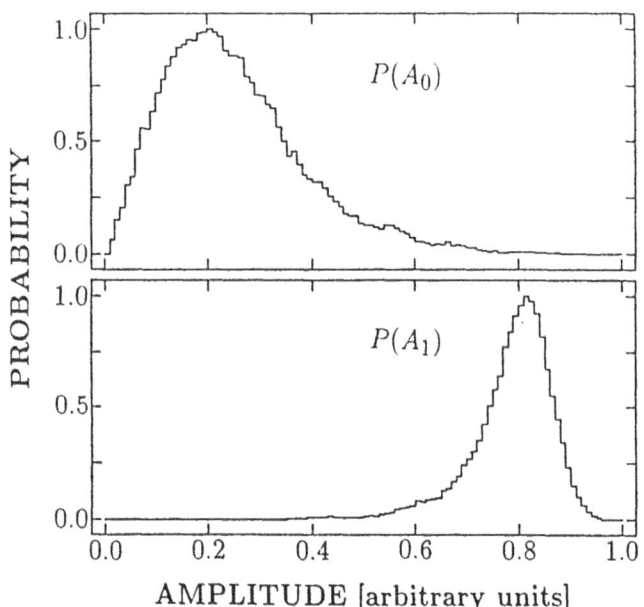

Fig. 1.   Empirical probability distribution functions of the amplitudes after integrating Eqs. (1)–(2). The parameters are: $\kappa_0 = 0.01$, $\kappa_1 = 0.06$, $Q_{00} = -10$, $Q_{01} = -15$, $Q_{10} = Q_{11} = -50$, $S_0 = S_1 = 10^{-6}$.

scatter. The details of this non-trivial problem are given by Buchler and Kovács (1992). Here we just mention the highlights of the solution.

Without jeopardizing generality, we assume that there are three modes only, that the first two are linearly excited and that the parameters correspond to a 'first overtone only' situation. Also, to obtain some analytically tractable expression, the noise is omitted in the first two modes ($S_0 = S_1 = 0$). The switching rates in the respective fundamental and first overtone limit cycles are

$$\bar{\kappa}_1 = \kappa_1 + Q_{10}A_0^2 + Q_{12}A_{20}^2, \tag{3}$$

$$\bar{\kappa}_0 = \kappa_0 + Q_{01}A_1^2 + Q_{02}A_{21}^2, \tag{4}$$

where A_{2j} is the amplitude of the noise excited third mode in the presence of the limit cycle j. The equilibrium solutions are given by

$$\kappa_j + Q_{jj}A_j^2 + Q_{j2}A_{2j}^2 = 0, \tag{5}$$

$$\kappa_2 + Q_{2j}A_j^2 + Q_{22}A_{2j}^2 + \frac{1}{2}\frac{S_2}{A_{2j}^2} = 0, \quad j = 0, 1. \tag{6}$$

Because $Q_{ij} < 0$ for all (i, j), it follows from Eq. (5) that the amplitudes of the principal modes *decrease* in the presence of this type of noise. Then, according to Eq. (3) and (4), the limit cycles may become less stable if the stabilizing effect of the $Q_{j2}A_{2j}^2$ terms does not overwhelm the opposite trend caused by the decrease of the limit cycle amplitudes. It is an important and interesting fact that one can readily find some parameters for which one of the switching rates decreases, while the other one increases (it is easy to show that the two switching rates cannot increase simultaneously). It means that this type of interaction cannot cause double-mode solution if the noiseless system is in the 'ether-or region' ($\bar{\kappa}_0 < 0$, $\bar{\kappa}_1 < 0$). It is also clear that the coupling with the principal modes in Eq. (6) is very important, since otherwise only the effective linear growth rates are decreased, which does not alter the stability of the system.

Once a proper set of parameters is found with $\bar{\kappa}_0 > 0$, $\bar{\kappa}_1 > 0$, we can tune the noise and let the system bifurcate (in this specific case for example) from the first overtone (low noise, $A_0 = 0$, $A_1 \neq 0$, $A_{j1} \neq 0$) through double-mode (medium noise, $A_0 \neq 0$, $A_1 \neq 0$, $A_j \neq 0$) to fundamental (high noise, $A_0 \neq 0$, $A_1 = 0$, $A_{j0} \neq 0$).

In the case of many high-order modes we can redistribute the total high-order mode energy necessary to destabilize the quasi single-mode states among many modes. Then, since the individual mode energies become lower and their fluctuations are almost independent one can expect a lower dispersion for the principal modes. Some more detailed considerations do indeed show that this is the case. The r.m.s. scatter of the amplitudes of the principal modes changes as $1/\sqrt{N}$, where N is the total number of the high overtone modes.

To illustrate the existence of the true double-mode solution and the decrease of the amplitude dispersion as the number of the modes increase, we perform a similar numerical simulation as we have already done in the two-mode case. The results are shown in Fig. 2. The noiseless system corresponds to a 'first overtone only' case and the nonlinear parameters to an 'either-or' case (*i.e.* no double-mode solution is possible for the noiseless system).

The difference between this and the precursor-type double-mode system is clearly shown (compare with Fig. 1). The simulations with a large number of modes exhibit the decrease of the amplitude dispersion in a quantitative agreement with the theoretical $1/\sqrt{N}$ dependence.

An application of the above idea to hydrodynamical models is fairly straightforward. The coupling coefficients can be extracted from the limit cycle analysis and the result can be used directly to evaluate the analytical stability conditions. A large number of tests made with standard RR Lyrae models indicate that the required conditions up to the 6-th order radial modes are not satisfied (at least not with 'standard' artificial viscosity). We think that the mechanism of noise generated 'smooth' multimode states should be further studied by exploring the effects of resonances. In any case as a partially *ad hoc* idea we cannot exclude that the nonradial modes, which

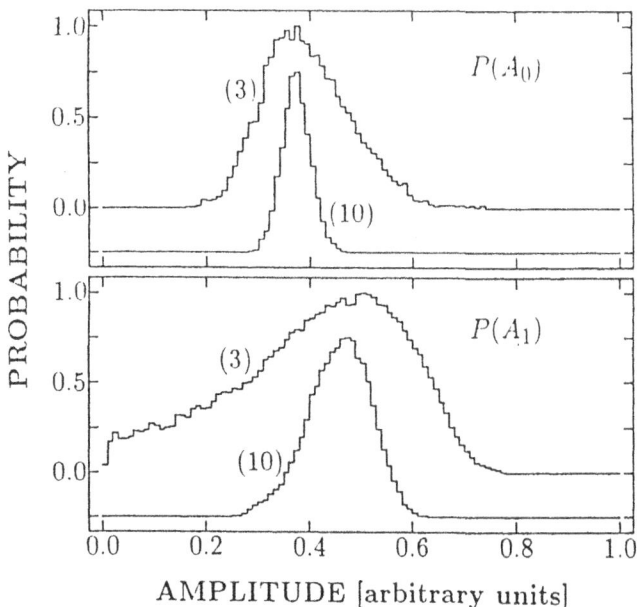

Fig. 2. Empirical probability distribution functions of the amplitudes of the principal modes in the case of multimode non-resonant interaction. The total number of modes are shown at the curves. The parameters are: $\kappa_0 = 0.01$, $\kappa_1 = 0.06$, $Q_{00} = -10$, $Q_{01} = -15$, $Q_{10} = Q_{11} = -50$, $\kappa_j = -0.12$, $Q_{0j} = Q_{j1} = -50$, $Q_{1j} = Q_{j0} = -500$. for $j \geq 2$ and all other Q_{ij} are zero. The intensities are: $S_0 = S_1 = 10^{-4} S_j$, $S_j = 2 \times 10^{-5}$ and 2.5×10^{-6} for the 3 and 10 mode case, respectively.

are most probably affected by turbulent convection, have proper couplings with the relevant radial modes. Then, a permanent double-mode state may exist, similarly as we have just described.

5. Hydrodynamical Modelling

Quite a few attempts have been made for numerical modelling of double-mode stars. After the early works (Stobie 1969; King *et al.* 1973), Stellingwerf (1975) was the first who, with the aid of the relaxation technique, performed the first systematic study in the limit cycle stability of the RR Lyrae stars. He found some cold models which study in the limit cycle stability of the RR Lyrae stars. He found some cold models which exhibited double-mode behavior. It is not only that those models have not much common with the observed double-mode stars, but also, subsequent simulations by Cox, Hodson and Davey (1976) could not confirm their double-mode nature. This is not very surprising in retrospect, since nonlinear behavior is very sensitive to some details of the model, which were certainly not identical in the two works.

Simon, Cox and Hodson (1980) tried to test the effect of three-mode resonance in supporting beat Cepheid pulsation. It turned out (in agreement with the subsequent analytical studies, as we have already mentioned) that this resonance is not likely to be important in double-mode pulsation.

The discovery of a large number of RRd stars further stimulated the search for double-mode pulsation models. Cox (1982) and Hodson and Cox (1982) made a moderate survey of RR Lyrae pulsation. They were unable to construct any RRd models. Re-investigating the problem, Kovács and Buchler (1988) performed a more extensive survey. They indeed found, in agreement with the analytical prediction of Dziembowski and Kovács (1984), that models which are in the proximity of the 2:1 resonance between the fundamental and the third overtone, exhibit less stable fundamental limit cycles compared to the ones outside of this resonance. This is a general property and is independent of the numerical details of the model. For some specific parameters, the resonance is able to destabilize the limit cycle and, if the first overtone (non-resonant) limit cycle is also unstable, lead to a double-mode model. The destabilizing effect of the 2:1 resonance has manifested itself in other types of models too, in particular, in Cepheid models (Buchler, Moskalik and Kovács 1990). Except for the fact that these models are the first, well established simulations showing permanent double-mode behavior, they are not relevant from practical point of view, because their periods are very different from the observed ones.

Convection, of course, may play an important role in the limit cycle properties of the models. Unfortunately, there has been no systematic and accurate work done in this area, except maybe for that of Ostlie (1990). He found that some of this RR Lyrae models indicate simultaneously unstable limit cycles. This observation was, however, not conclusive because the analysis was made on the perturbed, not perfectly settled limit cycles, and there was no subsequent direct time-integration performed to prove that the model really settles down on a double-mode state. Until the numerical problems are solved and more systematic studies are made on convective models, the role of convection in double-mode pulsation remains open.

Turning back to the purely radiative models, we would like to examine the effect of the artificial viscosity on the stability of the limit cycles. According to Stellingwerf (1975), the viscous pressure is given by

$$p_v(i) = C_Q P_g(i)[(u(i-1) - u(i))/c(i) - \alpha]^2,$$
$$\text{if} \quad (u(i-1) - u(i))/c(i) > \alpha > 0$$
$$= 0, \quad \text{otherwise.} \tag{7}$$

Here $c(i) = \sqrt{p_g(i)v(i)}$; $p_g(i)$, $v(i)$, $u(i)$ are the gas pressure, the specific volume and the velocity of the i-th mass shell. It is well known that the artificial viscosity has an important role in determining the limit cycle properties. Since there is not much of a restriction on the viscosity parameters

(C_Q, α), one cannot justify to fix them to any particular value. One possible methodology is that we are interested in the behavior of the pure radiative model and therefore, we should avoid 'unphysical' dissipation as much as possible. The problems which arise here are that: (1) amplitudes might become too high, (2) numerical instabilities might get amplified. To handle these problems, previous nonlinear works employed a numerical viscosity, which seemed to be a fair trade between numerical stability, size of amplitudes and avoiding very large dissipation. Here we would like to relax the restriction regarding the size of the amplitudes, while maintaining numerical stability. The main goal of this test is to *indicate* some type of behavior of the models which could be observed in the simulations with future less dissipative and more accurate codes. A very detailed discussion of our tests is given in Kovács and Buchler (1992). Additional tests regarding the artificial viscosity are presented in Kovács (1990).

To illustrate the dramatic effect of decreasing the viscous dissipation we compute a sequence of RR Lyrae models with two different artificial viscosity parameters. The sequence has the following parameters: $M = 0.75 M_\odot$, $L = 40 L_\odot$, $X = 0.7$, $Z = 0.0001$, $T_{\text{eff}} = 6000–6400$. We use the opacities published by Rogers and Iglesias (1992). Unless stated otherwise, all of our models contain 60 mass shells. The results are shown in Fig. 3.

The following important observations can be made: (1) The switching rate from the fundamental limit cycle toward the first overtone ($\bar{\eta}_1$) is strongly affected by the artificial viscosity. For the less dissipative ($C_Q, \alpha) = (4, 0.07)$ models $\bar{\eta}_1$ has a non-monotonic behavior and a very large positive value in an extended region. (2) The stability of the first overtone limit cycle is not much affected; the switching rate toward the fundamental ($\bar{\eta}_0$) does not change appreciably. (3) There is a region in which both switching rates are simultaneously positive. (4) The amplitudes of both limit cycles increase considerably for the less dissipative models.

We see that a completely different behavior is obtained for lower artificial dissipation. The 'standard' viscosity parameters $(C_Q, \alpha) = (4, 0.01)$ would predict a transition from an 'either - or' region ($T_{\text{eff}} > 6250$) to a 'fundamental only' one ($T_{\text{eff}} < 6250$), whereas for lower viscosity we get a 'double-mode only' regime ($T_{\text{eff}} = 6250–6080$) sandwiched between the two different singe-mode regimes. This behavior survives for many other (C_Q, α) combinations, when the resulting viscous dissipation is small enough. It is also comforting to know that the periods and period ratios in the double-mode region overlap with those of the Oo II RRd variables, namely: $P_0 = 0.51–0.56$ day, $P_1/P_0 = 0.746–0.747$ for our double-mode models.

The puzzling question regarding the behavior of $\bar{\eta}_1$ is that it has a non-monotonic variation and that it has a much larger value than that of the corresponding linear growth rate. None of these properties are easily explained within the framework of non-resonant pulsation (see *e.g.* Buchler,

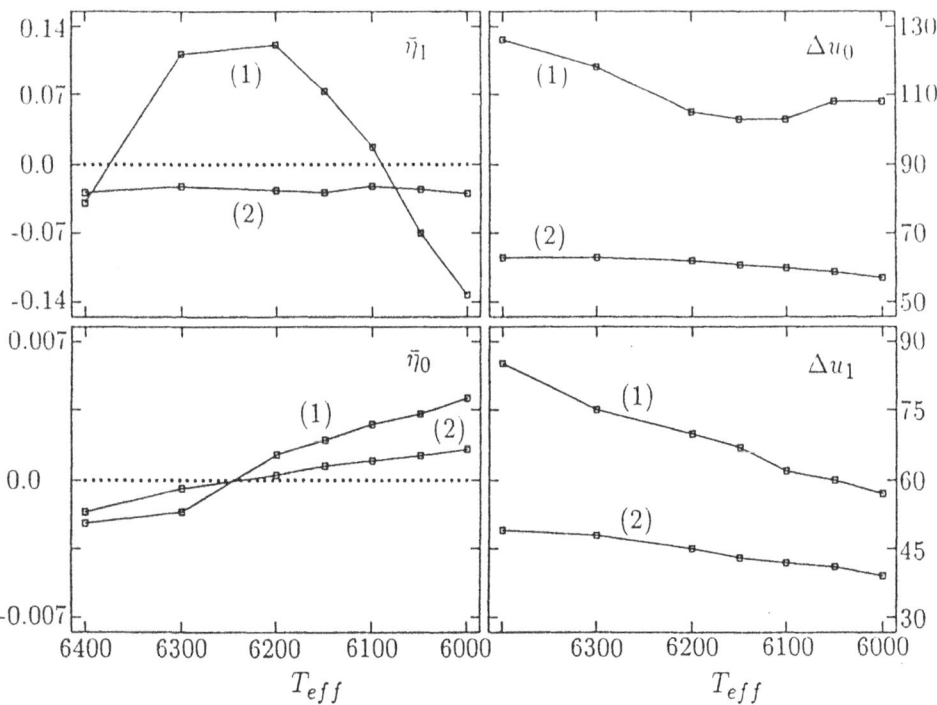

Fig. 3. Variation of the switching rates (*e.g.* Stellingwerf 1975) and velocity amplitudes (in [kms^{-1}]) along a model sequence with $M = 0.75M_\odot$, $L = 40L_\odot$, $Z = 0.0001$. Numbers at the lines denote the various values of the artificial viscosity parameters: (1): $C_Q = 4$, $\alpha = 0.07$; (2) $C_Q = 4, \alpha = 0.01$.

Moskalik and Kovács 1991). Therefore, we look for some type of resonance which leaves the system intact, except for the stability of the fundamental limit cycle. There are two different resonances in the proximity of the hump of $\bar\eta_1$, *viz.* $3\omega_0 \approx \omega_1 + \omega_2$ and $2\omega_1 \approx \omega_0 + \omega_2$, where ω_0, ω_1 and ω_2 are the fundamental, first and second overtone frequencies, respectively. The resonance centers are at $T_{\text{eff}} \approx 6300$ and $T_{\text{eff}} \approx 6100$ for the first and second resonance, respectively. Other more 'standard' resonances (two- and three-modes) are much 'weaker' in a numerological sense.

Which of these two resonances might cause the variation of $\bar\eta_1$? It is easy to see that the amplitude equation for A_1 is modified by the second resonance through a term which is proportional to $A_0 A_1 A_2$, consequently, the stability of the fundamental limit cycle will not be affected by this resonance, since $A_0 A_1 A_2$ is a higher order term, and therefore is negligible in the computation of $\eta\bar\eta_1$.

On the other hand, the first resonance $3\omega_0 \approx \omega_1 + \omega_2$ contributes to the amplitude equations for A_1 and A_2 by terms proportional to $A_0^3 A_2$ and $A_0^3 A_1$, respectively. Therefore, this resonance will affect the stability of the fundamental limit cycle. As we see, the perturbations with the first and

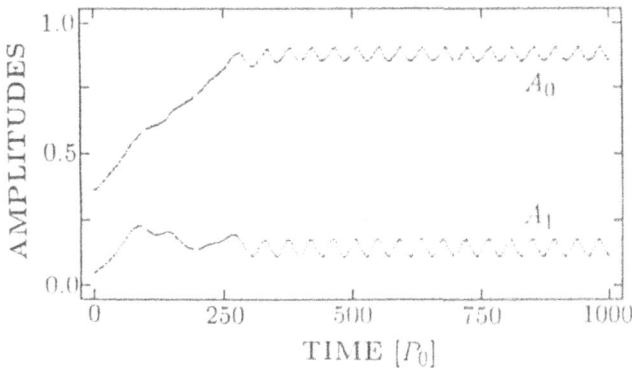

Fig. 4.   Amplitude evolution of the radius variation of the $P_0 = 0.529$ day, $M = 0.75 M_\odot$, $L = 40 L_\odot$, $T_{\text{eff}} = 6200$ K, $Z = 0.0001$ model with $C_Q = 4$, $\alpha = 0.07$. The amplitudes are given in arbitrary units.

second overtones are combined, a property which is also clearly observable in the Floquet analysis of the numerical models. The limit cycle amplitudes are not affected by this resonance in agreement with the data plotted in Fig. 3.

We now turn to the direct time integrations of the models. Each model is started from the static solution perturbed with a velocity distribution containing a mixture of first overtone (10%) and fundamental (90%) eigenmodes. The modal content of the radius or light variation is monitored by time-dependent Fourier analysis. One typical amplitude evolution is shown in Fig. 4. The double-mode state is independent of the initial conditions as is indicated by some supplementary tests.

It is interesting to note the periodic oscillations of the amplitudes, which is not an artifact of data analysis, but an inherent property of most of the models. This behavior is again an indication of the resonance discussed above.

Comparing with the observations one notes that the models have inverted amplitude ratios. The question is how we can reverse it? At the moment our guess is that some fine tuning of the dissipation together with a better spatial resolution might cause the desired effect.

In Fig. 5 we show the integrations of a model with $M = 0.75 M_\odot$, $L = 35 L_\odot$, $T_{\text{eff}} = 5900$K, $Z = 0.0001$. In both cases the single-mode states are unstable, and therefore, the final states are surely double-mode states. We see that in the more dissipative case (lower panel), despite of the long integration, the system has still not reached equilibrium, but that the amplitude ratio is reversed. Except for the imperfectly matched periods ($P_0 = 0.56$ day, $P_1/P_0 = 0.744$) the more dissipative model reproduces all directly observable main features of all Oo II RRd star. (Although our values for L and

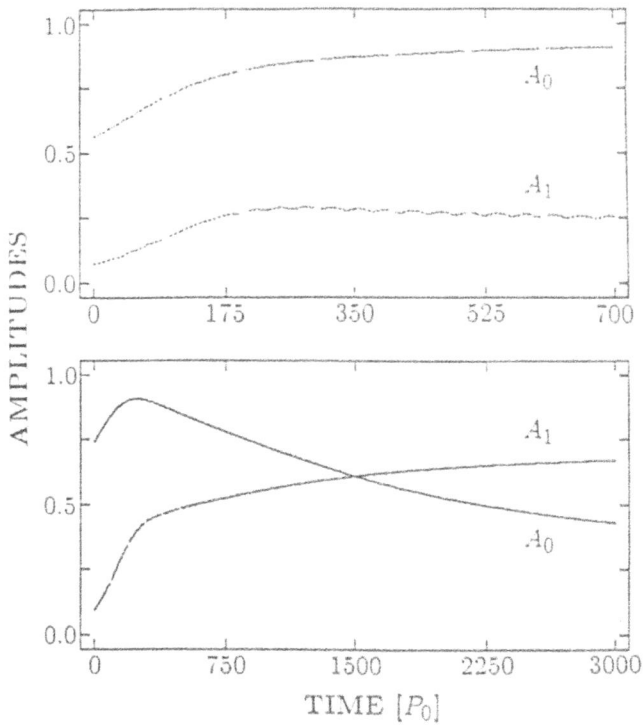

Fig. 5.   Amplitude evolution of the radius variation of the $P_0 = 0.561$ day, $M = 0.75 M_\odot$, $L = 35 L_\odot$, $T_{\text{eff}} = 5900$ K, $Z = 0.0001$ model with $C_Q = 1$, $\alpha = 0.004$ (upper panel) and $C_Q = 1$, $\alpha = 0.0$ (lower panel). The amplitudes are given in arbitrary units.

T_{eff} are certainly lower for this model than the 'standard' one, we consider this discrepancy of a secondary importance. These non-direct observables contain large uncertainties, therefore, they enter with lower weights in a comparison with the observations.)

   To illustrate the persistence of the double-mode behavior in the case of higher spatial resolution, we integrate a 90 zone model. The parameters are: $M = 0.75 M_\odot$, $L = 40 L_\odot$, $T_{\text{eff}} = 6200$ K, $Z = 0.0$; *i.e.* except for Z, this is the same model as shown in Fig. 4. The artificial viscosity parameters are also different, namely $(C_Q, \alpha) = (2, 0.006)$ in contrast to $(4, 0.07)$ used in the simulation shown in Fig. 4. We note that in the present form of the artificial viscosity, its effect on the behavior of the model depends on the number of zones. The same viscosity parameters for a finer zoned model cause larger amplitudes and possibly more violent behavior than for the coarse zoned model. The amplitude evolution shown in Fig. 6 proves that the double-mode behavior is maintained for higher zoned models too, without destroying the numerical stability and even improving the amplitude ratio.

   We conclude this section by emphasizing that the latest simulations with

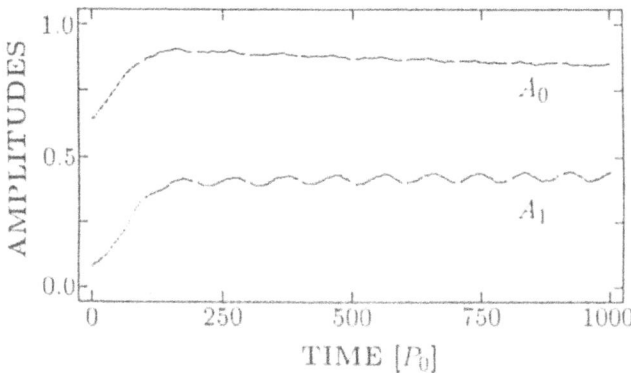

Fig. 6.   Amplitude evolution of the light variation of a 90 zone $P_0 = 0.529$ day, $M = 0.75 M_\odot$, $L = 40 L_\odot$, $T_{\text{eff}} = 6200$ K, $Z = 0.0$ model with $C_Q = 2$, $\alpha = 0.006$. The amplitudes are given in arbitrary units.

a less dissipative purely radiative code strongly indicate that permanent double-mode behavior is possible at least for RRd models. These simulations are by no means devoted to modelling the detailed nature of double-mode variables, but rather are meant to stimulate further investigations with other, more accurate and less dissipative codes.

6. Conclusions (and Questions)

Understanding double-mode variables is one of the main objectives of the theory of nonlinear stellar pulsations. The simple and indisputable observational constraints (periods and amplitudes) put clear criteria on the acceptance of different models. The reasonable assumption of purely radial pulsation gives us the hope to model this simplest type of multimode pulsation with 1-D hydrodynamical codes. The achievement of this goal is still in the more distant future for the vast majority of multimode pulsators which are oscillating in non-radial modes.

Though the observations are fairly extensive in some cases, there are many unanswered questions, that could fairly simply be answered by employing more accurate (but still standard) observational techniques. More specifically, some of the questions to be addressed are as follows.

(1) Stability of the light variation. Especially for globular cluster RRd stars the accuracy of the photographic observations do not allow us to make any strict statement about this problem. If there were sizable fluctuations in the amplitudes. the theoretical interpretation of the double-mode variables could be quite different from the one we have for a pulsation of constant amplitude.

(2) Are there any other modes besides the two principal ones excited in

double-mode variables? The beautiful new analysis of AI Vel by Walraven *et al.* (1992) shows that other modes at much lower amplitudes may well be excited. We think it is needless to emphasize the theoretical importance of the discovery of these low-amplitude modes. If some of them are radial modes then simple linear theory puts very strict restrictions on the parameters of the model. Excitation of these modes should tell us also something about the mechanism of double- (now multi-) mode pulsation.

(3) Statistics of the double-mode stars. What is the relation between cluster properties and the presence (or absence) of RRd stars? Why do we have only one RRd star in our galaxy? How many dwarf and beat Cepheids are in extragalactic objects?

(4) Is there any relation between some basic physical properties (like chemical composition, rotational velocity) and double-mode pulsation?

From the theoretical side, there is certainly a lot more to do in finding the cause of double-mode behavior and construct physically sound models. There are, however, two encouraging developments which may lead to some progress in the very near future.

(a) Turbulent convection may drive some modes which are marginally stable in some limiting state (either single- or double-mode). This leads to the possibility of observing a few additional modes (including radial ones) besides the large-amplitude principal modes. The dwarf Cepheid AI Vel could be just the first example for this type of mode excitation.

(b) Decreased artificial viscosity dissipation leads to double-mode pulsations in current purely radiative RR Lyrae models. The periods of the nonlinear models now fit the observed values with reasonable stellar parameters, but the amplitude ratio depends sensitively on the specific choice of the artificial viscosity parameters.

It is clear, there is now an even more urgent need to develop less dissipative, more accurate nonlinear codes. A progress has recently been made in this field (Dorfi and Feuchtinger 1991; Cox, Deupree and Gehmeyr 1991; Gehmeyr 1991) and there are certainly more results to be expected in the near future (Buchler and Marom 1992). In addition, because of the excessive amplitudes, there is not doubt that the unphysical artificial viscosity should be replaced by (or supplemented with) some physical mechanism, such as turbulent viscous dissipation. Though the correct treatment of turbulent convection in 1- D will obviously remain a problem for many years, some approximate parametrized model might be enough to cure the problem of the amplitudes.

Acknowledgements

The author is grateful to the LOC of this conference, and especially to Dr. Mine Takeuti for the hospitality and for the financial support. The travel grant from the Institute for Fundamental Theory in the Physics Department of the University of Florida is also very much acknowledged. Fruitful discussions with Robert Buchler and Ariel Marom were very useful during the preparation of this paper. This work was supported by NSF (AST 89-14425) and by an RCI grant through IBM and the NER Data Center at the University of Florida.

References

Andreasen, G. K.: 1988, *Astronomy and Astrophysics* **191**, 71.
Balona, L. A.: 1985, in *Cepheids: Theory and Observations*, Proc. IAU Coll. No 82, ed. B. F. Madore (Cambridge University Press), p. 17.
Buchler, J. R.: 1985, in *Chaos in Astrophysics*, NATO ASI Ser. C, Vol. 161, eds. J. R. Buchler, J. M. Perdang and E. A. Spiegel (Dordrecht: Reidel), p. 137.
Buchler, J. R.: 1992, *these proceedings*.
Buchler, J. R. and Kovács, G.: 1986, *Astrophysical Journal* **308**, 661.
Buchler, J. R. Kovács, G.: 1992, *Physica D*, submitted.
Buchler, J. R., Goupil, M.-J. and Kovács, G.: 1992, *Astronomy and Astrophysics*, submitted.
Buchler, J. R. and Marom, A.: 1992, in preparation.
Buchler, J. R., Moskalik, P., and Kovács, G.: 1990, *Astrophysical Journal* **351**, 617.
Buchler, J. R., Moskalik, P., and Kovács, G.: 1991, *Astrophysical Journal* **380**, 185.
Christensen-Dalsgaard, J. and Frandsen, S.: 1983, *Solar Phys.*, **82**, 469.
Clement, C. M. and Walker, I. R.: 1991, *Astronomical Journal* **101**, 1352.
Cox, A. N.: 1982, in *Pulsations in Classical and Cataclysmic Variable Stars*, eds. J. P. Cox and C. J. Hansen (Boulder: JILA), p. 157.
Cox, A. N.: 1987, in *The Second Conference on Faint Blue Stars*, eds. A. G. D. Philip, D. S. Hayes and J. W. L. Liebert (L. Davis Press, Inc., Schenectady, New York), p. 161.
Cox, A. N.: 1991, *Astrophysical Journal, Letters to the Editor* **381**, L71.
Cox, A. N., Deupree, R. G. and Gehmeyr, M.: 1991, in *Experimental Mathematics: Computational Issues in Nonlinear Science*, Proc. of the CNLS 11th Annual Conference, held in Los Alamos, May 20–24, 1991 (preprint).
Cox, A.N., Hodson, S. W. and Clancy, S. P.: 1983, *Astrophysical Journal* **266**, 94.
Cox, A. N. , Hodson, S. W. and Davey, W. R.: 1976, in *Solar and Stellar Pulsation*, eds. A. N. Cox and R. G. Deupree (Los Alamos: LA-6544-C), p. 188.
Dorfi, E. A. and Feuchtinger, M. U.: 1991, *Astronomy and Astrophysics* **249**, 417.
Dziembowski, W. and Kovács, G.: 1984, *Monthly Notices of the RAS* **206**, 497.
Dziembowski, W. and Krolikowska, M.: 1985, *Acta Astr.* **35**, 5.
Dziembowski, W., Krolikowska, M. and Kosovitchev, A.: 1988, *Acta Astr.* **38**, 61.
Fernley, J. A., Jameson, R. F., Sherrington, M. R. and Skillen, I.: 1987, *Monthly Notices of the RAS* **225**, 451.
Fitch, W.: 1980, *Lecture Notes in Physics* **125**, p. 7.
Gehmeyr, M.: 1991, *On Non-Lagrangian Computations of Convective RR Lyrae Stars*, Ph. D. Thesis, University of New Mexico.
Goldreich, P. and Keeley, D. A.: 1977, *Astrophysical Journal* **212**, 243.
Goranskij, V. P.: 1981. *Inf. Bull. Var. Stars*, No. 2007.
Hodson, S. W. and Cox, A. N.: 1982, in *Pulsations in Classical and Cataclysmic Variable Stars*, eds. J. P. Cox and C. J. Hansen (Boulder: JILA), p. 201.

Iglesias, C. A. and Rogers, F. J.: 1991, *Astrophysical Journal, Letters to the Editor* **371**. L73.

Jerzykiewicz, M.: 1988, in *Multimode Stellar Pulsations*, eds. g. Kovács, L. Szabados and B. Szeidl, Konkoly Observatory (Kultura: Budapest), p. 19.

Jerzykiewicz, M. and Wenzel, W.: 1977, *Acta Astr.* **27**. 35.

Jurcsik, J. and Barlai, K.: 1990, in *Confrontation Between Stellar Pulsation and Evolution*, Astron. Society of the Pacific. Conf. Ser., vol. 11, eds. C. Cacciari and G. Clementini, p. 112.

King, D. S., Cox, J. P., Eilers, D. D. and Davey, W. R.: 1973, *Astrophysical Journal* **182**, 859.

Kovács, G.: 1990, in *The Numerical Modelling of Nonlinear Stellar Pulsations; Problems and Prospects*, ed. J. R. Buchler (Kluwer, Dordrecht), p. 73.

Kovács, G. and Buchler, J. R.: 1988, *Astrophysical Journal* **324**, 1026.

Kovács, G. and Buchler, J. R.: 1992, *Astrophysical Journal*, submitted.

Kovács, G., Shlosman, I., and Buchler, J. R.: 1986, *Astrophysical Journal*, **307**, 593.

Kovács, G., Buchler, J. R., and Marom, A.: 1991, *Astronomy and Astrophysics (Letters)* **252**, L27.

Kovács, G., Buchler, J. R., Marom, A., Iglesias, C. A. and Rogers, F. J.: 1992, *Astronomy and Astrophysics (Letters)*, submitted.

Kovács, G. and Kolláth, Z.: 1988, in *Multimode Stellar Pulsations*, eds. G. Kovács, L. Szabados and B. Szeidl, Konkoly Observatory (Kultura: Budapest), p. 33.

Moskalik, P.: 1985, *Acta Astr.* **35**, 229.

Moskalik, P., Buchler, J. R. and Marom, A.: 1991, *Astrophysical Journal* **385**, 685.

Nemec, J. M. and Clement, C. M.: 1989, *Astronomical Journal* **98**, 960.

Nemec, J. M. and Mateo, M.: 1990, in *Confrontation Between Stellar Pulsation and Evolution*, Astron. Society of the Pacific. Conf. Ser., Vol. 11, eds.: C. Cacciari and G. Clementini, p. 64.

Ostlie, D. A.: 1990, in *The Numerical Modelling of Nonlinear Stellar Pulsations; Problems and Prospects*, ed. J. R. Buchler (Kluwer, Dordrecht), p. 89.

Petersen, J. O.: 1973, *Astronomy and Astrophysics* **27**, 89.

Rogers, F. J. and Iglesias, C. A.: 1992. *Astrophysical Journal, Supplement Series* **79** (in press, April).

Simon, N. R.: 1979, *Astronomy and Astrophysics* **75**, 140.

Simon, N. R.: 1982, *Astrophysical Journal, Letters to the Editor* **260**, L87.

Simon, N. R. Cox, A. N. and Hodson. S. W.: 1980, *Astrophysical Journal* **237**, 550.

Simon, N. R. and Schmidt, E. G.: 1976, *Astrophysical Journal* **205**, 162.

Stellingwerf, R. F.: 1975, *Astrophysical Journal* **195**, 441.

Stellingwerf, R. F.: 1984, *Astrophysical Journal* **284**. 712.

Stobie, R. S.: 1969, *Monthly Notices of the RAS* **144**, 511.

Szabados, L.: 1988, in *Multimode Stellar Pulsations*, eds. G. Kovács, L. Szabados and B. Szeidl, Konkoly Observatory (Kultura: Budapest), p. 1.

Szeidl, B.: 1988, in *Multimode Stellar Pulsations*, eds. G. Kovács, L. Szabados and B. Szeidl, Konkoly Observatory (Kultura: Budapest), p. 45.

Walraven, Th., Walraven, J. and Balona, L. A.: 1992, *Monthly Notices of the RAS* **254**, 59.

Wheeler, J. C., Sneden, C., and Truran, J. W.: 1989, *Annual Review of Astronomy and Astrophysics* **27**, 279.

Zalewski, J.: 1992, *these proceedings*.

PERIOD DOUBLING WITH HYSTERESIS IN
BL HER-TYPE MODELS

P. MOSKALIK

Copernicus Astronomical Center, ul. Bartycka 18,00-716 Warsaw, Poland

and

J. R. BUCHLER

University of Florida, Department of Physics, Gainesville, FL 32611, USA

1. Hydrodynamics

We have performed recently a survey of the nonlinear hydrodynamical models of the BL Her-type variables (Buchler & Moskalik 1992). Within this project we have studied several sequences of models, *i.e.*, families in which *only* T_{eff} has been varied from model to model, while all other stellar parameters have been kept constant. The fundamental mode pulsations of each model have been converged to strict periodicity with the relaxation code (Stellingwerf 1974). Such approach speeds up the calculations and simultaneously yields information about the stability properties of the resulting limit cycles. In all studied sequences except one, we have found a narrow range of T_{eff} (typically 100–150K), in which regular solution becomes unstable towards a period doubling bifurcation. The instability has its origin in a half-integer resonance, namely the 3:2 coupling between the fundamental mode and the first overtone (*cf.* Moskalik & Buchler 1990; hereafter MB90). This is the same resonance, which also causes period doubling in the models of classical Cepheids (Moskalik & Buchler 1991). The bifurcation leads to stable *period-two* oscillations, characterized by an RV Tau-like, albeit *strictly periodic* behavior of all variables. In other words, the pulsation light curves and velocity curves will exhibit two alternating minima (as well as maxima) of different values.

In Fig. 1 we show the bifurcation for one of our sequences, by plotting minimum pulsational velocities V_{min} versus T_{eff}. Filled (open) circles correspond to stable (unstable) period-one limit cycles, while asterisks represent differing minima of the period-two solutions. The latter are always stable. The alternations are very pronounced and can reach up to 23.4 km/s. Their size gradually decreases as the edges of the period doubling window are approached. On the low temperature edge of the instability domain, though, the alternations clearly *do not vanish*. This curious behavior prompted us to perform a detailed study of this temperature range.

In Fig. 2 we display the behavior of the models with T_{eff} around 5800K.

Astrophysics and Space Science **210**: 301–305, 1993.

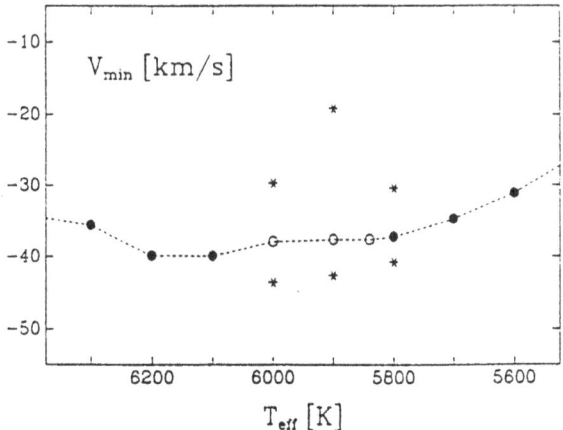

Fig. 1.   Minimum pulsational velocities (corrected for the limb darkening) *vs.* $T_{\rm eff}$ for sequence of models with $M = 0.55 M_\odot$, $L = 125 L_\odot$, $X = 0.7$ and $Z = 0.001$. Circles correspond to period-one limit cycles. Solutions, which are unstable towards period doubling are marked with open symbols. Asterisks represent (alternating) minima of the stable period-two cycles.

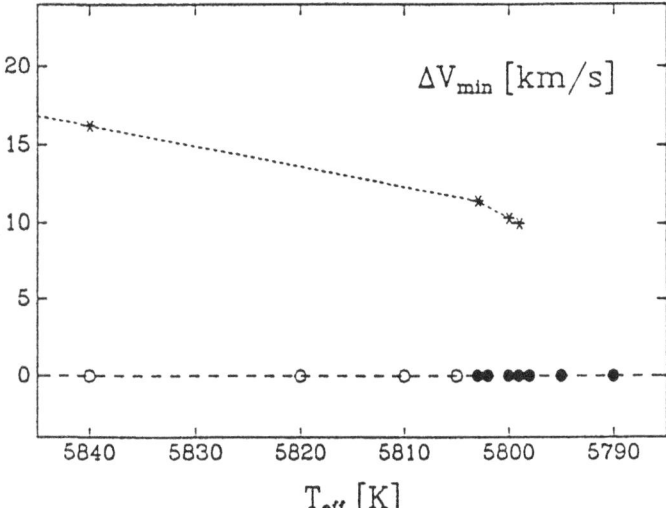

Fig. 2.   Differences between velocity minima (corrected for the limb darkening) *vs.* $T_{\rm eff}$ for sequence of Fig. 1. All symbols have the same meaning as in Fig. 1. The period-one limit cycles change their stability at $T_{\rm eff} \simeq 5805$K. For $T_{\rm eff} \in (5799{\rm K}{-}5805{\rm K})$ a hysteresis occurs.

For better visualization we plot here the *differences* between consecutive velocity minima, $\Delta V_{\rm min}$, again *versus* $T_{\rm eff}$ of the model. The period-one pulsations (circles), correspond now to $\Delta V_{\rm min} = 0$, since their minima are all equal. The period-two limit cycles (asterisks), on the other hand, correspond to $\Delta V_{\rm min} > 0$. We see that at the point of the stability exchange ($T_{\rm eff} \simeq 5805$K) the alternations indeed do not decrease to zero. Instead, the branch of the period-two solutions extends beyond that point and for $5799{\rm K} \leq T_{\rm eff} <$

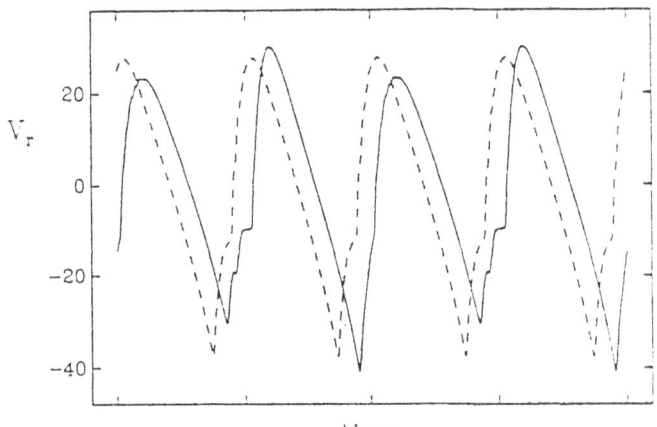

time

Fig. 3.   Pulsational velocity curves (corrected for the limb darkening) for the model with $M = 0.55M_\odot$, $L = 125L_\odot$, $X = 0.7$, $Z = 0.001$ and $T_{\text{eff}} = 5800$K; *solid line*: the period-two limit cycle; *dashed line*: the period-one limit cycle. Both solutions are stable.

5805K two different stable solutions can *coexits*: the period-one limit cycle and the period-two limit cycle. That means that a *hysteresis* occurs over that narrow interval of T_{eff} and that the pulsational state of the star will depend here on its evolutionary history. We have checked that this hysteresis is robust with respect to the numerical parameters (*i.e.*, number of timesteps) and it is certainly not a computational artifact. In Fig. 3 we plot the velocity curves of the two different stable solutions for the model of $T_{\text{eff}} = 5800$K. Both curves are quite similar, nevertheless the alternations in one of them are clearly visible, whereas the other one repeats after each cycle.

2. Amplitude Equations

The behavior of the hydrodynamical stellar models and their bifurcations can be captured qualitatively and quantitatively by the amplitude equations (*cf.* Buchler 1992). The apposite set of equations for the 3:2 resonance case has the form (MB90)

$$\frac{dA}{dt} = \kappa_0 A + ReQ_0 A^3 + ReT_0 AB^2 + Re(\Pi_0 e^{i\Gamma})A^2 B^2$$

$$\frac{dB}{dt} = \kappa_1 B + ReQ_1 B^3 + ReT_1 A^2 B + Re(\Pi_1 e^{-i\Gamma})A^3 B$$

$$\frac{d\Gamma}{dt} = 2\Delta\omega + Im(2T_1 - 3Q_0)A^2 + Im(2Q_1 - 3T_0)B^2 +$$

$$+2Im(\Pi_1 e^{-i\Gamma})A^3 - 3Im(\Pi_0 e^{i\Gamma})AB^2 \qquad (1)$$

where A and B are the (real) amplitudes of the fundamental mode and of the first overtone, and the quantity Γ is a linear combination of the respective modal phases: $\Gamma = 2\phi_1 - 3\phi_0$. The coefficients Q_j and T_j describe the

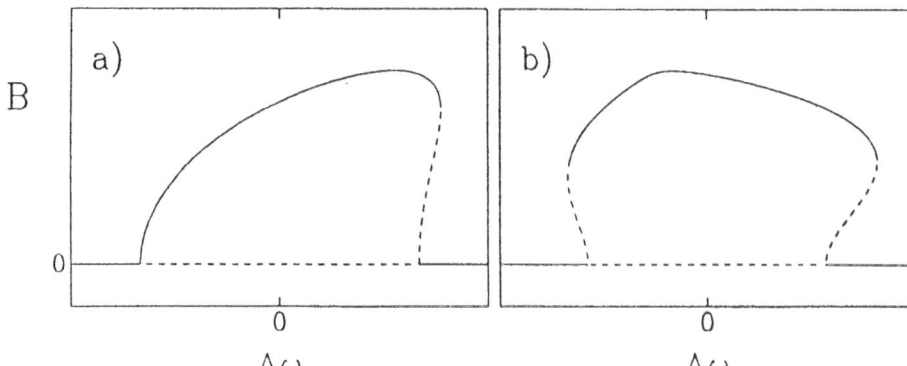

Fig. 4.   Amplitude of the first overtone, B, for the fixed points of Eqs.(1) $vs.$ resonance distance parameter $\Delta\omega$. Adopted parameters are: (a)$\Pi_0 = 0$, $\kappa_1 = \kappa_0$, $ReQ_0 = -2^{-1/3}\kappa_0^{1/3}|\Pi_1|^{2/3}$, $ReT_0 = 0$, $ReT_1 = -2^{2/3}\kappa_0^{1/3}|\Pi_1|^{2/3}$, $Im(2T_1 - 3Q_0) = 0$, $Im(2Q_1 - 3T_0) = 4ReQ_1$ and (b) $\kappa_1 = 3\kappa_0$, $ReQ_0 = -2^{-1/3}\kappa_0^{1/3}|\Pi_1|^{2/3}$, $ReT_0 = 0$, $ReQ_1 = -5 \times 2^{2/3}\kappa_0^{1/3}|\Pi_0||\Pi_1|^{-1/3}$, $ReT_1 = -2^{2/3}\kappa_0^{1/3}|\Pi_1|^{2/3}$, $Im(2T_1 - 3Q_0) = 0$, $Im(2Q_1 - 3T_0) = 0$, $\arg(\Pi_0\Pi_1) = 2.5$.

nonresonant nonlinear effects (namely saturation of the driving mechanism) and the coupling coefficients Π_j measure the strength of the resonant interaction. The driving rates κ_j as well as $\Delta\omega = \omega_1 - \frac{3}{2}\omega_0$ are given by the linear pulsation theory. The last parameter measures the distance to the resonance center. Within the framework of the amplitude equation formalism, the surface radius displacement can be expressed as

$$\delta R(t) = Ae^{i\omega_0 t} + Be^{i\frac{\Gamma}{2}}e^{\frac{3}{2}i\omega_0 t} + higher\ order\ terms. \qquad (2)$$

According to this formula, a fixed point of Eqs. (1) corresponds to a limit cycle oscillation of the stellar model. In particular, the fixed point with $B \neq 0$ represents the period-two cycle, since $\delta R(t)$ varies then with the period of $2P_0(P_0 = 2\pi/\omega_0)$. To the lowest order, the amplitude B measures the size of alternations in such a cycle. The solution with $B = 0$ corresponds to the period-one oscillations.

The fixed points of Eqs.(1) and their stability has been discussed in MB90. The authors have shown that the equations can reproduce very well the period doubling behavior found in the classical Cepheid models. In those models, though, the bifurcation is of the supercritical type, and the alternations in the period-two solutions vanish at the points of the stability exchange. It is interesting to check whether Eqs.(1) can also capture the subcritical bifurcation (*i.e.*, hysteresis) encountered in the BL Her models.

Fig. 4 shows that Eqs.(1) are indeed capable of reproducing such a behavior. The amplitude of the first overtone, B, has been calculated and plotted here as a function of $\Delta\omega$, with all other coefficients in the equations kept constant. Solutions displayed in Fig. 4a have been obtained for the simplified system with $ReT_0 = 0$ and $\Pi_0 = 0$, although in contrast to MB90 we have assumed that $Im(2T_1 - 3Q_0)$ and $Im(2Q_1 - 3T_0)$ are in general nonzero. The unstable fixed points are marked with the dashed lines. We see, that for $\Delta\omega > 0$, which corresponds to the low temperature side of the period doubling window, the period-two solution ($B \neq 0$) extends beyond the bifurcation point, bending back subsequently to join that point through the unstable branch. Thus, over the narrow interval of $\Delta\omega$ the stable period-two and period-one solutions coexist, just like in the hydrodynamical models. According to the amplitude equations, they are accompanied by the *unstable* period-two solution. In the simplified case considered in Fig. 4a it can be shown analytically, that the hysteresis will occur only if

$$|Im(2Q_1 - 3T_0)| > 2ReQ_1 \frac{\kappa_1 + ReT_1 A_0^2}{\sqrt{|\Pi_1|^2 A_0^6 - (\kappa_1 + ReT_1 A_0^2)^2}} \tag{3}$$

where $A_0^2 = -\kappa_0/ReQ_0$. Depending on the sign of $Im(2Q_1 - 3T_0)$, it will occur either for positive for negative $\Delta\omega$, but never for both. This latter property does not hold in the general case (*i.e.*, $ReT_0 \neq 0$, $\Pi_0 \neq 0$), when the hysteresis can be found on both sides of the period doubling window. An example of such a behavior is presented in Fig. 4b.

References

Buchler, J. R.: 1992, this conference.
Buchler, J. R. and Moskalik, P.: 1992, *Astrophysical Journal*, in press.
Moskalik, P. and Buchler, J. R.: 1990, *Astrophysical Journal* **355**, 590 (MB90).
Moskalik, P. and Buchler, J. R.: 1991, *Astrophysical Journal* **366**, 300.
Stellingwerf, R. A.: 1974, *Astrophysical Journal* **192**, 139.

NONLINEAR RR LYRAE MODELS WITH NEW

LIVERMORE OPACITIES

J. A. GUZIK and A. N. COX
Los Alamos National Laboratory

Abstract. A. N. Cox recently showed that a 20% opacity decrease in the 20,000-30.000 K region as indicated by the new Livermore OPAL opacities reconciles the discrepancy between pulsation and evolution masses of double-mode RR Lyrae variables. Nonlinear hydrodynamic calculations were performed for RR Lyrae models of mass $0.75M_\odot$, $51L_\odot$. and $Z = 0.0001$ (Osterhoff II type) including this opacity decrease. The Stellingwerf periodic relaxation method was used to converge the models to a limit cycle, and the Floquet matrix eigenvalues calculated to search for a tendency of the fundamental mode to grow from the full-amplitude overtone solution, and the overtone mode to grow from the full-amplitude fundamental solution, thereby predicting double-mode behavior. Models of $T_{eff} < 7000$ K with the opacity decrease have positive fundamental-mode growth rates in the overtone solution, in contrast to earlier results by Hodson and Cox (1982), and models with $T_{eff} > 7000$ have positive 1st overtone growth rates in the fundamental-mode solution, but double-mode behavior was not found.

A. N. Cox (1991) recently showed that the 20% opacity decrease of the new Livermore (1991) opacities compared to the Los Alamos opacities for $T < 100,000$ K removes the discrepancy between pulsation and evolution masses of double-mode RR Lyrae variables. This conclusion was verified by Petersen (1992). This paper presents nonlinear calculations for RR Lyrae models of $0.75M_\odot$, $51L_\odot$, and $Z = 0.0001$ (Osterhoff II type) with this opacity decrease to search for a tendency of the fundamental mode to grow from the full-amplitude overtone solution, and the overtone mode to grow from the full-amplitude fundamental solution, thereby predicting double-mode behavior.

The 60-zone radiative models have envelope masses 4 to 8×10^{30} g, with temperature at the base $\sim 600,000$ K, and initial radial velocity amplitudes 20–40 km/s. The Stellingwerf (1975a, b) analytical fit to the Cox-Tabor (1976) opacity tables is modified by an opacity ramp decreasing from unity to 0.8 between 10,000 and 20,000 K, remaining constant at 0.8 between 20.000 and 30,000 K, and increasing to unity again by 100,000 K. Comparisons of the Stellingwerf fit to the OPAL opacities show that the fit is adequate for other temperatures (Iglesias and Rogers 1991). The nonlinear code described by Ostlie (1990, see also review by Cox and Ostlie 1992) is used to run several hydrodynamic cycles, to converge to a periodic limit cycle via the Stellingwerf periodic relaxation method, and to check for instability of other modes by analysing the Floquet matrix. The models use artificial viscosity parameters $C_Q = 4$, and $\alpha = 0.02$. The results do not seem to be sensitive to the choice of these parameters, in contrast to the findings of Kovacs

Astrophysics and Space Science 210: 307–309, 1993.

TABLE I

Period, radial velocity amplitudes, and growth rates from Floquet matrix for
RR Lyrae models.

1st Overtone Solutions

$T_{\rm eff}$ (K)	Period (d)	$V_{\rm min}$ (km/s)	$V_{\rm max}$ (km/s)	F in 1H (per period)
6800	0.3487	−21.4	20.6	0.000955
6900	0.3309	−20.7	20.5	0.000687
7000	0.3183	−20.9	24.1	−0.000476
7100	0.3009	−20.5	21.6	−0.000990

Fundamental Mode Solutions

$T_{\rm eff}$ (K)	Period (d)	$V_{\rm min}$ (km/s)	$V_{\rm max}$ (km/s)	1H in F (per period)
6800	0.4721	−34.1	41.4	−0.060858
6900	0.4487	−31.3	38.0	−0.035650
7000	0.4273	−27.8	31.8	−0.015758
7100	0.4063	−20.2	23.2	0.004022
7150	0.3963	−2.9	2.8	0.015353

(1992) and Kovacs and Buchler (1988). Table I summarizes the results for
the fundamental and 1st overtone limit cycles for a grid of models with
$T_{\rm eff} = 6800$–7150 K.

In their nonlinear investigation of RR Lyrae models, Hodson and Cox
(1982) did not find an effective temperature region with positive fundamental growth rates in the first overtone solution. We find that the 20% opacity
decrease for $T < 100,000$ K indicated by the OPAL opacities causes the
fundamental mode growth rates in the first overtone solutions to becomes
positive for models with $T_{\rm eff}$ less than 7000 K. Still, we find no overlapping
effective temperature region where the overtone mode is predicted to grow
from the fundamental mode solution. Opacity adjustments such as widening or deepening the ramp on the Stellingwerf fit do change the growth
rates somewhat. For example, widening the temperature range with the full
20% decrease to between 10,000 and 40,000, makes the 1H in F growth
rate/period at 7000 K slightly less negative (-0.00037), and the F in 1H
growth rate/period positive (0.0023). Additional studies are in progress.

In this study and in the analysis of Cox (1991), it appears most promising
to search for double-mode RR Lyrae models just redward of fundamental
blue edge, at $T_{\rm eff} \sim 7000$ K, whereas Kovacs and Buchler (1988) and Kovacs
(1992) find double-mode behavior for much cooler models ($T_{\rm eff} = 6100$–
6500 K).

References

Cox, A. N.: 1991, *Astrophysical Journal* **381**, 171.

Cox, A. N. and Ostlie, D. A.: 1992, "A Linear and Nonlinear Study of Mira," *Nonlinear Phenomena in Stellar Variability*, IAU 134, Mito, Japan, Jan. 7–10.

Cox, A. N. and Tabor, J. E.: 1976, *Astrophysical Journal, Supplement Series* **31**, 271.

Iglesias, C. A. and Rogers, F. J.: 1991, *Astrophysical Journal* **371**, L73

Kovacs, G.: 1992, "Double-Mode Stellar Pulsation," *Nonlinear Phenomena in Stellar Variability*, IAU 134, Mito, Japan. Jan. 7–10.

Kovacs, G. and Buchler, J. R.: 1988, *Astrophysical Journal* **324**, 1026.

Ostlie, D. A.: 1990, "Time-Dependent Convection in Stellar Pulsation," in *proceedings of conference* in Les Arcs, France.

Petersen, J. O.: 1992, "Kappa-Effect Functions for Period Ratios," preprint.

Stellingwerf, R. F.: 1975a, *Astrophysical Journal* **195**, 441.

Stellingwerf, R. F.: 1975b, *Astrophysical Journal* **199**, 705.

A LINEAR AND NONLINEAR STUDY OF MIRA

A. N. COX

Theoretical Division, Los Alamos National Laboratory

and

D. A. OSTLIE

Weber State University

Abstract. Both linear and nonlinear calculations of the 331 day, long period variable star Mira have been undertaken to see what radial pulsation mode is naturally selected. Models are similar to those considered in the linear nonadiabatic stellar pulsation study of Ostlie and Cox (1986). Models are considered with masses near one solar mass, luminosities between 4000 and 5000 solar luminosities, and effective temperatures of approximately 3000 K. These models have fundamental mode periods that closely match the pulsation period of Mira. The equation of state for the stellar material is given by the Stellingwerf (1975ab) procedure, and the opacity is obtained from a fit by Cahn that matches the low temperature molecular absorption data for the Population I Ross-Aller 1 mixture calculated from the Los Alamos Astrophysical Opacity Library. For the linear study, the Cox, Brownlee, and Eilers (1966) approximation is used for the linear theory variation of the convection luminosity. For the nonlinear work, the method described by Ostlie (1990) and Cox (1990) is followed. Results showing internal details of the radial fundamental and first overtone modes behavior in linear theory are presented. Preliminary radial fundamental mode nonlinear calculations are discussed. The very tentative conclusion is that neither the fundamental or first overtone mode is excluded from being the actual observed one.

1. Background

Mira variables have been studied observationally for hundreds of years, because many of them are bright enough to be seen with the naked eye or with small telescopes, and their periods are so long that constant attention is not needed. Theoretical studies, however have been far fewer, mostly because the extensive and deep convection is difficult to cope with in the context of stellar pulsation. In this paper we present new theoretical results, but a definitive conclusion for which radial mode of pulsation is actually occurring still eludes us.

The basic parameters of Mira variables are their pulsation periods (or many periods for the red semiregular variables), their radii, and their luminosities. Surface compositions seem to be at least hydrogen rich as for most stars, even though these stars are obviously remnants of more massive stars that have undergone significant mass loss. Thus the composition of the pulsating envelopes. while surely homogeneous, may not be that for normal solar type stars. Numerous composition anomalies are known. For example, these red giants are often in spectral classes R, N, and S. Mira masses are unknown and disputed, but most seem to come from stars with original mass of less than 2 solar masses. More massive progenitors become red supergiants

Astrophysics and Space Science **210**: 311–319, 1993.
© 1993 *Kluwer Academic Publishers.*

and even Mira variables, but their red supergiant lifetimes must be short
as they rapidly lose mass to create planetary nebulae and ultimately white
dwarfs. Most Miras have currently near only one solar mass.

The evolution of the Mira stars is just at the Hayashi line on the Hertzsprung-Russell diagram, because their luminosity is very large for their mass.
The vary low mass envelopes result from the high luminosity blowing out
the matter to large radii. The low temperature of the matter gives it a high
opacity, and consequently strong convection to carry the high luminosity.

Nonlinear calculations to study the behavior of the Mira pulsations have
been made by Wood (1974), by Tuchman, Sack, and Barkat (1978, 1979),
and by Perl and Tuchman (1990). We at Los Alamos have been trying to
do these calculations for 10 years, but we have not been able to be satisfied
that our results are correct. We believe that the neglect of turbulent pressure
in all previous calculations is a significant deficiency, and the instantaneous
adaptation of convection by the Israel authors may be a bad approximation.
Studies of the upper atmospheres of Miras have also been carried out by
numerous groups using a driving piston boundary condition near the photosphere, with the latest papers being by Bowen (1988) and Beach (1990).
The results of these various studies give conflicting conclusions as to the
pulsation mode of Mira. Our conclusion is that either the fundamental or
the first overtone mode is allowed, but our demonstration is not yet secure.

2. Time-Dependent Convection Procedure

In both our linear and nonlinear studies we have adopted the approximate
time-dependent theory of Cox, Brownlee, and Eilers (CBE, 1966). The concept is that the instantaneous conditions suggest a convection luminosity
that cannot be realized because the turbulent eddies have an inertia that
the hydrodynamic forces cannot overcome rapidly. Thus there is a lag behind the desired luminosity that can be represented by the formula for the
convection luminosity increment:

$$dL_c = \frac{dt}{\tau}[\bar{L}_c(t) - L_c(t)],$$

where $\tau(t)$ is the mean eddy lifetime at time t at the level of interest in the
convection zone, and $\bar{L}_c$ is the instantaneous convection luminosity desired.
If τ is small, convection can adapt well, but for Mira variables it is often
comparable to the pulsation period Π. The instantaneous luminosity is given
by

$$\bar{L}_c = Ae^{i\omega t},$$

where ω is $2\pi/\Pi$, and A is the amplitude for completely adapting convection.

The solution for such a model with time-dependent convection is that the
convection luminosity increment, with the τ taken constant in time, is

$$\delta L_c(t) = B \; e^{i(\omega t - \theta)},$$

where

$$B = \frac{A}{\sqrt{(1 + (2\pi\tau/\Pi)^2}},$$

and

$$\theta = \tan^{-1}(2\pi\tau/\Pi).$$

The amplitude B actually realized is often much smaller than the amplitude A. This is because as the convection is struggling to reach its amplitude, the pulsation configuration changes sign rapidly, overcoming any convection luminosity changes before they can be attained.

Implementation of this time-dependent convection involves first calculating the partial derivatives of the convection luminosity with respect to the temperature and density on both sides of an interface between Lagrangian mass shells in the stellar model. These are calculated during model construction. They are then converted to derivatives with respect to the perturbations of our linear theory, that is δr and $T\delta S$. Then these terms are added to the existing matrix elements (coefficients of the perturbation variables) that represent the linearized momentum and energy equations for the stellar mass elements. Allowance is made for the fraction of the total luminosity due to convection, the actual reduced amplitude from the CBE procedure, and the cosine of the lag angle θ. Additional matrix elements are necessary for the imaginary parts of the convection luminosity variations also, and they use the sine of the lag angle instead of the cosine.

For many stellar models, the logarithmic derivatives of the convection luminosity are huge, maybe over 200! For radiation they rarely get about 15 for the temperature derivative and much smaller for the density derivative. Thus matrix elements are sometimes greatly increased, and the nonadiabatic eigenvector and eigensolution can be considerably different from the adiabatic one.

3. Mira Models

For this work we have constructed many models. The one discussed in detail here has one solar mass, a luminosity of 4000 solar luminosities, and an effective surface temperature of 3000 K. The opacity used is given by a fit by Cahn to the Ross-Aller 1 solar composition with $X = 0.70$ and $Z = 0.02$. The Stellingwerf (1975ab) analytic equation of state is used throughout. For both linear and nonlinear studies 60 mass shells have been used. Table I gives additional interior details.

The construction of our envelope models necessarily includes turbulent pressure using only the convective velocity at the exterior interface of each

TABLE I
Mira Variable Star Model

Mass: $1.0 M_\odot$ ($1.989 \times 10^{33} g$)
Luminosity (constant through model): 1.53×10^{37} erg/s ($4000 L_\odot$)
Effective Temperature: 3000 K (1.628×10^{13} cm)
Composition (constant through model): $X = 0.70$ $Z = 0.02$
Mass Shells: 60
Last Shell Mass: $7 \times 10^{28} g$ ($\tau = 3 \times 10^{-2}$)
Central Ball Mass: $9.2 \times 10^{32} g$ ($q = 0.46$)
Central Ball Radius Fraction: 0.04
Mass Ratios: range from 1.3 to 1.0 to 1.1 to 0.965
Convection Zone Temperature Range: 2.6×10^3 K to over 2.8×10^5 K
Radiative Luminosity Minimum: 2.0×10^{-5} at 15,800 K
Mixing Length/Pressure Scale Height: 2.60
Turbulent Pressure
Time Dependent Convection
Fundamental Mode Period and Kinetic Energy Growth Rate: 330 days, 1.23
First Overtone Mode Period and Kinetic Energy Growth Rate: 152 days, 0.73

mass shell, because as the integration proceeds, the interior interface convection is not yet known. When this model and its linear theory eigenvector is used to start the hydrodynamic calculations, turbulent pressure from both sides of a mass shell is needed. This is produced linearly over the first hydro period, and it affects the entire mean structure as well as the nonlinear solution. The relaxation can be done rather easily, because the time scale of the layers where the turbulent pressure is significant is typically a small number of pulsation periods.

4. Linear Results

Plots of the work per pulsational cycle to drive or damp pulsations are given in the five figures. The first one shows the hydrogen ionization driving that has been considered for years as the sole cause of the instability. This driving occurs just exterior to the strong convection zone top near 9000 K, because at the usual hydrogen driving temperature of 11,000 K, convection is too strong to allow the κ effect periodic radiation blocking. Figure 1 presents the strong fundamental and weaker overtone net driving, with extremely small damping.

Figure 2 shows a very different situation where our time-dependent convection is allowed to adapt completely to the current configuration. This means that θ is zero at all times for all mass zones. Again the fundamental mode is more strongly driven. The overtone has two peaks in its driving be-

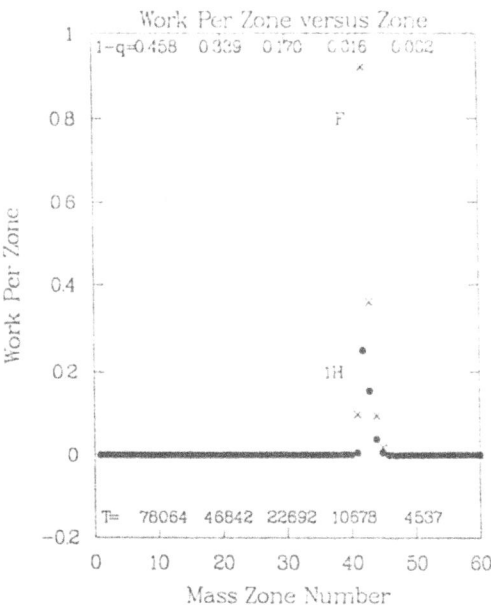

Fig. 1. The work per zone each pulsation cycle versus zone number and external mass fraction for a linear solution with frozen-in convection.

cause it has a node in its oscillation eigenvector at a temperature just over 11,000 K. The driving for both modes occurs between about one and ten percent of the mass of the star, but the damping extends deeply to almost half of the mass.

Figure 3 presents our best linear interpretation of what occurs for driving and damping in Mira stars. Here θ is allowed to be greater than zero, as appropriate for convection lagging. Note first that there are two fundamental mode peaks and three overtone peaks in the driving, whereas completely adapting convection gave only one for the fundamental and two for the overtone.

Figure 4 shows the time-dependent convection case when turbulent pressure is ignored. Again significant differences can be seen. The fundamental mode period is changed from 330 days to 343 days, but the growth rate is unchanged. For the overtone, the period is changed from 152 days to 153 days, and the growth rate is decreased from 0.73 to 0.44 per cycle relative to our best case in figure 3.

An interpretation of these time-dependent cases is displayed as Figure 5. Here $\Gamma_3 - 1$ is plotted versus the same zone numbers. This quantity reveals the relation between the temperature, density, and entropy variations as

$$\delta T/T = (\Gamma_3 - 1)\delta\rho/\rho + \delta S/c_v.$$

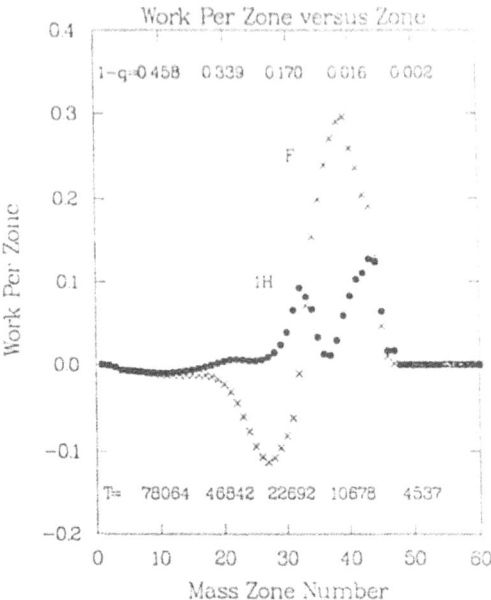

Fig. 2.   The work per zone each pulsation cycle versus zone number and external mass fraction for a linear solution with completely adaptive convection.

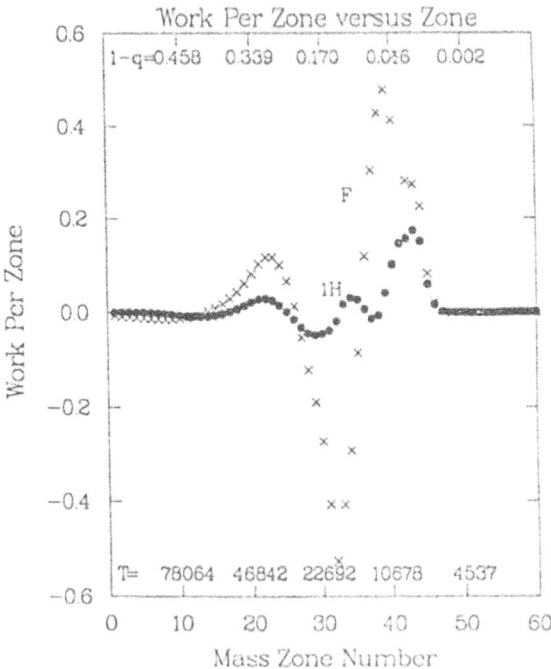

Fig. 3.   The work per zone each pulsation cycle versus zone number and external mass fraction for a linear solution with the properly lagging adaptive convection.

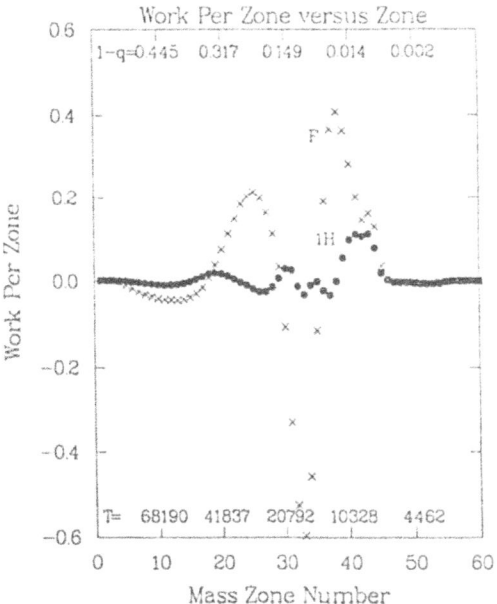

Fig. 4.   The work per zone each pulsation cycle versus zone number and external mass fraction for a linear solution with the properly lagging adaptive convection, but no turbulent pressure.

Places where $(\Gamma_3 - 1)$ is small are where the temperature fluctuations are small and where convection cannot be modulated so much. These then are places where the pulsation damping is smaller. Convection seems to drive when the $\Gamma_3 - 1$ is rapidly changing giving a rapid change in space of the temperature eigenvector. Note that the driving comes largely from the spatial variations of δr and $T\delta S$ in the eigenvector. The other factors in the convection luminosity variations, such as the fraction of the luminosity being carried by convection, the CBE amplitude, and the lag angle are only very slowly varying throughout the model.

5. Nonlinear Results

Our procedures for nonlinear calculations were described by Ostlie (1990) and Cox (1990). The main features are: Turbulent pressure is included. Turbulent viscosity, while small, is included. There is weighted spatial averaging of convective velocities from neighboring interfaces at the previous time step, with an adjustment for their velocities relative to the interface of interest. This spatially averaged velocity for an interface is time lagged according to the CBE procedure, and this velocity is then used in the mixing length luminosity formula. Finally, for an interface with a subadiabatic gradient,

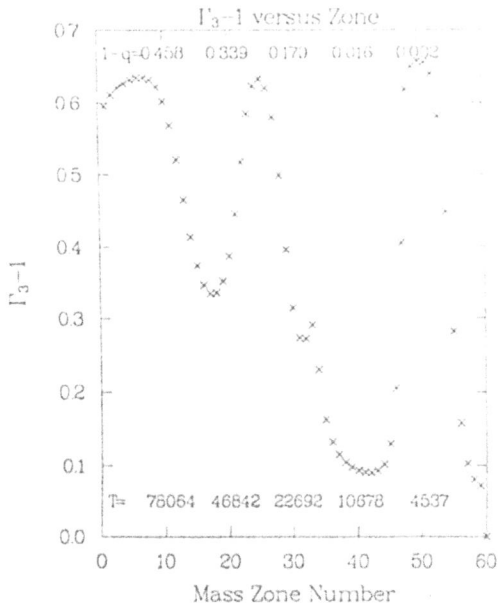

Fig. 5. The $\Gamma_3 - 1$ versus mass zone number and external mass fraction for the 3000 K, 5000 solar luminosity model.

the spatially average convective velocity is decreased in time by the mixing length theory dragging formula, and the luminosity at this subadiabatic interface is taken to be negative.

Current results for our nonlinear work include a fundamental mode case at the very long period of 998 days with an amplitude of about one kilometer per second. We have found that most models tend to decay in amplitude as they expand and cool from the hydrostatic models. Many more calculations are in progress.

6. Current Conclusions

Since turbulent pressure plays a significant role in limiting Mira pulsation amplitudes, hydrodynamic calculations without this real pressure, such as those by all earlier calculations, are probably not realistic.

Completely adapting convection produces a different phasing of the convection luminosity than that for the correct lagging, and thus results using completely adapting convection are not highly accurate.

Our use of masses at and above one solar mass produces unobserved low pulsation amplitudes, indicating that Mira masses are small.

For masses as low at 0.9 solar mass, both fundamental and overtone modes grow rapidly in amplitude and do not reach realistic limiting values.

Most important. fundamental mode pulsation for Mira variables is not excluded.

References

Beach, T. E.: 1990, thesis, Iowa State University.

Bowen, G. H.: 1988, *Astrophysical Journal* **329**, 299.

Cox, A. N.: 1990, Proceedings of the Ninth Florida Workshop in Nonlinear Astronomy.

Cox. A. N., Brownlee. R. R. and Eilers, D. D.: 1966, *Astrophysical Journal* **144**, 1024.

Ostlie, D. A.: 1990, Proceedings of the NATO Advanced Research Workshop on *The Numerical Modelling of Nonlinear Stellar Pulsations: Problems and Prospects*, ed. J. R. Buchler (Kluwer Academic Publishers, Dordrecht), p. 89.

Ostlie, D. A. and Cox. A. N., 1986, *Astrophysical Journal* **311**, 864.

Perl, M. and Tuchman. Y.: 1990, *Astrophysical Journal* **360**, 554.

Stellingwerf, R. F.: 1975a, *Astrophysical Journal* **195**, 441.

Stellingwerf, R. F.: 1975b, *Astrophysical Journal* **199**, 705.

Tuchman, Y., Sack, N., and Barkat, Z.: 1978, *Astrophysical Journal* **219**, 183.

Tuchman, Y., Sack, N., and Barkat, Z.: 1979, *Astrophysical Journal* **234**, 217.

Tuchman, Y., 1991, *Astrophysical Journal* **383**, 779.

RADIAL AND NONRADIAL PERIODS AND GROWTH

RATES OF AN AI VELORUM MODEL

J. A. GUZIK

Applied Theoretical Physics Division, Los Alamos National Laboratory

Abstract. Walraven, Walraven and Balona recently discovered several new periodicities in addition to the well-known fundamental and first overtone periods of the high-amplitude δ Scuti star AI Velorum. Linear nonadiabatic pulsation calculations were performed for an AI Velorum model of mass $1.96 M_\odot$, $24.05 L_\odot$, and T_{eff} 7566 K for the radial and low-degree nonradial modes to help verify the tentative identifications made by Walraven, et al. Comparison of the calculated periods with the observations suggests some alternatives to the identifications proposed by Walraven, et al.

Walraven, and Balona (1991) analyzed photometric observations of the high-amplitude δ Scuti star AI Velorum made in 1951–53, 1979, 1987, and 1989, and discovered a number of new periodicities. In addition to the radial fundamental ($P_0 = 0.1115740$ d) and first overtone ($P_1 = 0.0862086$ d) modes, periods $P_2 = 0.0444014$ d, $P_3 = 0.0626077$ d, $P_4 = 0.109438$ d, and $P_5 = 0.091575$ d were found. Walraven et al. make the following tentative mode identifications: P_2 is identified as the fifth radial overtone. If an aliasing problem occurred, and $F_3 = 16.973$ d^{-1} instead of 15.973 d^{-1}, P_3 could be 0.058917 d, which is near the third radial overtone. P_4 cannot be a radial mode, and is proposed to be the $n = 1$, $l = 1$ p-mode. P_5 is interpreted as a nonlinear interaction between P_2 and P_1, since the frequency difference $F_2 - F_1$ is very close to F_5, and the amplitude of P_5 is proportional to the amplitude of P_2.

This paper presents the radial and low-l nonradial linear nonadiabatic periods and growth rates for an AI Velorum envelope model with fundamental and first overtone periods that closely match the observed periods. The 250-zone model has $M = 1.96 M_\odot$, $L = 24.05 L_\odot$, $T_{\text{eff}} = 7566$ K, mixing length/pressure scale height $\alpha = 1.5$, and fixed composition Y=0.29, Z=0.01, comprises 74% of the total mass, and uses the Stellingwerf (1975a, 1975b) analytical fit to the Cox-Tabor (1976) opacities. The period ratios and growth rates were found to be sensitive to the depth of the envelope, but were not affected significantly by modest changes in opacity or helium abundance, or by finer zoning.

The static model and linear radial nonadiabatic periods and growth rates were calculated using a code developed at Los Alamos, and described by Cox (1983). Linear nonradial nonadiabatic periods and growth rates were calculated using a code developed by Pesnell (1990). Tables I–IV summarize the periods and growth rates for this AI Vel model.

Astrophysics and Space Science **210**: 321–323, 1993.

TABLE I

Linear Radial Periods and Growth Rates for
AI Vel Model

Mode	Period	Growth Rate/Period
F	0.111573	5.4e-06
1H	0.086197	7.9e-05
2H	0.070316	4.4e-04
3H	0.058872	1.4e-03
4H	0.050293	3.2e-03
5H	0.043882	5.6e-03
6H	0.038934	5.3e-03
7H	0.034899	−3.8e-04

TABLE II

Linear Nonradial Periods and Growth Rates
for AI Vel Model, $l = 1$

Order n	Period	Growth Rate/Period
1	0.107402	8.5e-06
2	0.082299	1.2e-04
3	0.067036	6.1e-04
4	0.056228	1.8e-03
5	0.048310	3.9e-03
6	0.042426	6.1e-03
7	0.037834	5.0e-03
8	0.034055	−1.0e-03

TABLE III

Linear Nonradial Periods and Growth Rates
for AI Vel Model, $l = 2$

Order n	Period	Growth Rate/Period
0	0.111460	3.7e-06
1	0.091683	3.3e-05
2	0.075178	2.4e-04
3	0.062506	9.3e-04
4	0.053011	2.4e-03
5	0.045967	4.9e-03
6	0.040639	6.2e-03
7	0.036382	3.3e-03
8	0.032835	−4.0e-03

TABLE IV
Linear nonradial periods and growth rates for
AI Vel Model, $l = 3$

Order n	Period	Growth Rate/Period
0	0.107304	3.6e-06
1	0.087681	6.0e-05
2	0.071041	4.0e-04
3	0.059152	1.3e-03
4	0.050421	3.2e-03
5	0.043987	5.7e-03
6	0.039065	5.8e-03
7	0.035061	1.1e-03
8	0.031710	−7.2e-03

It is interesting that many modes that have not been observed are calculated to have large growth rates. The $P_2 = 0.0444014$ d periodicity agrees well with the 5th radial overtone, as proposed by Walraven, but also is close to the $l = 3$, $n = 5$ p-mode period. $P_3 = 0.0626077$ d agrees well with the $l = 2$, $n = 3$ mode period, but its alias of 0.058917 d also matches the 3rd radial overtone period. $P_4 = 0.109438$ d cannot be a radial mode, but does not match closely any low-l nonradial period either; the closest matches are $l = 1$, $n = 1$ (0.107402 d), $l = 2$, $n = 0$ (0.111460 d), and $l = 3$, $n = 0$ (0.107304 d). The interpretation of the $P_5 = 0.091575$ d periodicity as an interaction between P_2 and P_1 is most plausible, but P_5 also is close to the $l = 2$, $n = 1$ period.

References

Cox, A. N. and Tabor, J. E.: 1976, *Astrophysical Journal, Supplement Series* **31**, 271.
Cox, A. N.: 1983, in *Astrophysical Processes in Upper Main-Sequence Stars*, Thirteenth Advanced Course of the Swiss Society of Astronomy and Astrophysics, eds. A. N. Cox, S. Vauclair, and J. -P. Zahn (Switzerland, Geneva Observatory).
Pesnell, W. D.: 1990, *Astrophysical Journal* **363**, 227.
Stellingwerf, R. F.: 1975a, *Astrophysical Journal* **195**, 441.
Stellingwerf, R. F.: 1975b, *Astrophysical Journal* **199**, 705.
Walraven, Th., Walraven, J., and Balona, L. A.: 1992, "Discovery of additional pulsation modes in AI Velorum," *Monthly Notices of the RAS*, in press.

THE IMPORTANCE OF RADIATIVE TRANSFER IN STELLAR PULSATION MODELS

C. G. DAVIS

Los Alamos National Laboratory, Physics Division, P. O. Box 1663,
Los Alamos, NM 87545, USA

Abstract. With the advent of new astrophysical opacities it seems appropriate to discuss the need for a full radiative transfer (RT) theory instead of the usual equilibrium diffusion theory used in most nonlinear pulsation codes. Early studies on the importance of RT in the calculation of light curves for Cepheid models showed little effect over diffusion theory. The new opacities though may help to explain the "bump" mass discrepancy problem. For RR Lyrae models the use of RT theory causes some effects both in the color differences (U-B) as well as in the light curves. New opacities help to explain the period ratios for double mode RR Lyrae and beat Cepheids. A new area of research is in the modeling of stars with high luminosity to mass ratios that show tendencies for doubling and transitions to chaos, such as W Virginis and RV Tauri stars. For these stars it has been shown that RT is necessary in calculating their light curves and that the understanding of the shock dynamics depends on the transfer of lines in the pulsating RT dependent atmospheres (Fokin 1991).

1. Introduction

Considerations of the effect of RT as opposed to equilibrium diffusion theory on the light curves of pulsating stars has been studied since the early 60s. The early nonlinear pulsation models used the simpler approximation of equilibrium diffusion to calculate light curves. Because color bands were being used extensively (for example, U, V, B) in observing pulsating stars and the evidence of shocks from line spectra were being observed in RR Lyrae and W Virginis stars, it seemed appropriate to consider multigroup RT in our nonlinear stellar pulsation models. At this time (1965), the computers were also becoming more capable of handling these larger sets of equations with finer zoning and multigroup opacity tables necessary for RT calculations.

2. Radiative Transfer Models of Cepheids RR Lyrae and W Virginis Stars

While Castor and Christy were doing RT in RR Lyrae models, we concentrated initially on Cepheids and the importance of RT in the calculation of their light curves. We increased the normal zoning (usually around 60 zones, also used by Christy) to 72 zones and retained 7–11 zones in the optically thin region. The results confirmed our expectation that the effects would be small. Only small variations in the light curve around the region where the shock transits the atmosphere (near the so called pseudo-viscosity dip) occurred. The static structure would then be used with a snapshot analysis at

Astrophysics and Space Science **210**: 325–327, 1993.
© 1993 *Kluwer Academic Publishers.*

a finer frequency grouping to obtain the colors. All these frequency grouped opacities were formed by using Rosseland means from the Cox-Stewart opacity tables.

For our RR Lyrae studies we concentrated on developing improved colors and the proper estimate of the mean color relation. The importance of RT on the light curve was only studied in one case (for a model of SW Andromadae)(Davis 1976). In this study there were indications that RT was important, not only from the effect on the excess U-B, but also on the limiting effect on the magnitude of the variation of M_{bol}. Following this work, Simon and Aikawa (1986) used models from Hill and Castor and showed that improved Fourier coefficients of $\phi 21$ and $\phi 31$ could be obtained from an RT model as compared with the usual diffusion or a convection model. There is a need in RR Lyrae modeling to do a more careful RT calculation with extended atmospheres and detailed zoning. In this regard, a recent paper (Ishida and Takeuti 1992) on zoning improvements to better understand the physics in Cepheid atmospheres should be consulted.

Finally we are convinced that RT is necessary in calculating the light curves for the high luminosity to mass stars such as W Virginis and RV Tauri. Our paper on a W Virginis model (Davis 1972) attest to this fact as well as the recent work of Fokin (1991).

3. The New Opacity Tables

With the advent of new astrophysical opacities (Iglesias et al. 1987, Daeppen 1991), it seems prudent to consider improved ways of handling these opacities in our RT coupled hydrodynamic codes. We should consider new methods to include these results such as "transmission means" (Freeman 1965) or possibly "multiband" methods (Cullen and Pomraning 1980) for further calculations.

4. Conclusions

There is the possibility that time-dependent convection will be added to our models in the near future. There is still a need for an improved radiative transfer coupled hydrodynamic code using the new opacities, and with a dynamic zoning algorithm to resolve the time-dependent structure of the atmospheres of these pulsating stars.

References

Cullen, D. and Pomraning, G.: 1980, *JQSRT* No.24, 29.
Daeppen, W.: 1991, *Present Status of the Opacity Project*, presented at IAU commission 35 meeting, Buenos Aires, July 23–Aug. 1.
Davis, C. G.: 1972, *Astrophysical Journal* 172, 419.

Davis, C. G.: 1976, in *The Proceedings of IAU Colloquium No.29*, ed. W. S. Fitch (Publishing House of the Hungarian Academy of Sciences, Budapest, Hungary).

Fokin, A. B.: 1991, *Monthly Notices of the RAS* **250**, 258.

Freeman, B.: 1965, *General Atomic Report* GAMD 6446.

Iglesias, C. A., Rogers. F. J., and Wilson, B. G.: 1987, *Astrophysical Journal* **322**, L45.

Ishida, T. and Takeuti, M.: 1992, *Publications of the ASJ*, to be published.

Simon N. and Aikawa, T.: 1986, *Astrophysical Journal* **304**, 249.

HYDRODYNAMICS AND MULTI-LEVEL NON-LTE RADIATIVE TRANSFER IN PULSATING ATMOSPHERES: CEPHEIDS

D. D. SASSELOV

Harvard-Smithsonian Center for Astrophysics, Cambridge, MA 02138, U.S.A.

The atmospheres of classical Cepheids cannot be represented adequately by a sequence of hydrostatic equilibrium atmospheric models (see Sasselov and Lester 1990 and references therein). Observational evidence for shocks is also available at least since Kraft (1956). Both the disequilibrium and the non-linear phenomena affect the emergent spectra and spectral line profiles of the Cepheids. In the lower amplitude variables the effects may be subtle, yet still lead to significant systematic discrepancies.

We have developed the code HERMES for time-dependent treatment of hydrodynamics and non-LTE radiative transfer in the *atmospheres* of classical Cepheids with periods shorter than 12 days. Our approach is applicable to stars in which the formed shock waves do not dominate the energy balance of the atmosphere. Consecutive detailed calculations are performed for several multi-level model atoms, including H, Ca II, Mg II, He I and II. We use a 1-D explicit conservative upwind monotonic second-order (in time and space) Godunov-type Lagrangian hydrodynamic scheme. Being a characteristics based scheme it allows natural handling of boundary conditions. The scheme is stable and without artificial dissipation, a crucial necessity for physically meaningful radiative transfer solutions (see Roe 1990, Sasselov and Raga 1991 for more details).

We present time-dependent models of Cepheid atmospheres, built in a semi-empirical way and with forms of the driving piston of the pulsation taken from interior envelope computations. At present we have studied the ways in which the non-linear phenomena in a pulsating atmosphere affect different spectral lines, and the effect on measured observables (*e.g.*, velocities, temperatures, etc.). A few brief notes can be made at this time. The major effect to the velocities measured from Doppler shifts of spectral lines is due to shock waves in a Cepheid atmosphere. These are, however, of no big concern because of their transient nature (less than 10% of the cycle length) and usually obvious impact on the line profile (line doubling, emission). Much more subtle, but by no means small, are the effects due to velocity gradients in a Cepheid atmosphere. These gradients stretch the contribution function of a photospheric line over a large interval of velocities, even for narrow depth-constrained transitions. The resulting profile is asymmetric, but smooth, and not mutilated. The asymmetry due to a velocity

Astrophysics and Space Science **210**: 329–330, 1993.
© 1993 *Kluwer Academic Publishers.*

gradient couples with the asymmetry due to the geometric projection effect, and produces a composite line profile, which is rarely strongly asymmetric, yet has a "wrong" Doppler shift. This nonlinear phenomenon, unlike shocks, is not transient, and may affect the photosphere for a quarter and more of the cycle length (longer for longer-period Cepheids). And, unlike geometric projection effects, it cannot be treated in a prescribed way – every Cepheid is affected according to its specific atmospheric structure and dynamics (*i.e.* the phases at which velocity gradient asymmetry and projection asymmetry couple differ). Finally, its effect on the velocity measured from the spectral lines is the same as that from the projection effect, for Cepheids of about 10 days period, *i.e.* a 20% effect. Both the frequency and the magnitude of the velocity gradient increase with period (luminosity) and amplitude.

The propagation of even weak shocks in a Cepheid atmosphere is related to the development of strong velocity gradients. The lack of shocks does not exclude velocity gradients. Most classical Cepheids exhibit strong symptoms of transient shock waves, therefore it should be expected that velocity gradients affect the profiles of their spectral lines. Semi-empirical consistent modelling of hydrodynamics and non-LTE radiative transfer of Cepheids is already feasible and should be used to improve the accuracy of the basic parameters of these stars as inferred from observations of their atmosphere (radii and distances, modes, etc.).

References

Kraft, R.: 1956, *Publications of the ASP* **68**, 137.

Roe, P.: 1990, in *Numerical Modelling of Nonlinear Stellar Pulsations*, ed. R. Buchler (Kluwer), p. 183.

Sasselov, D., and Lester, J.: 1990, *Astrophysical Journal* **362**, 333.

Sasselov, D., and Raga, A.: 1991, in *7th Cambridge Workshop on Cool Stars, Stellar Systems, and the Sun*, eds. M. Giampapa and J. A. Bookbinder, (A. S. P. Conference Ser.), in press.

NEW OPACITIES AND FIRST OVERTONE

MODE CEPHEIDS

E. ANTONELLO

Osservatorio Astronomico di Brera, Via E. Bianchi 46, 22055 Merate, Italy

Abstract. Some results of linear adiabatic and nonlinear pulsation models of first overtone mode Cepheids are discussed. New and augmented opacities and a nonstandard mass-luminosity relation have been taken into account. The models indicate the possible importance of the resonance $P1/P4 = 2$ near $P1 = 3$ days. The resonance could explain the observed characteristics of the light curve shape of first overtone mode Cepheids.

There are several first overtone mode pulsators among short period Cepheids (Antonello et al., 1990a, 1990b; Simon, 1990). The analysis of their light curves has yielded Fourier parameters showing trends and a discontinuity which suggest that a resonance phenomenon between pulsation modes must be present in a way similar in part to that found in classical Cepheids with period near 10 days. The resonance for the first overtone mode Cepheids should be of the type 2:1 between the first and the fourth overtone, and should be centered at $P1 \sim 3$ days. Up to now, however, there was no clear evidence of it from theoretical models. Petersen (1989) suggested that the possible solution of his problem should be given by the use of increased metal opacity.

The recent work by Moskalik et al. (1991) indicated that the new, increased opacities (Iglesias and Rogers, 1991) are able to reproduce the period ratios of double-mode Cepheids, and these opacities along with a non standard mass-luminosity relation (Chiosi, 1990) are able to reproduce very well the observed pulsational characteristics of fundamental mode Cepheids. We have introduced the new opacities in a LNA code of T. Aikawa in order to check the possible presence of the resonance in the first overtone mode pulsators, and we have adopted the mass-luminosity relation for full overshoot models. The linear adiabatic results for the fundamental mode pulsators are comparable with those obtained by Moskalik et al.; moreover, there is evidence of the $P1/P4 = 2$ resonance for models with $4M_\odot$ and $P1 \sim 3$ days. In other words, a comparison of linear adiabatic results with the observations shows that the models with luminosities and effective temperatures comparable with those of observed stars predict the resonance for fundamental mode pulsators with P0 in the range between 9.2 and 10.8 days, and the resonance for first overtone mode pulsators with P1 in the range between 3.2 and 3.5 days.

A further test has been done by Aikawa (these proceedings) with nonlinear models. In this test, augmented old opacities by a factor of 5 instead of the new opacities have been adopted. The effects of the augmented opac-

Astrophysics and Space Science **210**: 331–332, 1993.

ities by this factor should be analogous to those of new opacities (see e.g. Renzini, 1990). The model sequence consisted of some models with $4M_\odot$, with the same luminosity, $1316L_\odot$, and different effective temperatures in the range 5400–5900 K. The results of this test are very interesting, because for the first time it has been possible to obtain realistic light curves with small amplitude (0.4 mag) of first overtone mode Cepheid models. Moreover, the light curves suggest the presence of a Hertzsprung-like progression which is related to the resonance P1/P4. The preliminary results indicate that the Fourier parameter R_{21} is low, less than 0.16, in agreement with the observations, while the parameter o_{21} is generally larger by 0.5–1 rad than the observed one; moreover, the theoretical discontinuity in ϕ_{21} is small, 0.25 rad, in comparison with the observed one (1.5 rad). The theoretical resonance is centered at $P \sim 3.5$ days, while the observed center locates at about 3.2 days. There is an apparent clustering of the points in a narrow period range (2.9–3.9 days), which is explained by the use of a sequence of models with the same luminosity; we expect an improvement by using a sequence of models which runs approximately parallel to the blue edge (a similar sequence is also suggested by the observations; Antonello et al., 1991).

The results summarized in the present note will be discussed in detail elsewhere (Antonello and Aikawa, in preparation).

References

Antonello E., Poretti E., and Reduzzi L.: 1990a, *Astronomy and Astrophysics* **236**, 138.

Antonello E., Poretti E., and Reduzzi L.: 1990b, in *Confrontation between Stellar Pulsation and Evolution*, eds. C. Cacciari and G. Clementini (Astron. Soc. of the Pacific Conf. Series) 11, p. 209.

Antonello E., Poretti E., and Reduzzi L.: 1991, *Memorie. Soc. Astron. Italiana*, in press.

Chiosi C.: 1990, in *Confrontation between Stellar Pulsation and Evolution*, eds. C. Cacciari and G. Clementini (Astron. Soc. of the Pacific Conf. Series) 11, p. 158.

Iglesias C. A., and Rogers F. J.: 1991, *Astrophysical Journal* **371**, L73.

Moskalik P., Buchler J. R., and Marom A.: 1991, preprint.

Petersen J. O.: 1989, *Astronomy and Astrophysics* **226**, 151.

Renzini A.: 1990, in *Confrontation between Stellar Pulsation and Evolution*, eds. C. Cacciari and G. Clementini (Astron. Soc. of the Pacific Conf. Series) 11, p. 421.

Simon N. R.: 1990, in *Confrontation between Stellar Pulsation and Evolution*, eds. C. Cacciari and G. Clementini (Astron. Soc. of the Pacific Conf. Series) 11, p. 103.

COUPLED-OSCILLATORS

Y. TANAKA

Faculty of Education, Ibaraki University, Bunkyo, Mito 310, Japan

Abstract. Modal coupling oscillation models for the stellar radial pulsation and coupled-oscillators are reviewed. Coupled-oscillators with the second-order and third-order terms seemed to behave non-systematically. Using the equation by Schwarzschild and Savedoff (1949) with the dissipation term of van del Pol's type which is third-order, we demonstrate the effect of each term. The effects can be understood by the terms of the nonlinear dynamics, which is recently developing, that is, phase-locking, quasi-periodicity, period doubling, and chaos. As the problem of stellar pulsation, especially of double-mode cepheids on the period-ratio, we examine the dependence on the stellar structure from which the coupling constants in the second-order terms are derived. Eigen functions for adiabatic pulsations had been used for the calculation of the constants. It is noted that only two set of the constants are available, that is, for the polytrope model with $n = 3$ and a cepheid model without convection. Some examples of nonlinear dynamical effects will be shown.

It is shown that if the constants were suitable values, the period-ratio of double-mode cepheids is probably realized. The possibility is briefly suggested.

1. Introduction

Studies of the influence of second and higher order terms in the stellar pulsation have been continued for several decades. Woltjer (1935) shows equations of adiabatic stellar pulsation with a second order term can be expressed as the differential equation of time-dependent coefficients that the solution of stellar pulsation is expanded in terms of eigen functions of linear pulsation. He also applies it to Cepheid-variation with nonadiabatic perturbation and discuss the stationary state of periodic pulsation (Woltjer, 1937).

Resseland (1943) developed the mathematical theory of anharmonically pulsating gas spheres in the same way as Woltjer (1935, 1937) and applies it to the pulsation of homogeneous star, only taking the single mode into account. Bhatnager and Kothari (1944) give an exact solution for the pulsation of homogeneous modes for unperturbed stars.

Schwarzschild and Savedoff (1949) study anharmonic pulsation of the standard model which is more realistic model than the homogeneous model. The standard model is constructed by Schwarzschild (1941) with polytropic index 3 and various values of ratio of specific heats. Following Rosseland, they derive the equation of time-dependent amplitude for the fundamental and first overtone modes and compute it numerically. They result in that for the amplitude of characteristic cepheids, anharmonic pulsations yields the same period as harmonic pulsations. Anharmonics shows an appreciable skewness in the velocity-curve, but is still smaller than that observed.

Prasad (1949a, b) studies the interaction of multi-modes with calculations of coupling constants. The solutions are obtained by the use of Fourier series.

Astrophysics and Space Science **210**: 333–341, 1993.

After her, there are many studies of multi-mode coupled oscillator models, including the modes of non-radical oscillations.

Recently, the effects of nonadiabatic term and more realistic coupling constants of modes are studied by Takeuti and Aikawa (1981). Their work is mainly two properties. Following the mode coupling models, they use more realistic cepheid model and obtain the coupling constants. For studying to nonadiabatic effects, they introduce nonlinear damping terms of the van der Pol or the Krogdahl's type, which is introduced by Krogdahl (1955) to the stellar pulsation theory.

In the famous review of Ledoux and Walraven (1958, Fig. 46), the result by Krogdahl (1955) can be seen in comparison with the observational data and with the computed result by Schwarzschild and Savedoff (1949). The orbit in the radial-velocity space by Krogdahl does not fit to that observed at his period. However, this term is very compact and seen convenient for describing the effect of nonlinear dissipation, which may consist of many complicated terms of the nonlinear nonadiabatic effect. Using the equations which is the same ones as we will use later, Takeuti and Aikawa (1981) studied the behaviors of mode-coupling model analytically.

After them, analytical studies are developed by many authors such as Aikawa (1983, 1984), Dziembowski and Kovacs (1984), and Takeuti (1984, 1985, 1986).

2. Structure of Equations of Modal Coupling Model

The motion of simple pendulum is represented by

$$\frac{dx}{dt} = y, \quad \frac{dy}{dt} = -x \tag{1}$$

where x and y is the displacement from the equilibrium and the velocity, respectively. The solution of Eq. (1) is the sinusoidal functions such as $x = A\sin(t)$ which depends on the initial values. It should be noted that the equation of motion for the forced oscillation is written as follows:

$$\frac{dx}{dt} = y, \quad \frac{dy}{dt} = -x^3 + \mu(1 - x^2)y + B\cos(\sigma t), \tag{2}$$

where the second term is so-called van der Pol's damping term and the third is the external force. The system of Eq. (2) is given by Ueda and Akamatsu (1981), which shows the Japanese attractor. This is a hybrid of the Duffing and van der Pol oscillator. If we rewrite Eq. (2) as

$$\frac{dx}{dt} = y, \quad \frac{dy}{dt} = -x^3 + \mu(1 - x^2)y + z, \tag{3}$$

$$z = B\cos(\sigma t), \tag{4}$$

or

$$\frac{dx}{dt} = y, \quad \frac{dy}{dt} = -x^3 + \mu(1 - x^2)y + z, \tag{5}$$

$$\frac{dz}{dt} = a, \quad \frac{da}{dt} = -\sigma^2 z, \tag{6}$$

we can see the system is a coupled oscillator that there is no reaction from the first one to the second. The coefficient B in Eq. (2) is dependent on the initial values of Eq. (6) which lead chaotic behaviors of the system.

Now we discuss the equations of the mode coupled oscillator by Takeuti and Aikawa (1981). That is written as follows:

$$\frac{dx_1}{dt} = y_1, \tag{7}$$

$$\frac{dy_1}{dt} = -\sigma_1^2 x_1 + \sigma_1^2 \{[\frac{1}{2}C_{111}x_1 + C_{112}x_2]x_1 + \frac{\mu_1}{\sigma_1^2}(1 - \alpha_1^2 x_1^2)y_1 + \frac{1}{2}C_{122}x_2^2\}, \tag{8}$$

$$\frac{dx_2}{dt} = y_2, \tag{9}$$

$$\frac{dy_2}{dt} = -\sigma_2^2 x_2 + \sigma_2^2 \{[\frac{1}{2}C_{222}x_2 + C_{212}x_1]x_2 + \frac{\mu_2}{\sigma_2^2}(1 - \alpha_2^2 x_2^2)y_2 + \frac{1}{2}C_{211}x_1^2\}, \tag{10}$$

where C_{ijk} are the coupling constants between the modes 1 and 2.

If we suppose μ_2 and all C_{2jk} equal to zero, and C_{111} and C_{112} also zero, the system becomes the nonlinearly forced van der Pol oscillator. That is, the mode 1 with van der Pol's damping is forced to oscillate by mode 2 through the term of $C_{122}(A\sin\sigma_2 t + B\cos\sigma_2 t)^2$. Two sets of coupling constants between the fundamental and first overtone for stellar models have been evaluated with linear adiabatic eigen functions (Takeuti, 1985). Before analyzing the coupled oscillator for stellar models, we study oscillators with rather simple method.

It should be worthwhile to note that the forced oscillation can be reduced into a sine circle map (Chirikov, 1979), which shows the Arnold's tongues for phase locking and the devil's staircase. The discrete system seems rather convenient to study the rough behavior of differential equation system. The two-dimensional maps such as Henon's and Mira's attractors may also be useful (Gumowski and Mira, 1980).

3. Rough Behaviors of the Equations

The values of C_{1ij} and C_{2ij} obtained by Takeuti (1985) for cepheid models nearly equals 2~4, and 6, respectively. If we simplify constant values for coupling constants such as $C_{111} = C_{112} = C_{121} = C_1$ and so on, we can decrease the number of parameters whose dependence should be examined. The brief method and results are given by Nakahara and Tanaka (1992) and more detailed discussion can be seen in Tanaka et al. (1992 a, b). Here we sketch their results.

First, fixing the values of $C_2 = 6$ and $\mu_1/\sigma_1^2 = 0.1$, they examine the dependence of solutions on other parameters. They show various phase locking of period ratios from 2/3 to 1/1 (Tanaka et al., 1992a). Between them, we can see the phase locking of period ratios in Farey series where the ratio of 5/7 is included. The ratio of 5/7, however, can be seen in very narrow region of parameter space. The regions between the phase locking are not distinguished whether higher ratios of phase locking, chaotic or quasi-periodic states. They also compute the Lyapunov exponents for examining quasi-periodicity, phase locking and chaos (Nakahara and Tanaka, 1992). But they do not recognize the chaotic states within their computed results.

Tanaka et al. (1992b) show the dependence of C_1 and C_2 in the manner of the Mandelbrot set. They select the color for each point in (C_1, C_2) plane when the solution has diverged (Fig. 1). In Fig. 2, the enlargement of Fig. 1 in the first quarter is shown. We can see complicated patterns in the divergent region and the complex boundary. In Fig. 3, we see the structure of the non-divergent region. The phase locking of lower period ratios such as 3/4, 5/7 etc. are shown. The shape of typical phase locked regions seems similar to so-called Arnold's tongues in the sine maps. It is noted that the non-divergent region is located $C_2 < -1.5C_1 + 15$, depending on the other parameters. As their results is very preliminary ones, further examples are desired.

4. Oscillation of the "Standard Model"

Equations used by Schwarzschild and Savedoff (1949) are the same as those of us without nonadiabatic term and give values of coupling constants between the fundamental and first overtone modes for the standard model. It is noted that their coupling constants are the first one for realistic stellar models. Following Takeuti and Aikawa (1981), Nakahara and Tanaka (1993) introduce the van der Pol's term as the nonadiabatic effect to the equations by Schwarzchild and Savedoff (1949) and demonstrate the behavior of the oscillation of polytropic stars.

They report that in the polytropic stellar pulsation the fundamental and first overtone mode synchronize each other for most values of parameters of the damping terms, while phase locking of 4/5 and 3/4 rarely appears. It is found that for the period ratio of 1/1 which located near the boundary between the divergent and non-divergent regions, the period doubling of 1/1 is observed, which will lead to chaos. Since the mesh of parameters for

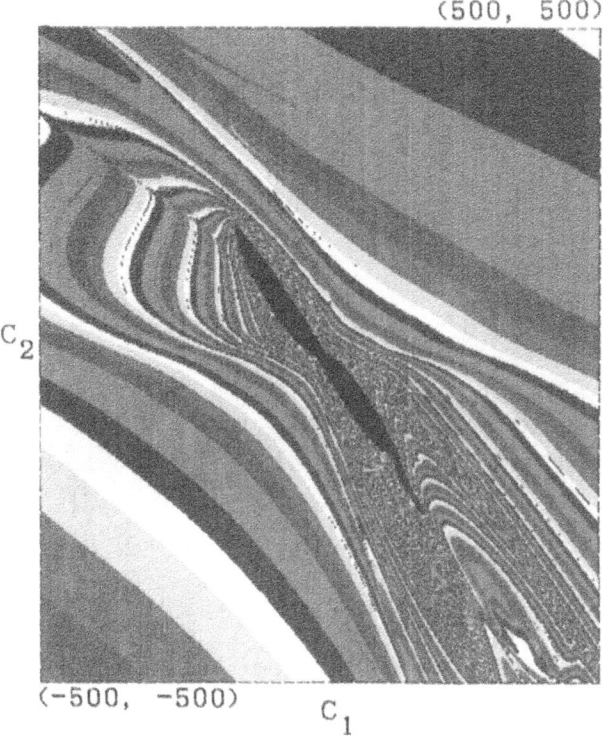

Fig. 1.   Pattern of divergence.

researching lower period ratios is rather coarse, more careful survey may be helpful for understanding period ratios of compact star's pulsation.

5. 5:7 Period Ratio by a Cepheid Model Without Convection

More realistic coupling constants of double-mode stellar pulsation are obtained by Takeuti and Aikawa (1981) and Takeuti (1984) as mentioned above. Following the mode coupling models, they use more realistic cepheid model ($M/M_\odot = 6.7$, $L/L_\odot = 2280$, $T_{\text{eff}} = 5850$ K, $X, Y, Z = 0.70, 0.28$, 0.02) and obtain the coupling constants. Using the constants, Seya et al. (1990) demonstrate the behavior of the solutions of Eq. (7)–(10). It should be noted that Moskalik and Buchler (1989) show period doubling bifurcation of coupled oscillators with van der Pol's term, but C_{1ij} equal to zero.

Seya et al. (1990) obtain time developments of mode 1 (the fundamental mode) and 2 (the first overtone) and the orbits in the phase planes of $(x_1, dx_1/dt)$ etc. for various sets of coefficients of the damping terms. They find the phase locking of several ratios. As pointed out by Ishida and Takeuti

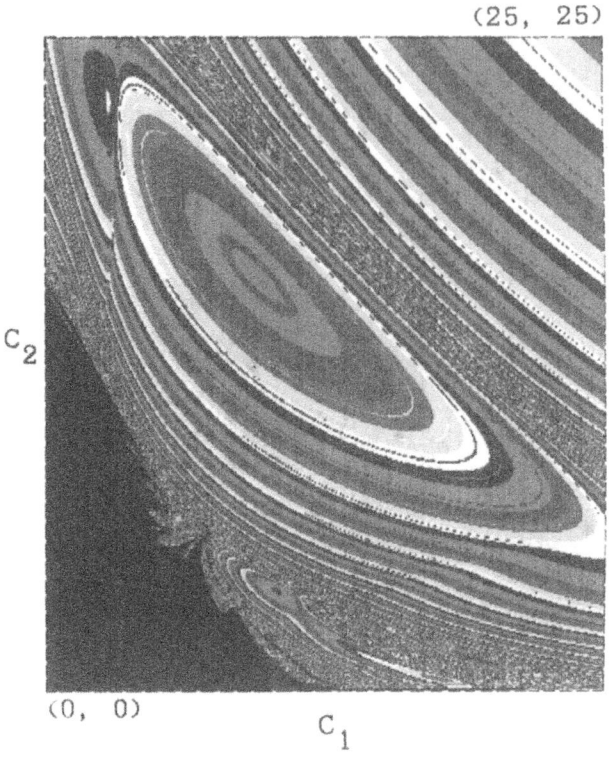

(25, 25)

C_2

(0, 0)

C_1

Fig. 2.   Enlargement of Fig. 1.

(1991), the ratio of 3/4 is dominant for the given coupling constants and the linear frequency ratio adopted by them. Seya et at. (1990) also show the phase locked orbits such as the period ratio of 3/4 will bifurcate in period doubling as the change of a_2^2. They use the first return map for showing the characteristics of the beat or the quasi-periodicity of the solution. The phenomena of quasi-periodicity may be important for studying the beat of variable stars.

Seya et al. (1991) demonstrate the behavior of solutions of Eqs. (7)–(10) in the Poincare section. On the section, the complex features can be understood as the chaotic states of solutions. Thus they conclude that the chaos of coupled oscillator are produced through folding the surface of torus. Ishida and Takeuti (1991) give the condition of synchronization of coupled oscillator with non-zero C_{ijk} of Takeuti (1985). They also find the phase locking of other ratios than 2/1.

The period ratio of 5/7 as phase locking is reported by Seya et al. (1990), Tanaka et al. (1990) and Ishida and Takeuti (1991). Ishida and Takeuti (1991) obtain it for fixed ratio of linear periods and fixed coupling constants,

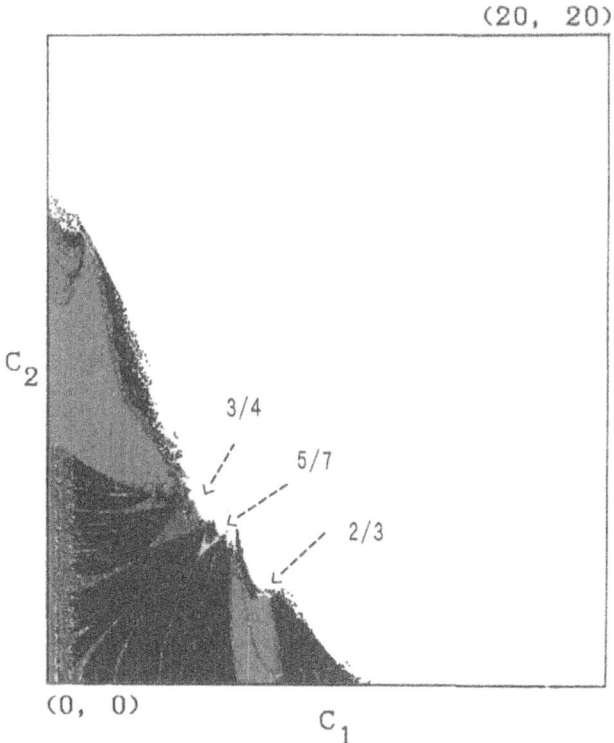

Fig. 3. Structure of non-divergent region. The phase locking can be seen as the Arnold's tongues.

while the others change the linear period ratios for showing the ratio of 5/7. The period ratio of 5/7 is important, because the double-mode cepheids seem have rather narrow period ratios 0.7 to 0.71. The standard stellar evolutionary theory derives the period ratios higher than those of observation. It is known that the mass of cepheids should be considerably reduced for fitting them. The review on this problem by Balona (1985) should be referred.

Tanaka et al. (1992c, d) try to search the condition of coupling constants for the period ratio of 5/7 with the realistic linear periods and the coefficients of the damping terms. Using the relation of $C_{112}/C_{211} = C_{122}/C_{212}$ and fixed C_{2jk}, they find out the relation of C_{111} and A where A is a multiplier of the original coupling constants of C_{1j2}.It is shown that the 5/7 phase locking is realized by the increase of self- and mutual-coupling constants of mode 1.

Takeuti et al. (1992) and Yamakawa et al. (1992) calculate the coupling constants, taking the nonadiabatic effect in radiative models. They obtain rather large values of the constants in order of 10–100. Such increases may make the oscillation to be synchronized or sometime diverge. Zalewski

(1992), however. show that the nonadiabatic constants become much smaller for the cepheid models if convection is taken into account. This may be the real model of double-mode cepheids which have the period ratio of 5/7.

It is noted that the present coupled oscillator model is still not very realistic. As the further correction. we should mention that Ishida et al. (1992) try to examine the effect of third order nonlinear terms. Their results will give us more information on the direction for studying double-mode cepheids and its evolutionary masses.

6. Summary

We have the constants of modal coupling for the standard model and some cepheid models. With these constants, the behaviors of the coupled oscillator model of stellar pulsation are analyzed by many authors. They find out the phase locking, quasi-periodicity, chaos and divergent states which depend on the constant and the coefficients of the van der Pol's damping term. The coupling constants for nonadiabatic models are now examined with and without convection. As the result, the period ratio of 5/7 and the mystery of mass reduction of cepheids are expected to be understood. Thus we may expect to resolve soon one of mysteries of multi-period variables, but no one can say the step might not be the devil's staircase. It is desired to work with further knowledge from observation, hydrodynamical simulations, and nonlinear dynamics.

Acknowledgements

The author express his heartful thanks to his students for their collaborations to numerical computations. He is also grateful to Prof. M. Takeuti for his kind discussion and encouragement.

References

Aikawa, T.: 1983, *Monthly Notices of the RAS* **204**, 1193.
Aikawa, T.: 1984, *Monthly Notices of the RAS* **206**, 833.
Bhatnagar, P. L. and Kothari, D. S.: 1944, *Monthly Notices of the RAS* **104**, 292.
Chirikov. B. V.: 1979, *Phys. Report* **52**, 263.
Dziembowski, W. and Kovacs, G.: 1984, *Monthly Notices of the RAS* **206**, 497.
Gumowski, I. and Mira. C.: 1980, *Recurrences and Discrete Dynamic Systems*, Lecture Note in Math. Sci.(Springer) , p. 809.
Ishida, T. Takano, R., Yamakawa, F. and Takeuti, M.: 1992, in these Proceedings.
Ishida, T. and Takeuti, M.: 1991, *Astrophysics and Space Science* **178**, 311.
Krogdahl, W. S.: 1955, *Astrophysical Journal* **122**, 43.
Ledoux, P. and Walraven, T.: 1958, *Handbuch der Physik* **51**, 549.
Nakahara, T. and Tanaka, Y.: 1992, in these Proceedings.
Nakahara, T. and Tanaka, Y.: 1993, *Bull. Fac. Education, Ibaraki Univ. (Nat. Sci.)*, **42**, 23.

Prasad, C.: 1949a, *Monthly Notices of the RAS* **109**, 528.

Prasad, C.: 1949b, *Monthly Notices of the RAS* **109**, 711.

Rosseland, S.: 1943, *Monthly Notices of the RAS* **103**, 233.

Schwarzschild, M.: 1941, *Astrophysical Journal* **94**, 245.

Schwarzschild, M. and Savedoff, M. P.: 1949, *Astrophysical Journal* **109**, 298.

Seya, K., Tanaka, Y. and Takeuti, M.: 1990, *Publications of the ASJ* **42**, 405.

Seya, K., Tanaka, Y. and Takeuti, M.: 1991, *Bull. Fac. Educ., Ibaraki Univ. (Nat. Sci.)* **40**, 1.

Takeuti, M.: 1984, in Non-Linear Phenomena in Stellar Outer-layers, ed. M. Takeuti, Astron. Institut., Tôhoku Univ., 1.

Takeuti, M.: 1985, *Astrophysics and Space Science* **109**, 99.

Takeuti, M.: 1986, *Astrophysics and Space Science* **119**, 37.

Takeuti, M. and Aikawa, T.: 1981. *Sci. Reports Tôhoku Univ., 8th Ser.* **2**, 106.

Takeuti, M. Yamakawa, F. and Ishida, T.: 1992, *Publications of the ASJ* **44**, 101.

Tanaka, Y., Nakahara, T., and Yokota, Y.: 1992a, *Bull. Fac. Education, Ibaraki Univ. (Nat. Sci.),* **41**, 41.

Tanaka, Y., Ogura, S., and Sekino, T.: 1992b, *Bull. Fac. Education, Ibaraki Univ. (Nat. Sci.)* **41**, 51.

Tanaka, Y., Seya, K. and Takeuti, M.: 1990, in *Confrontations between Stellar Pulsation and Evolution*, eds.Cacciari, C. and Clementini, G. (Astronomical Society of the Pacific), p. 145.

Tanaka, Y., Seya, K. and Takeuti, M.: 1992c, *Bull. Fac. Education, Ibaraki Univ. (Nat. Sci.)* **41**, 37.

Tanaka, Y., Seya, K. and Takeuti, M.: 1992d, *Publications of the ASJ* **44**, 331.

Ueda, Y. and Akamatsu, N.: 1981, *IEEE Trans. Circuits and Systems* **CAS-28**, 217.

Woltjer, J. Jr.: 1935, *Monthly Notices of the RAS* **95**, 260.

Woltjer, J. Jr.: 1937, *Bull. Astron. Inst. Netherlands* **8**, 193.

Yamakawa, F. Ishida, T. and Takeuti, M.: 1992, in these Proceedings.

Zalewski, J.: 1992, in these Proceedings.

AN OSCILLATOR MODEL FOR STELLAR VARIABILITY

T. NAKAHARA and Y. TANAKA
Faculty of Education, Ibaraki University, Bunkyo, Mito 310, Japan

Abstract. The dependence of coupling constants in a coupled oscillator model is examined with simplified methods. The Lyapunov exponents are preliminary introduced for the model. The behaviors of oscillator model are examined in a parameter plane. So-called the Arnold's tongues for phase-locking states are observed in fractal patterns.

1. Introduction

Two-mode coupling oscillator model for stellar variability has investigated in the view of nonlinear dynamics (Seya et al. 1989). The equations of coupled oscillator model are described as follows;

$$\frac{d^2x_1}{dt^2} = -\sigma_1^2 x_1 + \sigma_1^2 \{[(1/2)C_{111}x_1 + C_{112}x_2]x_1$$
$$+ \varepsilon_1(1 - \alpha_1^2 x_1^2)(dx_1/dt) + (1/2)C_{122}x_2^2\},$$
$$\frac{d^2x_2}{dt^2} = -\sigma_2^2 x_2 + \sigma_2^2 \{[(1/2)C_{222}x_2 + C_{212}x_1]x_2$$
$$+ \varepsilon_2(1 - \alpha_2^2 x_2^2)(dx_2/dt) + (1/2)C_{211}x_1^2\}, \qquad (1)$$

where the coupling constants C_{ijk} are evaluated by Takeuti (1985) for a classical cepheid model. It is featured that the van der Pol's damping term is introduced. Seya et al. (1989) show the complicated behaviors of Eq. (1) as the change of coefficients in damping terms. The behaviors seem to be phase-locking, quasi-periodicity or chaos. Using these equations, we examine the dependence of behaviors on coupling constants by rather simple manner. We simplify the constants as $C_{ijk} = C_i$ for all j, k. The evaluated values of C_{1jk} and C_{2jk} by Takeuti (1985) equal $2 \sim 4$ and 6, respectively. Thus the value of C_1 is varied from 2 to 6 and that of C_2 is fixed at 6.0. It should be worthwhile to research the tendency of solutions for wide values of C_1 and C_2. The set of angular frequencies is chosen as $\sigma_1^2 = 1.0$ and $\sigma_1^2 = 1.9$.

2. Results

First, we follow the results by Tanaka et al. (1991a) that the Lyapunov exponents are preliminary introduced in order to distinguish the solutions of Eq. (1). The computed results is shown in Table I. In these case, all λ_1 is nearly zero and λ_2 is zero and negative, which mean that quasi-periodicity or phase-locking occur. We also examined time developments and projected orbits on a phase plane for the same parameter and confirmed the state of

Astrophysics and Space Science **210**: 343–345, 1993.
© 1993 *Kluwer Academic Publishers.*

solutions in the Table. We should note that the other phase-locking states appear as the values of coupling constants and coefficients of damping terms are varied. The frequency ratios cover from 2/3 to 1/1 as Farey's series, while indistinguishable (quasi-periodic, chaotic and convergent) and divergence cases are also observed. These have been observed as the ratio of angular frequency is changed (Tanaka et al. 1990).

TABLE I

The samples of the Lyapunov exponents and the period ratios. The coupling constants and parameters of nonlinear damping term are also given. $\sigma_1^2 = 1.0$, $\sigma_2^2 = 1.9$, $\varepsilon_1 = 0.1$. Q-P means the quasi-periodicity.

Parameters					Lyapunov exponents				
C_1	C_2	ε_2	α_1^2	α_2^2	λ_1	λ_2	λ_3	λ_4	P_2/P_1
6.0	6.0	0.3	1600	1600	0.000	0.000	−0.148	−0.845	Q-P
6.0	6.0	0.3	1600	800	0.001	−0.016	−0.141	−0.872	7/10
6.0	6.0	0.4	1600	800	0.000	−0.021	−0.149	−1.195	5/7
6.0	6.0	0.3	1600	2400	0.000	−0.031	−0.113	−0.844	3/4
2.0	6.0	0.1	800	800	0.000	−0.063	−0.064	−0.329	4/5

Next, the behaviors of coupled oscillator are studied by Tanaka et al. (1991 b) on the parameters plane, (C_1, C_2) where we shall see fractal patterns. Numerical integration are carried out by the mixed Euler and Heun method with a personal computers. The first diagram is divided into two regions, divergent or non-divergent. The divergent region (D) is defined by $A = x_1^2 + x_2^2 > 10$. When A becomes greater than 10, the pixel at (C_1, C_2) is colored by the number of iteration. When A keeps to be small after the 500 iteration, the set of (C_1, C_2) is seemed as non-divergent region (ND) and colored black. In the diagram, D region is patterned by complex stripes. When α_1^2 becomes large and α_2^2 small, ND region tends to expand. They also illustrate the structure of ND region. Although phase-locking, quasi-periodicity and chaos are included in ND region, they only distinguish the phase-locking and color according to counted ratios. We seem that these regions are so-called the Arnold's tongues, which appear in the circle map. We can understand that the coupled oscillator model behaves as a simple sine map which is the simplified form of forced oscillator.

3. Concluding Remarks

We have shown the model coupling oscillator models for stellar variability by using simplified sets of constants. The route to chaos in the coupled oscillator are period doubling and quasi-periodicity which depends upon the control parameters.

References

Seya, K., Tanaka, Y., and Takeuti, M.: 1990, *Publications of the ASJ* **42**, 405.

Takeuti, M.: 1985, *Astrophysics and Space Science* **109**, 99.

Tanaka, Y., Nakahara, T., and Yokata, Y.: 1991a. *Bull. Fac. Educ., Ibaraki Univ.*, in press

Tanaka, Y., Ogura, S., and Sekino. T.: 1991b, *Bull. Fac. Educ., Ibaraki Univ.*, in press.

Tanaka, Y., Seya, K., and Takeuti, M.: 1990, in *Confrontations between Stellar Pulsation and Evolution*, eds. C. Cacciari and G. Clementini (Astronomical Society of Pacific), 145.

THE COUPLING COEFFICIENTS OF PULSATION FOR

RADIATIVE STELLAR MODELS

F. YAMAKAWA

Space System Dept., Fujitsu Ltd., Nakase 1-chome, Mihama-ku, Chiba 261, Japan

T. ISHIDA

Nishi-Harima Astronomical Observatory, Ohnadesan, Sayo-cho, Hyogo 679-53, Japan

and

M. TAKEUTI

Astronomical Institute, Tôhoku University, Sendai 980, Japan

Abstract. The second order theory of coupling is discussed regarding the radial pulsation of stellar models which are constructed ignoring convection. The formula including the nonadiabatic effect is presented. Numerical values given for model classical cepheids are considerably greater than the adiabatic values.

The interaction between different oscillation modes is important for studying the real stellar pulsation. The coupling is studied by Schwarzschild and Savedoff (1949) for polytropic gas spheres. Takeuti and Aikawa (1981) studied the coupling coefficients of realistic cepheid models. Their values are not so different with those of polytropic models. Both these papers treated only the coupling in adiabatic oscillation, so that only the effect of adiabatic changes of pressure of a mode on another mode is considered. In the real stellar pulsation, the pressure change is dominated not only by the adiabatic component, essentially the effect of density change, but also by the nonadiabatic change which is caused from the change of temperature or the entropy.

Complete study of coupling is performed by Buchler and Goupil (1984) including nonadiabatic effect. Their formulation is complete but is difficult to use for estimating the coupling before hydrodynamic simulation. So the nonadiabatic coupling only based on the linear calculation should be studied. We assumed a common time dependency for the density and the entropy change. Then we have the following expressions :

$$x(r_0, t) = \xi(r_0)q(t), \tag{1}$$

and

$$y(r_0, t) = \eta(r_0)q(t), \tag{2}$$

where

$$x(r_0, t) = \frac{\delta r}{r_0}, \tag{3}$$

Astrophysics and Space Science **210**: 347–348, 1993.

and

$$y(r_0, t) = \frac{\delta s}{c_{V0}}. \tag{4}$$

We denote here the change of radius as δr, and the change of entropy as δs. r_0 and c_{V0} are the radius and specific heat at the equilibrium state, respectively.

Since we use the linear pulsation for determining $x(r_0)$ and $y(r_0)$, the pulsation function ξ may similar to the eigenfunction of Sturm-Liouville problem. This is also crucial assumption in our study. When we assume such an idealized condition on ξ, a lot of higher order terms is canceled by its orthogonality. We may easily obtain the effect of the entropy change on the other mode. The most important results of this calculation is that the coupling coefficients become greater in the order of 2 than those of adiabatic approximation. This comes from the strong nonadiabatic effect in the critical zone of hydrogen ionization. This effect may be removed when we use the models including the effect of convection.

We wish to express our thanks to Drs. H. Saio and J. Zalewski for their discussion.

References

Buchler, J.-R. and Goupil, M.-J. : 1984, *Astrophysical Journal* **279**, 394.
Schwarzschild, M. and Savedoff,M. P. : 1949, *Astrophysical Journal* **109**, 298.
Takeuti, M. and Aikawa, T.: 1981, *Sci. Reports Tôhoku Univ., Eighth Ser.* **2**, 106.

THE COUPLING COEFFICIENTS OF RADIAL PULSATION
IN THIRD ORDER

T. ISHIDA

Nishi-Harima Astronomical Observatory, Ohnadesan, Sayo-cho, Hyogo 679-53, Japan

R. TAKANO

Astronomical Institute, Tôhoku University, Sendai 980, Japan

F. YAMAKAWA

Space System Dept., Fujitsu Ltd., Nakase 1-chome, Mihama-ku, Chiba 261, Japan

and

M. TAKEUTI

Astronomical Institute, Tôhoku University, Sendai 980, Japan

Abstract. The third order theory of coupling is discussed regarding the radial pulsation of stellar models.

The coupling constant of the stellar radial pulsation have been studied by various authors in the second order (Schwarzschild and Savedoff 1949, Takeuti and Aikawa 1981, and Takeuti *et al.* 1992). As demonstrated by Buchler and Goupil (1984), the third order terms play essential role in non-linear finite-amplitude oscillation. Therefore, the study of coupling in second order theory has its proper limit. Takeuti and Aikawa (1981) used an artificial third-order term for excitation in the form of van der Pol. We derived third order terms of coupling coefficients along the line of Schwarzschild-Savedoff (1949) and Takeuti *et al.* The expressions are not so complicated at least for adiabatic pulsation, but not so easy to make overview in the nonadiabatic case.

References

Buchler, J.-R. and Goupil, M.-J. : 1984, *Astrophysical Journal* **279**, 394.
Schwarzschild, M. and Savedoff, M. P. : 1949, *Astrophysical Journal* **109**, 298.
Takeuti, M. and Aikawa, T.: 1981, *Sci. Reports Tôhoku Univ., Eighth Ser.* **2**, 106.
Takeuti, M. Yamakawa, F., and Ishida, T. M. : 1992, *Publications of the ASJ to be published.*

Astrophysics and Space Science **210**: 349, 1993.

THE ROLE OF CONVECTION IN REDUCING NONADIABATICITY AND MODE COUPLING IN CEPHEIDS

J. ZALEWSKI*

Astronomical Institute, Tôhoku University, Aoba-ku, Sendai 980, Japan

Abstract. The effect of convection on the strength of coupling is examined. It is found that in Cepheid models the inclusion of convection smooths the sharp peak of entropy perturbation in the ionization region and reduces significantly the coupling.

1. Introduction

The nonadiabatic amplitude equations formalism has been extensively studied in the case of Cepheids and other stars populating the instability strip (see e.g. Buchler, Moskalik and Kovács, 1991). In radiative models, for which the formalism was developed (Buchler and Goupil, 1984, hereinafter BG) it is found that perturbations of thermodynamic quantities in the ionization zones become very large and numerical problems appear in the calculation of coupling coefficients (Pesnell and Buchler, 1986). Buchler and Kovács (1987, hereinafter BK, see also Kovács and Buchler, 1989) have devised a method in which the coupling coefficients are derived from a fit to the results of nonlinear calculations, however in this approach one has to verify the applicability of amplitude equations by comparing their prediction with the behavior of nonlinear models. To avoid the difficulties of the BG formalism Takeuti, Yamakawa and Ishida (1991, hereinafter TYI) have introduced a modified nonadiabatic formalism for amplitude equations.

With TYI formalism it is found that in radiative Cepheid models the coupling is very strong — the nonadiabatic coupling coefficients are by about two orders of magnitude larger than those calculated from adiabatic eigenfunctions. Basing on these results it has been suggested by Tanaka et al. (1991) that the inclusion of convection may reduce the strength of coupling.

Here I describe the effect the convection has on linear pulsations and on the coupling coefficients. The convection in the background models was described by the mixing length theory, and in linear pulsation calculations the usual approximation viz. that $\nabla \cdot F_c' = 0$ (see Unno et al., 1989) has been made.

---

* On leave from N. Copernicus Astronomical Center, Warsaw, Poland.

Astrophysics and Space Science **210**: 351–354, 1993.

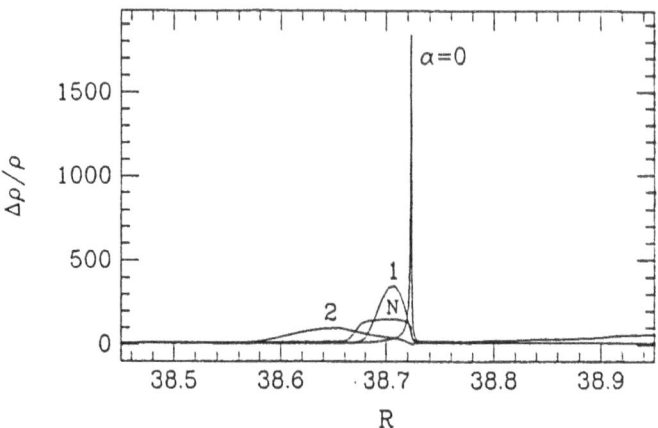

Fig. 1.   Dependence of $|\Delta\rho/\rho|$ on α in a Cepheid model. Curves are labeled with α. The curve labeled N represents the Lagrangean density fluctuation at the instant of maximum compression in the ionization region (see text).

2. The Effect of Convection on Linear Eigenvectors

In purely radiative models of Cepheids the amplitudes of density or temperature eigenfunctions (calculated from linear nonadiabatic equations) are sharply peaked in the ionization region of H/He (see e.g. Stellingwerf, 1990), partly due to strong nonlinearity of opacity and partly due to sharp gradients occurring in this region which are due to the lack of effective energy transport mechanism in these models.

To examine the effect of convection series of static models parametrized by $\alpha = l/H_p$ (see Paczynski, 1969), the ratio of mixing length to the pressure scale height, have been computed. The linear nonadiabatic pulsations have been calculated using Dziembowski's (1977) code. The dependence of the density eigenfunction on is shown in Fig. 1. The temperature perturbation behaves in a similar way. As it is seen the amplitude and steepness of the peak are strongly reduced relative to purely radiative model due to the decreased radiation flux. This also leads to the decrease of nonadiabaticity in the ionization zones because the total pressure perturbation is smooth across the ionization region. For efficient convection, then, it may be expected that, qualitatively, the nonadiabatic eigenfunctions will behave similarly to the adiabtic ones.

Also in Fig. 1 is shown the maximum lagrangean density variation obtained from nonlinear calculations (the density variation is scaled in such a way that the surface amplitude of relative radius variation is unity). As it is seen the linear $\Delta\rho/\rho$ for $\alpha \sim 1-2$ qualitatively resembles the nonlinear variation, what indicates that the effect of nonlinearity can be to a certain degree modeled by the inclusion of convection. Thus amplitude equations

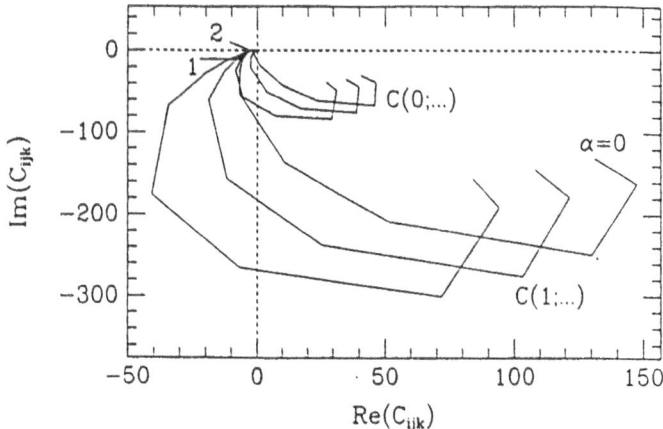

Fig. 2. Dependence of C on α for a model with $M = 6M_\odot$, $L = 2280L_\odot$, $\log(T_{\text{eff}}) = 3.767$.

calculated basing on the results of linear pulsations with $\alpha \sim 1.5$ may better represent the behavior of nonlinear models.

3. The Effect of Convection on Coupling in Cepheids

To examine qualitatively the effect of convection on the strength of coupling the nonadiabatic formalism of TYI is adopted. This formalism, although it contains several simplifying assumptions, is used in the present case to obtain qualitative estimate of the coupling coefficients because (as it uses the second law of thermodynamics in phase of energy equation) it does not require the explicit specification of the mode of energy transport (radiative vs. radiative/convective).

In the TYI formalism the (second order nonresonant) coupling coefficients are calculated from $C_{ijk} = Q_{ijk}/(\sigma_i^2 E K_i)$, where $E K_i$ and σ_i are the kinetic energy and nondimensional frequency of mode i, respectively, and the integral Q_{ijk} gives the strength of coupling between the three modes $\{i, j, k\}$ (TYI, Eq. (26)). The main contribution to this integral comes from the ionization region. With the increase of α the contribution from this region diminishes.

The dependence of C on α is shown in Fig. 2. From this figure it is seen that in radiative models the coupling is strong and strongly nonadiabatic, $|ReC| \sim |ImC| \sim 10^2$. For $\alpha = 1$ the coupling coefficients decrease to $|C| \sim 10$, and for larger α they become of comparable order to those calculated from adiabatic eigenfunctions, though they remain weakly nonadiabatic ($|ReC| \gg |ImC| > 0$).

For Cepheid models the coupling coefficients scale with model parameters approximately as $|C| \propto \lambda^2$, with $\lambda = (L/L_\odot)(M_\odot/M)^2$ (the fit is based on calculations for Cepheid models listed in Tables Ia,b in TYI). Hence with an increase of L/M, or a decrease of M they increase rapidly.

4. Conclusions

The inclusion of convection in the background models reduces the steep gradients in the ionization region and hence results in a decrease of the amplitudes of thermodynamic eigenfunctions in this region. The coupling which is determined by the behavior of eigenfunction in the ionization region is thus reduced.

For Cepheid models the investigations of the effect of the strength of coupling on the occurrence of double mode behavior (Dziembowski and Kovács, 1984, Kovács and Kolláth, 1988, Tanaka et al., 1991) have shown that the type of behavior is determined by the degree of nonadiabaticity and by the magnitude of coupling. Therefore in these stars the influence of convection should be included in the *ab initio* calculation of nonadiabatic coupling coefficients.

For higher L/M supergiants the convection becomes inefficient, so that it can not reduce the amplitudes of linear perturbations in the ionization region. Hence the coupling coefficients obtained from linear eigenvectors may lead to a significant overestimate of the strength of coupling. It would be interesting to see whether in extreme supergiant stars the approach of BK leads to amplitude equations which are compatible with the hydrodynamical calculations.

References

Buchler, J. R. and Goupil, M.: 1984, *Astrophysical Journal* **279**, 394, BG.
Buchler J. R. and Kovács, G.: 1987, *Astrophysical Journal* **318**, 232, BK.
Buchler, J. R., Moskalik, P. and Kovács, G.: 1991, *Astrophysical Journal* **380**, 185.
Dziembowski, W.: 1977, *Acta Astron.* **27**, 95.
Dziembowski, W. and Kovács, G.: 1984, *Monthly Notices of the RAS* **206**, 497.
Kovács, G. and Buchler, J. R., 1989, *Astrophysical Journal* **346**, 898.
Kovács, G. and Kolláth, Z.: 1988, in *Multimode Stellar Pulsations*, eds. G. Kovács, L. Szabados, and B. Szeidl (Kultura, Budapest), p. 33
Paczynski, B.: 1969, *Acta Astron.* **19**, 1.
Pesnell, D. W. and Buchler, J. R.: 1986, *Astrophysical Journal* **303**, 740.
Stellingwerf, R. F.: 1990, *The Numerical Modelling of Nonlinear Stellar Pulsations*, ed. by J. R. Buchler, NATO ASI Series, vol. 302 (Kluwer Academic Publishers, The Netherlands), p.27.
Takeuti, M., Yamakawa, F. and Ishida, T.: 1991, *Publications of the ASJ*, submitted.
Tanaka, Y., Seya, K. and Takeuti, M.: 1991, *Publications of the ASJ*, submitted.
Unno, W., Osaki, Y., Ando, H., Saio, H. and Shibahashi, H.: 1989, *Nonradial Oscilations of Stars*, 2nd ed. (University of Tokyo Press, Tokyo).

THE EFFECT OF CONVECTION IN NONLINEAR
ONE-ZONE STELLAR MODELS

M. SAITOU

Hitachi System Engineering. Ltd., Tokyo 143, Japan.

Abstract. We study the convective nonlinear one-zone models of the pulsating variable stars. In the small convective case, the solution shows chaotic behavior through the period doubling bifurcation as the temperature is lower although the temperature at which the chaos appears is lower than in the case of no convection. On the other hand, in the strong convective case, the solution only shows period-one limit cycle.

1. Introduction

The study of nonlinear simple one-zone models of the pulsating stars with the saturation effect of the kappa mechanism was performed by using numerical integration (Saitou et al., 1989). Their models were represented with the three ordinary differential equations and the solutions showed the chaotic behavior through the period-doubling bifurcation as the temperature is lower.

In this study, we examine the effect of the convection for the same one-zone models.

2. Model

We construct the pulsating stellar model from the Baker's (1966) one-zone stellar model and modify it adding Stellingwerf's (1986) equations for the convection.

The equations that describe our model are four ordinary differential equations as follows:

$$\frac{dx}{dt} = y, \tag{1}$$

$$\frac{dy}{dt} = (1+x)^2(1+z) - (1+x)^{-2}, \tag{2}$$

$$\frac{dz}{dt} = -3\gamma y(1+x)^{-1}(1+z)$$
$$-\varepsilon(1+x)^{-3}[(1-\eta)(1+x)^{\alpha}(1+z)^{\beta} + \eta(1+x)^{-1}(1+u)^3 - 1], \tag{3}$$

$$\frac{du}{dt} = \zeta[(1+x)^{3/2}(1+z)^{1/2} - (1+u)], \tag{4}$$

Astrophysics and Space Science **210**: 355–358, 1993.

where t is the time in the unit of the free-fall time, x, y, z, and u respectively, the stellar radius, the radial velocity, the pressure, and the mean convective velocity in the unit of the equilibrium values, γ the adiabatic exponent, ε the nonadiabaticity or the ratio of the free-fall time to the thermal time, η the ratio of the convective luminosity to the total luminosity, and ζ the convective efficiency or the ratio of the free-fall time to a convective adjustment time.

We used the saturation effect of the kappa mechanism, and put α and β artificially as follows:

$$\alpha = a[(1 + x)^3(1 + z) - 1.2] + 21.6, \tag{5}$$

$$\beta = 3.6(1 + x)^3(1 + z)[(1 + x)^3(1 + z) - 0.2], \tag{6}$$

where a is a parameter to control the kappa mechanism. Also the parameter a correspond to the temperature of the shell of the model star and the temperature is high when the parameter a is large (see Saitou et al., 1989).

3. Calculation

The equations are solved by the Runge-Kutta method with the initial values of $x = 0$, $y = 0.001$, $z = 0$, and $u = 0$ at $t = 0$. The calculation is performed changing the parameters a, η, and ζ, and Table I shows the summary of the solutions. Figure 1 shows also the phase map and return maps, which are plotted the values of successive maxima, of a selected solution ($t = 500$ to 600).

TABLE I
Summary of Solution

η	a							
	20	15	12	10	7	5	3	1
0.0	1	4	c					
0.1		1	2	8	c			
0.2			1	1	2	c		
0.3							1	2
0.4							1	1
0.6							1	1

$\zeta = 1.0$
1: period 1, 2: period 2, 4: period 4,
8: period 8, c: chaos

4. Discussion

From Figure 1 and Table I, we fine that in the case of small convective luminosity, η, the solution shows chaotic behavior. However, the trajectory

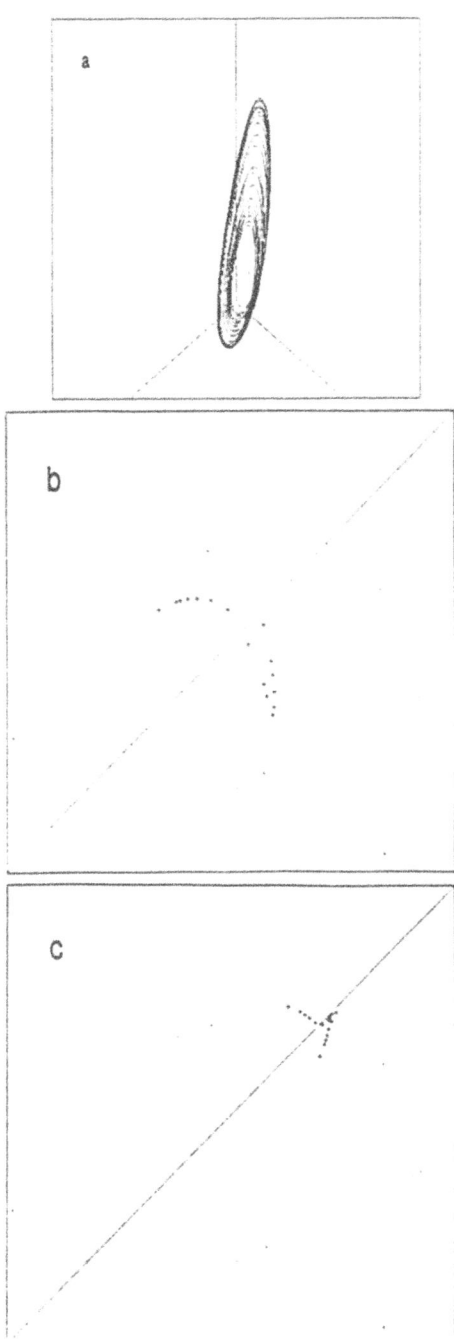

Fig. 1.   The phase map and return maps of the solution for $a = 12.0$, $\eta = 0.04$, and $\zeta = 1.0$. (a) shows orbit in the x-y-z phase space. (b) and (c) show return maps plotted values of successive maxima for x and for y, respectively.

of the larger convective luminosity is more regular and is especially only simple limit cycle at $\eta = 0.20$. This result can imply that the convective effect generally stabilizes the pulsation as well as the result of Stellingwerf (1986).

Table I shows that in the case of small convective luminosity the chaotic behavior is induced as the parameter a is small as well as in the case of no convection. However, the temperature at which the chaos appears is lower than no convection case.

5. Summary

We summarize our study for the convective one-zone model of the pulsating star as follows:

(1) Small convection induces the chaos as well as no convection but stabilizes the pulsation.

(2) Large convection products only the regular pulsation.

Acknowledgements

We would like to thank Professor M. Takeuti for this helpful advices and recommendation to me for participation in the IAU Colloquium.

References

Baker, N.: 1966, in *Stellar Evolution*, eds. R. F. Stein and A. G. W. Cameron (Plenum Press, New York), p. 333.
Saitou, M., Takeuti, M. and Tanaka, Y.: 1989, *Publications of the ASJ* **41**, 297.
Stellingwerf, R. F.: 1986, *Astrophysical Journal* **303**, 119.

A UNIFIED MODEL OF DWARF NOVA OUTBURSTS BASED ON THE DISK INSTABILITY

Y. OSAKI, M. HIROSE and S. ICHIKAWA
Department of Astronomy, Faculty of Science, The University of Tokyo,
Bunkyo-ku, Tokyo 113, Japan

Abstract. A unified model for outbursts of dwarf novae is proposed based on the disk instability model in cataclysmic variable stars. In this model, two different intrinsic instabilities (i.e., the thermal instability and the tidal instability) within accretion disks are considered in non-magnetic cataclysmic variable stars. It is suggested that all of three sub-classes of dwarf novae (i.e., U Gem-type, Z Cam-type and SU UMa-type dwarf novae) may be explained in terms of two model parameters of the orbital period of the binary and of the mass transfer rate within the framework of the disk instability model.

1. Introduction

Dwarf novae are eruptive variables showing repetitive outbursts of amplitudes of 2-6 magnitude. There are three sub-classes in dwarf novae: the U Gem-type showing only the ordinary outbursts, the Z Cam-type showing occasional standstill, and the SU UMa-type showing superoutbursts. Various outbursting phenomena in these stars are now believed to be caused by variable accretion onto the white dwarf component in the cataclysmic binary system. Two different models have been competing to explain the outbursts of dwarf novae: one model is the mass transfer burst model advocated by Bath (1973) and by his group and the other is the disk instability model proposed first by Osaki (1974) and extensively pursued by many groups (for discussion of these models, see a recent review by Osaki 1989b). Here we present a unified model of dwarf nova outbursts from the standpoint of the disk instability model.

2. A Unified Model of Dwarf Nova Outbursts

In the disk instability model for outbursts of dwarf novae, the mass transfer rate from the secondary star is supposed to be constant at all time and all time-dependent and outbursting phenomena are thought to be caused by intrinsic instabilities in accretion disks. In our unified model, different outbursting behaviors among non-magnetic cataclysmic binary systems are classified by two-parameters characterizing accretion disks in these systems; that is, the orbital period of the system and the mass transfer rate to the accretion disk from the secondary star. There exist two different instabilities in accretion disks relevant in cataclysmic variables: one is the thermal instability due to the hydrogen ionization-recombination phase transition

Astrophysics and Space Science **210**: 359–360, 1993.

and the other is the tidally driven eccentric instability (Whitehurst 1988; Hirose and Osaki 1990) by which instability an accretion disk is deformed into an eccentric form and its apsidal line slowly rotates in the inertial frame of reference. It has already been well known in the disk instability theory that difference between non-outbursting system and outbursting system are due to difference in the mass transfer rate from the secondary. Among outbursting systems, the ordinary U Gem stars and the "superoutbursting" SU UMa stars are distinguished by the difference in the orbital period of the binary. This is in turn due to the difference in the mass ratio of the binary because the tidally driven eccentric instability occurs only in those systems with low mass secondary. The superoutburst and superhump phenomena of SU UMa stars can be explained by the combined mechanism of thermal and tidal instability of accretion disks (see, Osaki 1989a). The Z Cam sub-type dwarf novae, which are characterized by the "standstill" in the outburst and outbursting stars in the orbital period and the mass transfer rate diagram. The possible causes of the standstill phenomenon are discussed. A possible model is suggested to explain the standstill in Z Cam stars within the disk instability model, in that the disk radius may exhibit a long term variation, leading these stars sometime in a cyclic outbursting state and sometime in the standstill.

More details on this model will be presented elsewhere.

References

Bath, G. T.: 1973, *Nature Phys. Sci.* **246**, 84.
Hirose, M. and Osaki, Y.: 1990, *Publications of the ASJ* **42**, 135.
Osaki, Y.: 1974, *Publications of the ASJ* **26**, 429.
Osaki, Y.: 1989a, *Publications of the ASJ* **41**, 1005.
Osaki, Y.: 1989b, in *Theory of Accretion Disks*, eds. F. Meyer, W. J. Duschl, J. Frank, and E. Meyer-Hofmeister (Kluwer Academic Publishers, Dordrecht), p. 183.
Whitehurst, R.: 1988, *Monthly Notices of the RAS* **232**, 35.

PULSATIONAL INSTABILITY OF ACCRETION DISKS
AROUND COMPACT OBJECTS

T. OKUDA

Hakodate College, Hokkaido Univ. of Education, Hakodate 040, Japan

and

S. MINESHIGE

Department of Physics, Ibaraki Univ., Mito 310, Japan

Abstract. Linear analysis shows that radial oscillations in accretion disks around compact object are overstable to axisymmetric perturbation under a variety of conditions. Furthermore, numerical simulations confirm that the radial oscillations induce quasi-periodic modulations of the disk luminosity. The disk oscillation model may be responsible for quasi-periodic oscillations (QPOs) observed in low mass X-ray binaries (LMXBs), cataclysmic variables (CVs), and other compact objects.

Pulsational instability of accretion disks around compact objects was first examined by Kato (1978) and later confirmed by Blumenthal, Yang, and Lin (1984) under a variety of conditions. They found that the accretion disks are unstable to axisymmetric radial perturbations and suggested that the radial oscillations with local Keplerian frequencies may explain QPOs in CVs. Papaloizou and Stanley (1986) treated analytically and numerically stability of an axisymmetric accretion disk which includes the disk-boundary layer. Okuda and Mineshige (1991) and Okuda et al. (1992) recently performed numerical simulations of the radial oscillations, with special attention to the QPOs in LMXBs and CVs, respectively.

We summarize these results briefly. Excitation mechanism of the radial oscillations can be understood in analogy with the mechanism in stellar pulsation. In the frame work of α-model as to viscous stress, the major thermal energy is supplied by viscous dissipation and the viscous stress $P_{r\phi}$ is assumed to be proportional to local pressure. The viscous energy generation rate increases in the compressed phase during the radial oscillations, leading to amplification of the oscillations.

Our numerical simulations start with initial conditions of the stationary solution by Shakura and Sunyaev (1973) corresponding to an input mass flow rate, $\dot{M}_0$, which is kept constant during the calculations. The radial pressure gradient force in the momentum equation, which is neglected in the stationary solution, works as initial small perturbations. The disks are also assumed to be Keplerian both at inner and outer edges.

As far as the input mass flow rate $\dot{M}_0$ is low in comparison with the critical accretion rate $\dot{M}_c$ corresponding to the Eddington luminosity, the

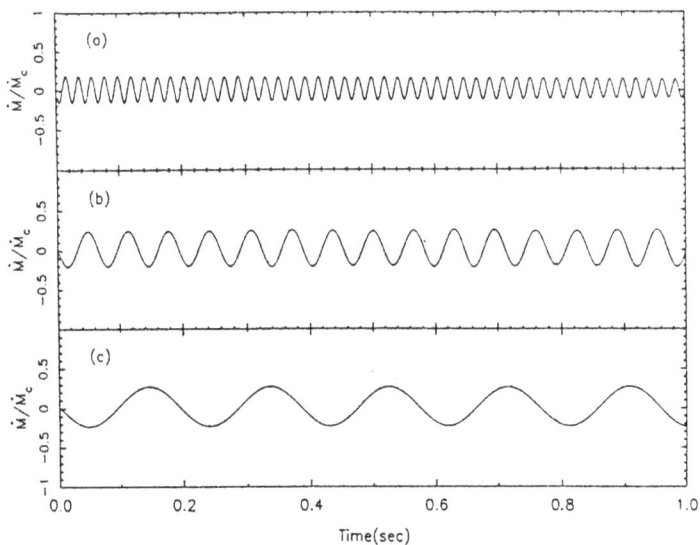

Fig. 1.   Modulations of mass flow rate, $\dot{M}$, at $r/r_{\rm in}$ = 12.9(a), 27.3(b), 56.2(c) for a neutron star with $\dot{M}_0 = 0.02\,\dot{M}_c$ and $\alpha = 0.1$.

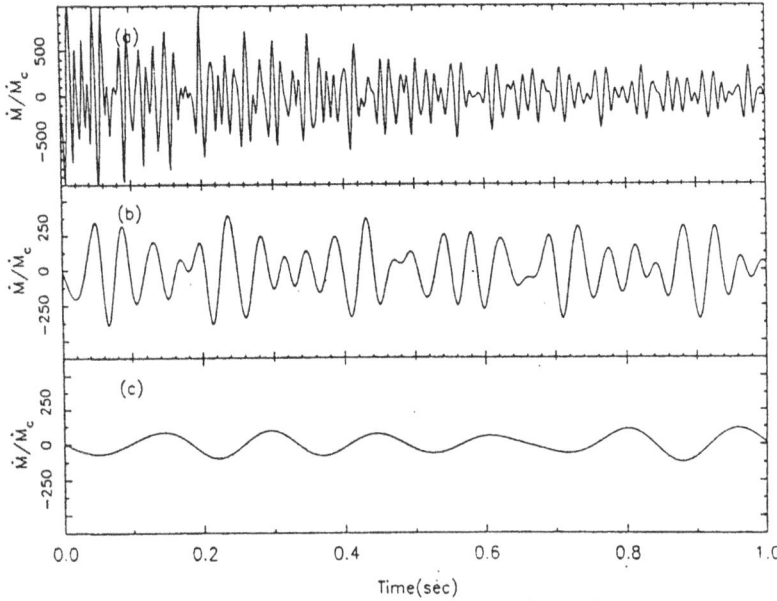

Fig. 2.   Same as in Fig. 1 but with $\dot{M}_0 = \dot{M}_c$.

flow variables modulate in good agreement with the result of the linear analysis. Fig. 1 shows modulations of the mass flow rate $\dot{M}$ at $r/r_{\rm in}$ = 12.9(a), 27.3(b), and 56.2(c) for an accretion disk around a neutron star with $\dot{M}_0/\dot{M}_c = 0.02$ and $\alpha = 0.1$, where $r_{\rm in}$ is the disk inner edge. The mass flow rate shows sinusoidal oscillations with Keplerian frequencies. When $\dot{M}_0$

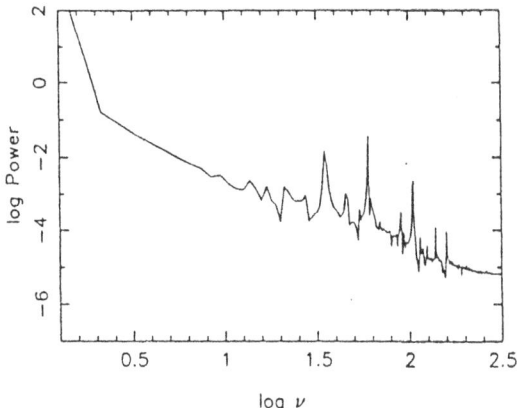

Fig. 3.   Power spectra of 1–3 kev disk luminosity oscillation in the same model as Fig. 2.

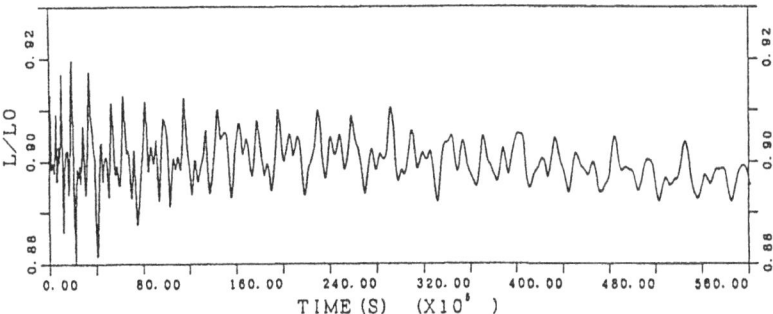

Fig. 4.   Disk luminosity modulations for an active galactic nucleus with $M = 10^8 M_\odot$, $\dot{M}_0 = \dot{M}_c$, and $\alpha = 0.1$.

is so high, overall features of the oscillations deviate from the sinusoidal form and become somewhat chaotic due to nonlinear effect in the disk equations. Fig. 2 depicts the time variation of $\dot{M}$ at three radii for the neutron star with $\dot{M}_0/\dot{M}_c = 1.0$ and $\alpha = 0.1$, where the viscous stress $P_{r\phi} = -\alpha P_g$ is adopted.

It should be noted that relative oscillation amplitudes of $\dot{M}$ are generally large, whereas oscillation amplitudes of the temperature T, the surface density Σ and the azimuthal velocity v_ϕ are as small as 0.1~10 percent, depending on the input mass flow rate $\dot{M}_0$. Large oscillation amplitudes of the flow variables are numerically found in the region of $r = 4 \sim 12 r_{\rm in}$.

As the results, the total disk luminosity, L_d, which is an integral of local emergent flux over the whole disk, is modulated quasi-periodically. The modulations are characterized by the power spectra of L_d, which is shown in Fig. 3.   Some discrete peaks are found in the power spectra. This may be attributed to the coarse mesh spacing of the disk model used here. Actual disk oscillations are represented as compound of an infinite number of oscillators, each having a different frequency near the Keplerian frequency and a different amplitude. Resultant power spectra should show a more smooth and broader peak (see Okuda et al. 1992). These quasi-periodic light variations are also expected in other class of compact objects such as white dwarfs, black holes, and active galactic nuclei. Fig. 4 shows luminosity modulations for an accretion disk around an active galactic nucleus with $M = 10^8 M_\odot$, $\dot{M}_0 = \dot{M}_c$, and $\alpha = 0.1$. The disk oscillation model may be promising for the QPOs observed in many species of compact objects.

References

Blumenthal, G. R., Yang, L. T. and Lin, D. N. C.: 1984, *Astrophysical Journal*, **287**, 774.
Kato, S.: 1978, *Monthly Notices of the RAS*, **185**, 629.
Okuda, T. and Mineshige, S.: 1991, *Monthly Notices of the RAS*, **249**, 684.
Okuda, T., Ono, K., Tabata, M. and Mineshige, S.: 1992, *Monthly Notices of the RAS*, **254**, 427.
Papaloizou, J. C. B. and Stanley, G. Q. G.: 1986, *Monthly Notices of the RAS*, **220**, 593.
Shakura, N. I. and Sunyaev, R. A.: 1973, *Astronomy and Astrophysics*, **24**, 337.

NUMERICAL SIMULATIONS OF PULSATIONALLY UNSTABLE ACCRETION DISKS AROUND SUPERMASSIVE BLACK HOLES

F. HONMA
*Department of Astronomy, Faculty of Science, Kyoto University, Sakyo-ku, Kyoto
606-01, Japan*

R. MATSUMOTO
College of Arts and Sciences, Chiba University, Yayoi-cho, Chiba 260

and

S. KATO
*Department of Astronomy, Faculty of Science, Kyoto University, Sakyo-ku, Kyoto
606-01, Japan*

Abstract. The innermost region of slim accretion disks with standard α viscosity is unstable against axisymmetric radial inertial acoustic perturbations under certain conditions. Numerical simulations are performed in order to demonstrate behaviors of such unstable disks. It is shown that oscillations with the period of $\sim 10^{-3}(M_{\rm BH}/M_\odot)$ s can be excited near the inner edge of the disks, where $M_{\rm BH}$ is the mass of the central object. This kind of unstable disks is a possible origin of the periodic X-ray time variabilities with period of $\sim 10^4$ s observed in a Seyfert galaxy NGC 6814.

1. Introduction

Accretion disks around a supermassive black hole are the most promising model of active galactic nuclei (AGN's). Various types of time variabilities observed from AGN's and their comparison with possible disk oscillations are important clues to diagnose the model.

Under certain conditions the innermost region of relativistic accretion disks is pulsationally unstable against axisymmetric radial oscillations, which propagate in the disk as inertial-acoustic waves (Matsumoto et al. 1988, 1989 and references therein). This paper presents briefly recent results of numerical simulations performed in order to investigate behaviors of such unsteady disks around a supermassive black hole. Detailed results will be published elsewhere (Honma et al. 1992). Similar investigations in the case of stellar mass black hole have been performed by Matsumoto et al. (1988, 1989).

2. Results

Slim, transonic disk models with standard α viscosity are used. The parameters involved in our simulations are the mass the central object, $M_{\rm BH}$, the accretion rate describing the initial steady state, $\dot{M}_{\rm init}$, and the viscosity parameter, α. Values of them are set to $M_{\rm BH} = 10^4 M_\odot$, $\alpha = 0.1$, and

Astrophysics and Space Science 210: 365–367, 1993.
© 1993 *Kluwer Academic Publishers.*

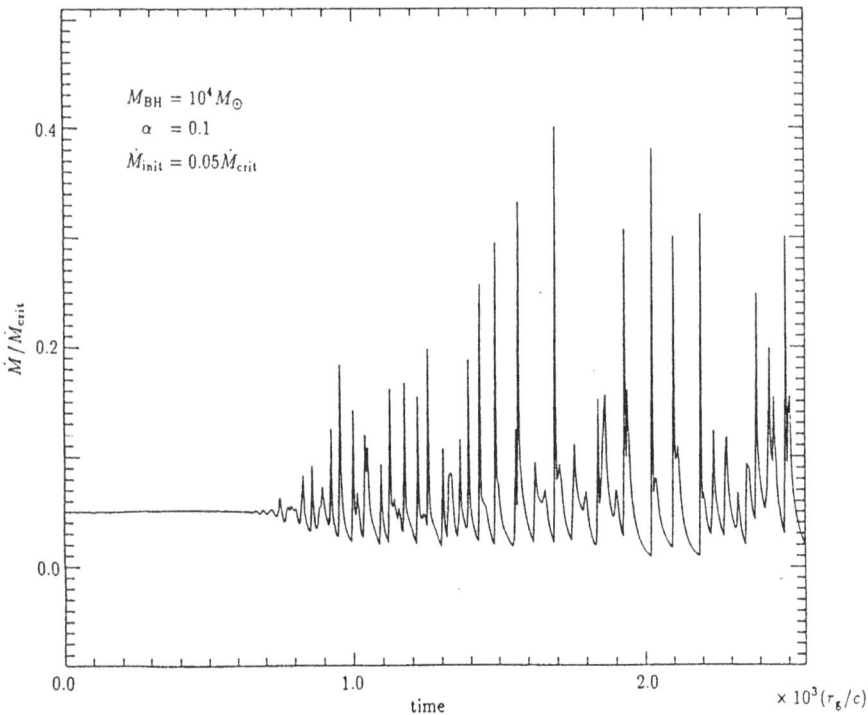

Fig. 1. Time variations of accretion rate at $\varpi = 2.8r_{\mathrm{g}}$ in the simulation with $M_{\mathrm{BH}} = 10^4 M_\odot$, $\alpha = 0.1$, and $\dot{M}_{\mathrm{init}} = 0.05\dot{M}_{\mathrm{crit}}$.

$\dot{M}_{\mathrm{init}} = 0.05\dot{M}_{\mathrm{crit}}$, where, $\dot{M}_{\mathrm{crit}}$ is the critical accretion rate, which is defined as $\dot{M}_{\mathrm{crit}} = 16L_{\mathrm{Edd}}/c^2$.

The overall feature of the disk behavior is essentially the same as that obtained by Matsumoto et al. (1988, 1989) to the case of stellar mass central objects: The innermost region of the disk (radius $\sim 3r_{\mathrm{g}} - 3.5r_{\mathrm{g}}$) oscillates quasi-periodically and generate waves, which propagate both outwards and inwards from this region. The waves immediately grow to shock waves. Figure 1 shows the time variation of the accretion rate through the inner boundary. The period of the oscillation is $\lesssim 100(r_{\mathrm{g}}/c)$.

3. Discussion

Recently, periodic variability of X-ray intensity with the period of $\sim 10^4$ s is discovered in a Seyfert galaxy NGC 6814 (Mittaz and Branduardi-Raymont 1989). This period of time variation of NGC 6814 can be explained as the oscillation of the innermost region of the accretion disk, if $M_{\mathrm{BH}} \sim 10^7 M_\odot$ is employed.

The time variability presented here as our simulation results is not exactly periodic; quasi-periodic and somewhat chaotic. The system of equations which we treat is similar with that of stellar pulsation, although boundary conditions are different. Hence. depending on values of parameters, semi-regular and chaotic variations will be expected even in the present case of disk oscillations, as in stellar pulsation. This is interesting theoretically and also when we want to explain observations.

Acknowledgements

Computations were performed on the FACOM M780/30 and the Fujitsu VP2600 at the Data Processing Center of Kyoto University. This work was supported in part by the Space Data Analysis Center of Institute of Space and Astronautical Science.

References

Honma, F., Matsumoto. R., and Kato, S: 1992, submitted to *Publications of the ASJ*
Matsumoto, R., Kato, S., and Honma, F.: 1988, in *Physics of Neutron Stars and Black Holes*, ed. Y. Tanaka (Universal Academy Press, Tokyo, Japan), p. 155.
Matsumoto, R., Kato, S., and Honma, F.: 1989, in *Theory of Accretion Disks*, eds. F. Meyer, W. J. Duschl, J. Frank, and E. Meyer-Hofmeister (Kluwer Academic Publishers, Dordrecht, Holland), p. 167.
Mittaz, J. P. D., and Branduradi-Raymont, G.: 1989, *Monthly Notices of the RAS*, **238**, 1029.

LONG-TERM V/R VARIATIONS OF BE STARS DUE TO GLOBAL ONE-ARMED OSCILLATIONS OF EQUATORIAL DISKS

ATSUO T. OKAZAKI

College of General Education, Hokkai-Gakuen University,
Toyohira-ku, Sapporo 062, Japan

Abstract. We study the long-term variations of Balmer line profiles due to global one-armed oscillations in Be-star disks. In order to examine the qualitative effects of oscillations on line profiles, we assume that the eigenfunctions of one-armed nonlinear oscillations are similar to those of linear oscillations. Computing the line profiles for various values of disk parameters, we find that in small disks or in disks with steep density gradients the one-armed fundamental modes cause remarkable variabilities similar to the observed V/R variations.

The long-term V/R variation is one of the most puzzling phenomena in Be stars. Periods of the variations range from years to decades; they are much longer than the dynamical time-scales in the central stars and the envelopes. In addition behaviors of the profile variations are bizarre: a profile as a whole shifts blueward (redward) when the red (violet) component is the stronger (e.g., McLaughlin 1961).

Recently, Okazaki (1991) proposed a model based on a theory of global oscillations in nearly Keplerian disks. According to this theory of oscillations (e.g., Kato 1983), the model suggests that the long-term V/R variations are phenomena caused by the global one-armed oscillations in the equatorial disks of Be stars. Studying the eigenmodes of linear one-armed isothermal oscillations in isothermal disks with finite radial sizes ($r_* \leq r \leq r_{\mathrm{out}}$), Okazaki (1991) found that the one-armed oscillation model naturally explains observed periods of the V/R variations.

In the present paper we examine the qualitative effects of the one-armed oscillations on line profiles. For this purpose we adopt a simplified treatment: we assume that the eigenfunctions of one-armed nonlinear oscillations are similar to those of linear oscillations found in Okazaki (1991); the eigenfunctions are normalized so that the maximum value of the perturbed part of the angular velocity is 5% of the unperturbed part. In addition we assume that the source function is constant over the entire disk region. On these assumptions we compute the optically-thick line profiles emitted from the entire disk by integrating fluxes along a bundle of line-of-sights penetrating the disk; neither the radiation from the central star nor the continuum radiation from the disk are included. The thermal broadening is taken into account as the line broadening mechanism. The density profile adopted is

Astrophysics and Space Science **210**: 369–370, 1993.
© 1993 *Kluwer Academic Publishers.*

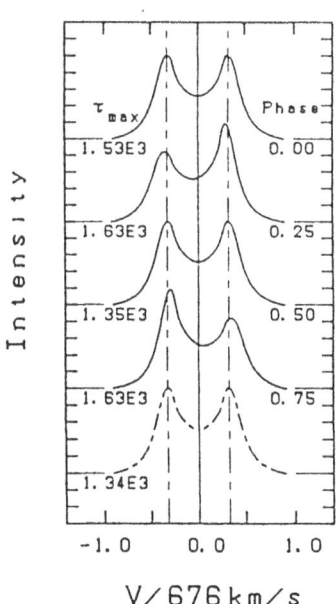

V/676 km/s

Fig. 1.   Variability caused by the one-armed fundamental mode for $r_{out}/r_* = 10$ and $\alpha = 3$. The central star is in B0 main-sequence and the inclination angle is $60°$. The four solid profiles denote the profiles at the different phases and the dash-dotted profile denote the profiles from the unperturbed disk. The line optical depth is given on the left side of each profile. The vertical dash-dotted lines represent the peak velocities of the unperturbed profiles.

a simple power-law form: the equatorial density of the unperturbed disk is proportional to $r^{-\alpha}$.

Examining the line profile variabilities for various values of disk parameters, we obtain the following conclusions: In small disks ($r_{out}/r_* \sim 2$) or in disks with steep ($\alpha \gtrsim 3$) density gradients (see figure 1), the one-armed fundamental modes cause remarkable variabilities similar to the observed V/R variations. These variabilities result mainly from the eccentric deformation of the region through which the optical depths along line-of-sights are of order unity.

References

Kato, S.: 1983, *Publications of the ASJ*, **35**, 249.
McLaughlin, D. B.: 1961, *J. Roy. Astron. Soc. Canada*, **55**, 73.
Okazaki, A. T.: 1991, *Publications of the ASJ*, **43**, 75.

A SIMPLIFIED MODEL FOR A NONLINEAR TIDAL EFFECT ON ACCRETION DISKS IN CVS

ZHONG-YONG ZHANG and JIAN-SHENG CHEN
*Beijing Astronomical Observatory, Chinese Academy of Sciences,
Beijing 100080, People's Republic of China*

Abstract

This paper investigates the tidal effect on accretion disk in CVs and sets up a simplified model in which the secondary's gravitation is substituted by a mean tidal torque. We find that a linear tidal torque will not be able to maintain an equilibrium disk. By using the result of the radius of the equilibrium disk approximately equals to the tidal radius, which was obtained by using the two dimensional numerical simulation invoking nonlinear tidal effect, we give the modified tidal dissipation function for our simplified model which could be used to interpret the outburst of the dwarf nova with tidal effect. The paper also shows that the radius of an equilibrium disk with a torus is slightly small than the Lubow-Shu radius, and the tidal effect may also cause the cycle of quiescence-superoutburst in addition to the cycle of quiescence-outbursts-superoutburst.

Key words: CVs – Accretion disk – Tidal effect

References

Chen J. S., Liu X. W., and Wei M. Z.: 1991, *Astronomy and Astrophysics* **242**, 397.
Frank J., King A. R., and Raine D. J.: 1985, *Accretion Power in Astrophysics* (Cambridge University Press, Cambridge), p. 60.
Lubow S. H. and Shu F. H.: 1975, *Astrophysical Journal* **198**, 383.
Mayer-Hofmeister and Ritter H.: 1991, preprint.
Osaki Y.: 1989a, in *Theory of Accretion Disks*, eds. F. Meyer et al. (Kluwer Academic Publishers), p. 183.
Osaki Y.: 1989b, *Publications of the ASJ* **41**, 1005.
Paczynski B.: 1977, *Astrophysical Journal* **216**, 822.
Papaloizou J., Pringle J. E.: 1977, *Monthly Notices of the RAS* **181**, 441.
Pringle J. E.: 1981, *Annual Review of Astronomy and Astrophysics* **19**, 137.
Ritter H.: 1990, *Catalogue of Cataclysmic Binaries, Low-Mass X-ray Binaries and Related Objects* (fifth edition).
Smak J.: 1983, *Astrophysical Journal* **272**, 234.
Smak J.: 1984, *Acta Astron.* **34**, 161.
Whitehurst R.: 1988, *Monthly Notices of the RAS* **232**, 35.
Whitehurst R.: 1989, in *Theory of Accretion Disks*, eds. F. Meyer et al., p. 213.
Whitehurst R.: 1991, *Monthly Notices of the RAS* **249**, 25.

LIST OF PARTICIPANTS

Name	Affiliation
Agu, M.	Ibaraki University, Japan
Aikawa, T.	Tohoku Gakuin University, Japan
Ando, H.	National Astronomical Observatory, Japan
Antonello, E.	Osservatorio Astronomico di Brera, Italy
Baranov, A. S.	Institute for Theoretical Astronomy, Saint Petersburg, Russia
Barthes, D.	Université de Montpellier I, France
Breger, M.	Institut für Astronomie, Wien, Austria
Buchler, R.	University of Florida, USA
Davis, C. G.	Los Alamos National Laboratory, USA
Dziembowski, W.	Copernicus Astronomical Center, Warsaw, Poland
Fadeyev, Yu. A.	Institute for Astronomy, Academy of Science, Russia
Foong, S. K.	Ibaraki University, Japan
Goupil, M.-J.	Observatoire de Meudon Paris, France
Guzik, J. A.	Los Alamos National Laboratory, USA
Hamada, T.	Ibaraki University, Japan
Honma, F.	Kyoto University, Japan
Ishida, T.	Nishi-harima Astronomical Observatory, Hyogo, Japan
Ishizuka, T.	Ibaraki University, Japan
Israelian, G.	Byurakan Astrophysical Observatory, Armenia
Itoh, Y.	Kakuda Girl's Senior High School, Miyagi, Japan
Jiang, S.-Y.	Beijing Astronomical Observatory, China
Kambe, E.	National Defense Academy, Japan
Kanetake, R.	Tohoku University, Japan
Kato, S.	Kyoto University, Japan
Kitamura M.	National Astronomical Observatory, Japan
Kobayashi, E.	Osaka Science Education Center, Japan

Kogure, T.	Kyoto University, Japan
Kolláth, Z.	Konkoly Observatory, Hungary
Kovács, G.	University of Florida, USA
Kozai, Y.	National Astronomical Observatory, Japan
Kubiak, M.	Warsaw University Observatory, Poland
Kurtz, D. W.	University of Cape Town, South Africa
Li, Z.-P.	Beijing Astronomical Observatory, China
Liu, Z.-L.	Beijing Astronomical Observatory, China
Makishima. K.	University of Tokyo, Japan
Matsuoka. M.	Institute of Physical and Chemical Research, Japan
Mineshige, S.	Ibaraki University, Japan
Mori, H.	Kyushu University, Japan
Moskalik. P.	Copernicus Astronomical Center, Warsaw, Poland
Nakahara, T.	Ibaraki University, Japan
Nakamura, Y.	Ibaraki University, Japan
Okazaki, A.	Gunma University, Japan
Okazaki, A. T.	Hokkai-Gakuen University, Japan
Okuda, T.	Hokkaido University of Education, Japan
Osaki, Y.	University of Tokyo, Japan
Pajdosz, G.	Pedagogical University, Cracow, Poland
Paparó, M.	Konkoly Observatory, Hungary
Percy, J. R.	University of Toronto, Canada
Petersen, J. O.	University Observatory, Copenhagen, Denmark
Pfeiffer, B.	Observatoire Midi-Pyrénées, France
Saijo, K.	National Science Museum, Tokyo, Japan
Saio, H.	Tohoku University, Japan
Saitou, M.	Hitachi System Engineering Ltd., Japan
Sano, M.	Tohoku University, Japan
Sasselov, D. D.	Harvard-Smithsonian Center for Astrophysics, USA
Sato, H.	University of Tokyo, Japan
Sawada, Y.	Tohoku University, Japan
Serre, T.	Observatoire de Paris, France
Shibahashi, H.	University of Tokyo, Japan
Shimada, M.	Kyoto University, Japan

Tainaka, K.	Ibaraki University, Japan
Takano, R.	Tohoku University, Japan
Takeuti, M.	Tohoku University, Japan
Tamura, S.	Tohoku University, Japan
Tanaka, Y.	Ibaraki University, Japan
Uji-iye, K.	Tohoku University, Japan
Unno, W.	Kinki University, Osaka, Japan
Vauclair, G.	Observatoire Midi-Pyrénées, France
Walton, S. J.	Queen Mary and Westfield, UK
Yamakawa F.	Fujitu, Ltd., Japan
Yanagida, T.	Institute of Statistical Mathematics, Tokyo, Japan
Yokosawa, M.	Ibaraki University, Japan
Yoshida, H.	National Astronomical Obseervatory, Japan
Yu, Z.-Y.	Shanghai Observatory, China
Zalewski, J.	Tohoku University, Japan
Zhang, C.-S.	Purple Mountain Observatory, China
Zhang, Z.-Y.	Beijing Astronomical Observatory, China

The manufacturer's authorised representative in the EU is Springer
Nature Customer Service Centre GmbH, Europaplatz 3, 69115 Heidelberg,
Germany. If you have any concerns regarding our products, please
contact ProductSafety@springernature.com

Printed and bound by CPI Group (UK) Ltd, Croydon, CR0 4YY

24/04/2026
02096308-0011